Nanotechnology: Nanoparticles and Nanostructures

Nanotechnology: Nanoparticles and Nanostructures

Edited by Hector Lamb

Clanrye International
www.clanryeinternational.com

Clanrye International,
750 Third Avenue, 9th Floor,
New York, NY 10017, USA

ISBN: 978-1-63240-717-7

Cataloging-in-Publication Data

Nanotechnology : nanoparticles and nanostructures / edited by Hector Lamb.
p. cm.
Includes bibliographical references and index.
ISBN 978-1-63240-717-7
1. Nanotechnology. 2. Nanoparticles. 3. Nanostructures. I. Lamb, Hector.

T174.7 .N36 2018
620.5--dc23

For information on all Clanrye International publications
visit our website at www.clanryeinternational.com

CLANRYE
INTERNATIONAL

Printed in China.

Contents

Preface

Nanotechnology is a new and advanced scientific and engineering study which deals with manipulation of matter of atomic, supramolecular, and molecular level. Nanotechnology further diversities into nanomedicine, nanomaterials, nanoelectronics, etc. This book is compiled in such a manner, that it will provide in-depth knowledge about the theory and practice of nanotechnology and the nanostructures and nanoparticles studied under it. Some of the diverse topics covered in it address the varied branches that fall under this category. Those in search of information to further their knowledge will be greatly assisted by this textbook.

A foreword of all chapters of the book is provided below:

Chapter 1 - Nanoparticles are particles that vary from 1 to 100 nanometers in size. Some of the methods of creating nanoparticles are attrition, gas condensation, hydrothermal synthesis, pyrolysis and chemical precipitation. This is an introductory chapter which will introduce briefly all the significant aspects of nanoparticles; **Chapter 2 -** Copper nanoparticles are particles that are formed by natural processes or by chemical synthesis. Silver nanoparticle, platinum nanoparticles, colloidal gold, cobalt oxide nanoparticles, gold nanobeacon, nanoscale iron particles, nanofiber, iron oxide nanoparticles, iron-platinum nanoparticles and nanoceramic are other classifications of nanoparticles by composition. This chapter has been carefully written to provide an easy understanding of the varied classifications of nanoparticles by composition; **Chapter 3 -** A nanostructure is any structure whose size varies between microscopic and molecular structure. Nanomesh, gradient multilayer nanofilm, nanohole, nanosheet, nanostructured film, sculptured thin film, icosahedral twins, nano flake and nanofoam are important topics related to the study of nanostructures. Nanotechnology is best understood in confluence with the major topics listed in the following chapter; **Chapter 4 -** Carbon nanotubes are carbon allotropes that have a cylindrical nanostructure. They posses exceptional stiffness and strength. The different types of carbon nanotubes that exist are single-walled, junctions and crosslinking, multi-walled, etc. The chapter strategically encompasses and incorporates the major components and key concepts of carbon nanotubes, providing a complete understanding; **Chapter 5 -** Nanotechnology is the technology that is used in the manipulation of matter that exist at the nanoscale. Nanotechnology is used in creating new material which can have various applications in fields such as tissue engineering and nanoelectronics. This chapter is an overview of the subject matter incorporating all the major aspects of nanotechnology.

At the end, I would like to thank all the people associated with this book devoting their precious time and providing their valuable contributions to this book. I would also like to express my gratitude to my fellow colleagues who encouraged me throughout the process.

Editor

1

Understanding Nanoparticles

Nanoparticles are particles that vary from 1 to 100 nanometers in size. Some of the methods of creating nanoparticles are attrition, gas condensation, hydrothermal synthesis, pyrolysis and chemical precipitation. This is an introductory chapter which will introduce briefly all the significant aspects of nanoparticles.

Nanoparticle

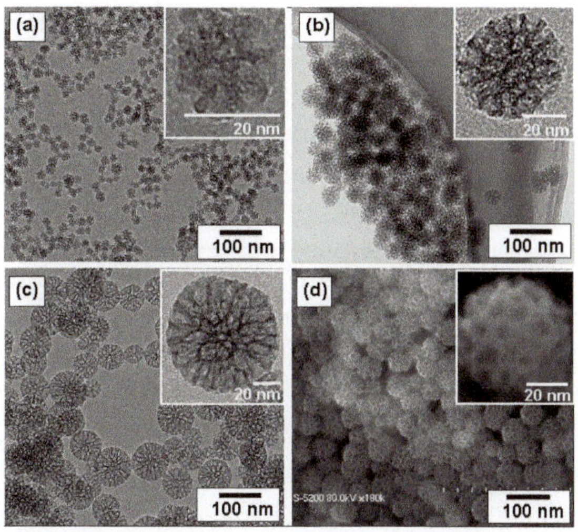

TEM (a, b, and c) images of prepared mesoporous silica nanoparticles with mean outer diameter: (a) 20nm, (b) 45nm, and (c) 80nm. SEM (d) image corresponding to (b). The insets are a high magnification of mesoporous silica particle.

Nanoparticles are particles between 1 and 100 nanometres (nm) in size. In nanotechnology, a particle is defined as a small object that behaves as a whole unit with respect to its transport and properties. Particles are further classified according to diameter. Ultrafine particles are the same as nanoparticles and between 1 and 100 nm in size, fine particles are sized between 100 and 2,500 nm, and coarse particles cover a range between 2,500 and 10,000 nm.

Scientific research on nanoparticles is intense as they have many potential applications in medicine, physics, optics, and electronics. The U.S. National Nanotechnology Initiative offers government funding focused on nanoparticle research.

Definition

The term "nanoparticle" is not usually applied to individual molecules; it usually refers to inorganic materials.

The reason for the synonymous definition of nanoparticles and ultrafine particles is that, during the 1970s and 80s, when the first thorough fundamental studies with "nanoparticles" were underway in the USA (by Granqvist and Buhrman) and Japan, (within an ERATO Project) they were called "ultrafine particles" (UFP). However, during the 1990s before the National Nanotechnology Initiative was launched in the USA, the new name, "nanoparticle," had become more common (for example, the same senior author's paper 20 years later addressing the same issue, lognormal distribution of sizes). Nanoparticles can exhibit size-related properties significantly different from those of either fine particles or bulk materials.

Nanoclusters have at least one dimension between 1 and 10 nanometers and a narrow size distribution. Nanopowders are agglomerates of ultrafine particles, nanoparticles, or nanoclusters. Nanometer-sized single crystals, or single-domain ultrafine particles, are often referred to as nanocrystals.

Background

Although nanoparticles are associated with modern science, they have a long history. Nanoparticles were used by artisans as far back as Rome in the fourth century in the famous Lycurgus cup made of dichroic glass as well as the ninth century in Mesopotamia for creating a glittering effect on the surface of pots. In modern times, pottery from the Middle Ages and Renaissance often retains a distinct gold- or copper-colored metallic glitter. This luster is caused by a metallic film that was applied to the transparent surface of a glazing. The luster can still be visible if the film has resisted atmospheric oxidation and other weathering.

The luster originates within the film itself, which contains silver and copper nanoparticles dispersed homogeneously in the glassy matrix of the ceramic glaze. These nanoparticles are created by the artisans by adding copper and silver salts and oxides together with vinegar, ochre, and clay on the surface of previously-glazed pottery. The object is then placed into a kiln and heated to about 600 °C in a reducing atmosphere. In heat the glaze softens, causing the copper and silver ions to migrate into the outer layers of the glaze. There the reducing atmosphere reduced the ions back to metals, which then came together forming the nanoparticles that give the color and optical effects. Luster technique showed that ancient craftsmen had a sophisticated empirical knowledge of materials. The technique originated in the Muslim world. As Muslims were not allowed to use gold in artistic representations, they sought a way to create a similar effect without using real gold. The solution they found was using luster.

Michael Faraday provided the first description, in scientific terms, of the optical properties of nanometer-scale metals in his classic 1857 paper. In a subsequent paper, the author (Turner) points out that: "It is well known that when thin leaves of gold or silver are mounted upon glass and heated to a temperature that is well below a red heat (~500 °C), a remarkable change of properties takes place, whereby the continuity of the metallic film is destroyed. The result is that white light is now freely transmitted, reflection is correspondingly diminished, while the electrical resistivity is enormously increased."

Uniformity

The chemical processing and synthesis of high-performance technological components for the private, industrial, and military sectors requires the use of high-purity ceramics (oxide ceramics, such as aluminium oxide or copper(II) oxide), polymers, glass-ceramics, and composite materials, as metal carbides (SiC), nitrides (Aluminum nitrides, Silicon nitride), metals (Al, Cu), non-metals (graphite, carbon nanotubes) and layered (Al + Aluminium carbonate, Cu + C). In condensed bodies formed from fine powders, the irregular particle sizes and shapes in a typical powder often lead to non-uniform packing morphologies that result in packing density variations in the powder compact.

Uncontrolled agglomeration of powders due to attractive van der Waals forces can also give rise to microstructural heterogeneity. Differential stresses that develop as a result of non-uniform drying shrinkage are directly related to the rate at which the solvent can be removed, and thus highly dependent upon the distribution of porosity. Such stresses have been associated with a plastic-to-brittle transition in consolidated bodies, and can yield to crack propagation in the unfired body if not relieved.

In addition, any fluctuations in packing density in the compact as it is prepared for the kiln are often amplified during the sintering process, yielding inhomogeneous densification. Some pores and other structural defects associated with density variations have been shown to play a detrimental role in the sintering process by growing and thus limiting end-point densities. Differential stresses arising from inhomogeneous densification have also been shown to result in the propagation of internal cracks, thus becoming the strength-controlling flaws.

Inert gas evaporation and inert gas deposition are free many of these defects due to the distillation (cf. purification) nature of the process and having enough time to form single crystal particles, however even their non-aggreated deposits have lognormal size distribution, which is typical with nanoparticles. The reason why modern gas evaporation techniques can produce a relatively narrow size distribution is that aggregation can be avoided. However, even in this case, random residence times in the growth zone, due to the combination of drift and diffusion, result in a size distribution appearing lognormal.

It would, therefore, appear desirable to process a material in such a way that it is physically uniform with regard to the distribution of components and porosity, rather than using particle size distributions that will maximize the green density. The containment of a uniformly dispersed assembly of strongly interacting particles in suspension requires total control over interparticle forces. Monodisperse nanoparticles and colloids provide this potential.

Monodisperse powders of colloidal silica, for example, may therefore be stabilized sufficiently to ensure a high degree of order in the colloidal crystal or polycrystalline colloidal solid that results from aggregation. The degree of order appears to be limited by the time and space allowed for longer-range correlations to be established. Such defective polycrystalline colloidal structures would appear to be the basic elements of submicrometer colloidal materials science and, therefore, provide the first step in developing a more rigorous understanding of the mechanisms involved in microstructural evolution in high performance materials and components.

Properties

Nanoparticles are of great scientific interest as they are, in effect, a bridge between bulk materials and atomic or molecular structures. A bulk material should have constant physical properties regardless of its size, but at the nano-scale size-dependent properties are often observed. Thus, the properties of materials change as their size approaches the nanoscale and as the percentage of the surface in relation to the percentage of the volume of a material becomes significant. For bulk materials larger than one micrometer (or micron), the percentage of the surface is insignificant in relation to the volume in the bulk of the material. *The interesting and sometimes unexpected properties of nanoparticles are therefore largely due to the large surface area of the material, which dominates the contributions made by the small bulk of the material.*

Silicon nanopowder

1 kg of particles of 1 mm^3 has the same surface area as 1 mg of particles of 1 nm^3

Nanoparticles often possess unexpected optical properties as they are small enough to confine their electrons and produce quantum effects. For example, gold nanoparticles appear deep-red to black in solution. Nanoparticles of yellow gold and grey silicon are red in color. Gold nanoparticles melt at much lower temperatures (~300 °C for 2.5 nm size) than the gold slabs (1064 °C);. Absorption of solar radiation is much higher in materials composed of nanoparticles than it is in thin films of continuous sheets of material. In both solar PV and solar thermal applications, controlling the size, shape, and material of the particles, it is possible to control solar absorption.

Other size-dependent property changes include quantum confinement in semiconductor particles, surface plasmon resonance in some metal particles and superparamagnetism in magnetic materials. What would appear ironic is that the changes in physical properties are not always desirable. Ferromagnetic materials smaller than 10 nm can switch their magnetisation direction using room temperature thermal energy, thus making them unsuitable for memory storage.

Suspensions of nanoparticles are possible since the interaction of the particle surface with the solvent is strong enough to overcome density differences, which otherwise usually result in a material either sinking or floating in a liquid.

The high surface area to volume ratio of nanoparticles provides a tremendous driving force for diffusion, especially at elevated temperatures. Sintering can take place at lower temperatures, over shorter time scales than for larger particles. In theory, this does not affect the density of the final product, though flow difficulties and the tendency of nanoparticles to agglomerate complicates matters. Moreover, nanoparticles have been found to impart some extra properties to various day to day products. For example, the presence of titanium dioxide nanoparticles imparts what we call the self-cleaning effect, and, the size being nano-range, the particles cannot be observed. Zinc oxide particles have been found to have superior UV blocking properties compared to its bulk substitute. This is one of the reasons why it is often used in the preparation of sunscreen lotions, is completely photostable and toxic.

Clay nanoparticles when incorporated into polymer matrices increase reinforcement, leading to stronger plastics, verifiable by a higher glass transition temperature and other mechanical property tests. These nanoparticles are hard, and impart their properties to the polymer (plastic). Nanoparticles have also been attached to textile fibers in order to create smart and functional clothing.

Metal, dielectric, and semiconductor nanoparticles have been formed, as well as hybrid structures (e.g., core–shell nanoparticles). Nanoparticles made of semiconducting material may also be labeled quantum dots if they are small enough (typically sub 10 nm) that quantization of electronic energy levels occurs. Such nanoscale particles are used in biomedical applications as drug carriers or imaging agents.

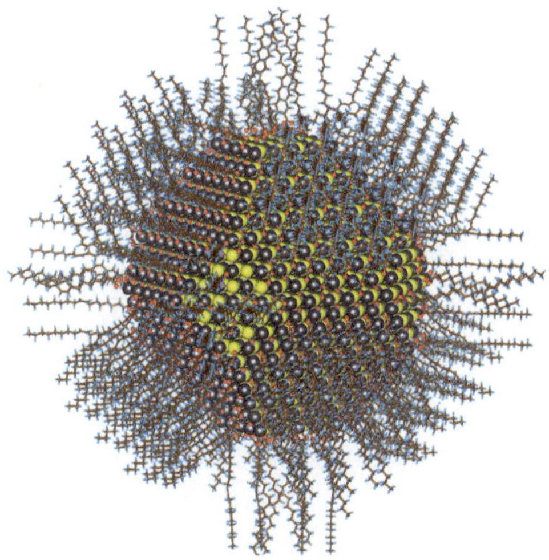

Semiconductor nanoparticle (quantum dot) of lead sulfide with complete passivation by oleic acid, oleyl amine and hydroxyl ligands (size ~5nm).

Semi-solid and soft nanoparticles have been manufactured. A prototype nanoparticle of semi-solid nature is the liposome. Various types of liposome nanoparticles are currently used clinically as delivery systems for anticancer drugs and vaccines.

Nanoparticles with one half hydrophilic and the other half hydrophobic are termed Janus particles and are particularly effective for stabilizing emulsions. They can self-assemble at water/oil interfaces and act as solid surfactants.

Hydrogel nanoparticles made of N-isopropylacrylamide hydrogel core shell can be dyed with affinity baits, internally. These affinity baits allow the nanoparticles to isolate and remove undesirable proteins while enhancing the target analytes.

Synthesis

There are several methods for creating nanoparticles, including gas condensation, attrition, chemical precipitation, pyrolysis and hydrothermal synthesis. In attrition, macro- or micro-scale particles are ground in a ball mill, a planetary ball mill, or other size-reducing mechanism. The resulting particles are air classified to recover nanoparticles. In pyrolysis, a vaporous precursor (liquid or gas) is forced through an orifice at high pressure and burned. The resulting solid (a version of soot) is air classified to recover oxide particles from by-product gases. Traditional pyrolysis often results in aggregates and agglomerates rather than single primary particles. Ultrasonic nozzle spray pyrolysis (USP) on the other hand aids in preventing agglomerates from forming.

A thermal plasma can deliver the energy to vaporize small micrometer-size particles. The thermal plasma temperatures are in the order of 10,000 K, so that solid powder easily evaporates. Nanoparticles are formed upon cooling while exiting the plasma region. The main types of the thermal plasma torches used to produce nanoparticles are dc plasma jet, dc arc plasma, and radio frequency (RF) induction plasmas. In the arc plasma reactors, the energy necessary for evaporation and reaction is provided by an electric arc formed between the anode and the cathode. For example, silica sand can be vaporized with an arc plasma at atmospheric pressure, or thin aluminum wires can be vaporized by exploding wire method. The resulting mixture of plasma gas and silica vapour can be rapidly cooled by quenching with oxygen, thus ensuring the quality of the fumed silica produced.

In RF induction plasma torches, energy coupling to the plasma is accomplished through the electromagnetic field generated by the induction coil. The plasma gas does not come in contact with electrodes, thus eliminating possible sources of contamination and allowing the operation of such plasma torches with a wide range of gases including inert, reducing, oxidizing, and other corrosive atmospheres. The working frequency is typically between 200 kHz and 40 MHz. Laboratory units run at power levels in the order of 30–50 kW, whereas the large-scale industrial units have been tested at power levels up to 1 MW. As the residence time of the injected feed droplets in the plasma is very short, it is important that the droplet sizes are small enough in order to obtain complete evaporation. The RF plasma method has been used to synthesize different nanoparticle materials, for example synthesis of various ceramic nanoparticles such as oxides, carbours/carbides, and nitrides of Ti and Si.

Inert-gas condensation is frequently used to make nanoparticles from metals with low melting points. The metal is vaporized in a vacuum chamber and then supercooled with an inert gas stream. The supercooled metal vapor condenses into nanometer-size particles, which can be entrained in the inert gas stream and deposited on a substrate or studied in situ.

Nanoparticles can also be formed using radiation chemistry. Radiolysis from gamma rays can create strongly active free radicals in solution. This relatively simple technique uses a minimum number of chemicals. These including water, a soluble metallic salt, a radical scavenger (often a secondary alcohol), and a surfactant (organic capping agent). High gamma doses on the order of 10^4 Gray are required. In this process, reducing radicals will drop metallic ions down to the zero-valence state. A scavenger chemical will preferentially interact with oxidizing radicals to prevent the re-oxidation of the metal. Once in the zero-valence state, metal atoms begin to coalesce into particles. A chemical surfactant surrounds the particle during formation and regulates its growth. In sufficient concentrations, the surfactant molecules stay attached to the particle. This prevents it from dissociating or forming clusters with other particles. Formation of nanoparticles using the radiolysis method allows for tailoring of particle size and shape by adjusting precursor concentrations and gamma dose.

Sol-gel

The sol-gel process is a wet-chemical technique (also known as chemical solution deposition) widely used recently in the fields of materials science and ceramic engineering. Such methods are used primarily for the fabrication of materials (typically a metal oxide) starting from a chemical solution (*sol*, short for solution), which acts as the precursor for an integrated network (or *gel*) of either discrete particles or network polymers.

Typical precursors are metal alkoxides and metal chlorides, which undergo hydrolysis and polycondensation reactions to form either a network "elastic solid" or a colloidal suspension (or dispersion) – a system composed of discrete (often amorphous) submicrometer particles dispersed to various degrees in a host fluid. Formation of a metal oxide involves connecting the metal centers with oxo (M-O-M) or hydroxo (M-OH-M) bridges, therefore generating metal-oxo or metal-hydroxo polymers in solution. Thus, the sol evolves toward the formation of a gel-like diphasic system containing both a liquid phase and solid phase whose morphologies range from discrete particles to continuous polymer networks.

In the case of the colloid, the volume fraction of particles (or particle density) may be so low that a significant amount of fluid may need to be removed initially for the gel-like properties to be recognized. This can be accomplished in a number of ways. The most simple method is to allow time for sedimentation to occur, and then pour off the remaining liquid. Centrifugation can also be used to accelerate the process of phase separation.

Removal of the remaining liquid (solvent) phase requires a drying process, which typically causes shrinkage and densification. The rate at which the solvent can be removed is ultimately determined by the distribution of porosity in the gel. The ultimate microstructure of the final component will clearly be strongly influenced by changes implemented during this phase of processing. Afterward, a thermal treatment, or firing process, is often necessary in order to favor further polycondensation and enhance mechanical properties and structural stability via final sintering, densification, and grain growth. One of the distinct advantages of using this methodology as opposed to the more traditional processing techniques is that densification is often achieved at a much lower temperature.

The precursor sol can be either deposited on a substrate to form a film (e.g., by dip-coating or spin-coating), cast into a suitable container with the desired shape (e.g., to obtain a monolithic ceramics, glasses, fibers, membranes, aerogels), or used to synthesize powders (e.g., microspheres, nanospheres). The sol-gel approach is a cheap and low-temperature technique that allows for the fine control of the product's chemical composition. Even small quantities of dopants, such as organic dyes and rare earth metals, can be introduced in the sol and end up uniformly dispersed in the final product. It can be used in ceramics processing and manufacturing as an investment casting material, or as a means of producing very thin films of metal oxides for various purposes. Sol-gel derived materials have diverse applications in optics, electronics, energy, space, (bio)sensors, medicine (e.g., controlled drug release) and separation (e.g., chromatography) technology.

Colloids

The term colloid is used primarily to describe a range of mixtures which have solid particles dispersed in a liquid medium. The term applies only if the particles are larger than atomic dimensions but small enough to exhibit Brownian motion. If the particles are large enough, their dynamic behavior in any given period of time in suspension would be governed by forces of gravity and sedimentation. If the particles are small enough, their irregular motion in suspension can be attributed to the collective bombardment of a myriad of thermally agitated molecules in the liquid suspending medium, as described originally by Albert Einstein in his dissertation. Einstein showed evidence that water was made up of discrete molecules by characterizing the observed erratic particle behavior using the theory of Brownian motion, with sedimentation being a possible result. The critical size range (or particle diameter) typically ranges from nanometers (10^{-9} m) to micrometers (10^{-6} m).

Morphology

Nanostars of vanadium(IV) oxide

Scientists have taken to naming their particles after the real-world shapes that they might represent. Nanospheres, nanochains, nanoreefs, nanoboxes and more have appeared in the literature. These morphologies sometimes arise spontaneously as an effect of a templating or directing agent present in the synthesis such as miscellar emulsions or anodized alumina pores, or from the innate crystallographic growth patterns of the materials themselves. Some of these morphologies may serve a purpose, such as long carbon nanotubes used to bridge an electrical junction, or just a scientific curiosity like the stars shown.

Amorphous particles usually adopt a spherical shape (due to their microstructural isotropy), whereas the shape of anisotropic microcrystalline whiskers corresponds to their particular crystal habit. At the small end of the size range, nanoparticles are often referred to as clusters. Spheres, rods, fibers, and cups are just a few of the shapes that have been grown. The study of fine particles is called micromeritics.

Characterization

The majority of nanoparticle characterization techniques are light-based, but a non-optical nanoparticle characterization technique called Tunable Resistive Pulse Sensing (TRPS) has been developed that enables the simultaneous measurement of size, concentration and surface charge for a wide variety of nanoparticles. This technique, which applies the Coulter Principle, allows for particle-by-particle quantification of these three nanoparticle characteristics with high resolution.

Functionalization

The surface coating of nanoparticles determines many of their properties, notably stability, solubility, and targeting. A coating that is multivalent or polymeric confers high stability. Functionalized nanomaterial-based catalysts can be used for catalysis of many known organic reactions.

Surface Coating for Biological Applications

For biological applications, the surface coating should be polar to give high aqueous solubility and prevent nanoparticle aggregation. In serum or on the cell surface, highly charged coatings promote non-specific binding, whereas polyethylene glycol linked to terminal hydroxyl or methoxy groups repel non-specific interactions. Nanoparticles can be linked to biological molecules that can act as address tags, to direct the nanoparticles to specific sites within the body, specific organelles within the cell, or to follow specifically the movement of individual protein or RNA molecules in living cells. Common address tags are monoclonal antibodies, aptamers, streptavidin or peptides. These targeting agents should ideally be covalently linked to the nanoparticle and should be present in a controlled number per nanoparticle. Multivalent nanoparticles, bearing multiple targeting groups, can cluster receptors, which can activate cellular signaling pathways, and give stronger anchoring. Monovalent nanoparticles, bearing a single binding site, avoid clustering and so are preferable for tracking the behavior of individual proteins.

Red blood cell coatings can help nanoparticles evade the immune system.

Health and Safety

Nanoparticles present possible dangers, both medically and environmentally. Most of these are due to the high surface to volume ratio, which can make the particles very reactive or catalytic. They are also able to pass through cell membranes in organisms, and their interactions with biological systems are relatively unknown. However, it is unlikely the particles would enter the cell nucleus, Golgi complex, endoplasmic reticulum or other internal cellular components due to the particle size and intercellular agglomeration. A recent study looking at the effects of ZnO nanoparticles on human immune cells has found varying levels of susceptibility to cytotoxicity. There are concerns that pharmaceutical companies, seeking regulatory approval for nano-reformulations of existing medicines, are relying on safety data produced during clinical studies of the earlier, pre-reformulation version of the medicine. This could result in regulatory bodies, such as the FDA, missing new side effects that are specific to the nano-reformulation.

Whether cosmetics and sunscreens containing nanomaterials pose health risks remains largely unknown at this stage. However considerable research has demonstrated that zinc nanoparticles are not absorbed into the bloodstream in vivo.

Concern has also been raised over the health effects of respirable nanoparticles from certain combustion processes. As of 2013 the U.S. Environmental Protection Agency was investigating the safety of the following nanoparticles:

- Carbon Nanotubes: Carbon materials have a wide range of uses, ranging from composites for use in vehicles and sports equipment to integrated circuits for electronic components. The interactions between nanomaterials such as carbon nanotubes and natural organic matter strongly influence both their aggregation and deposition, which strongly affects their transport, transformation, and exposure in aquatic environments. In past research, carbon nanotubes exhibited some toxicological impacts that will be evaluated in various environmental settings in current EPA chemical safety research. EPA research will provide data, models, test methods, and best practices to discover the acute health effects of carbon nanotubes and identify methods to predict them.
- Cerium oxide: Nanoscale cerium oxide is used in electronics, biomedical supplies, energy, and fuel additives. Many applications of engineered cerium oxide nanoparticles naturally disperse themselves into the environment, which increases the risk of exposure. There is ongoing exposure to new diesel emissions using fuel additives containing CeO_2 nanoparticles, and the environmental and public health impacts of this new technology are unknown. EPA's chemical safety research is assessing the environmental, ecological, and health implications of nanotechnology-enabled diesel fuel additives.
- Titanium dioxide: Nano titanium dioxide is currently used in many products. Depending on the type of particle, it may be found in sunscreens, cosmetics, and paints and coatings. It is also being investigated for use in removing contaminants from drinking water.
- Nano Silver: Nano silver is being incorporated into textiles, clothing, food packaging, and other materials to eliminate bacteria. EPA and the U.S. Consumer Product Safety Commission are studying certain products to see whether they transfer nano-size silver particles in real-world scenarios. EPA is researching this topic to better understand how much nano-silver children come in contact with in their environments.
- Iron: While nano-scale iron is being investigated for many uses, including "smart fluids" for uses such as optics polishing and as a better-absorbed iron nutrient supplement, one of its more prominent current uses is to remove contamination from groundwater. This use, supported by EPA research, is being piloted at a number of sites across the country.

Laser Applications

The use of nanoparticles in laser dye-doped poly(methyl methacrylate) (PMMA) laser gain media was demonstrated in 2003 and it has been shown to improve conversion efficiencies and to decrease laser beam divergence. Researchers attribute the reduction in beam divergence to improved dn/dT characteristics of the organic-inorganic dye-doped nanocomposite. The optimum composition reported by these researchers is 30% w/w of SiO_2 (~ 12 nm) in dye-doped PMMA.

Medicinal Applications

- Liposome
- Dendrimer
- Iron oxide nanoparticles
- Nanomedicine
- Polymer-drug conjugate
- Polymeric nanoparticle

Magnetic Nanoparticles

Magnetic nanoparticles are a class of nanoparticle that can be manipulated using magnetic fields. Such particles commonly consist of two components, a magnetic material, often iron, nickel and cobalt, and a chemical component that has functionality. While nanoparticles are smaller than 1 micrometer in diameter (typically 5–500 nanometers), the larger microbeads are 0.5–500 micrometer in diameter. Magnetic nanoparticle clusters that are composed of a number of individual magnetic nanoparticles are known as magnetic nanobeads with a diameter of 50–200 nanometers. Magnetic nanoparticle clusters are a basis for their further magnetic assembly into magnetic nanochains. The magnetic nanoparticles have been the focus of much research recently because they possess attractive properties which could see potential use in catalysis including nanomaterial-based catalysts, biomedicine and tissue specific targeting, magnetically tunable colloidal photonic crystals, microfluidics, magnetic resonance imaging, magnetic particle imaging, data storage, environmental remediation, nanofluids, and optical filters, defect sensor and cation sensors.

Properties

The physical and chemical properties of magnetic nanoparticles largely depend on the synthesis method and chemical structure. In most cases, the particles range from 1 to 100 nm in size and may display superparamagnetism.

Types of Magnetic Nanoparticles

Oxides: Ferrites

Ferrite nanoparticles or iron oxide nanoparticles (iron oxides in crystal structure of maghemite or magnetite) are the most explored magnetic nanoparticles up to date. Once the ferrite particles become smaller than 128 nm they become superparamagnetic which prevents self agglomeration since they exhibit their magnetic behavior only when an external magnetic field is applied. The magnetic moment of ferrite nanoparticles can be greatly increased by controlled clustering of a number of individual superparamagnetic nanoparticles into superparamagnetic nanoparticle clusters, namely magnetic nanobeads. With the external magnetic field switched off, the remanence falls back to zero. Just like non-magnetic oxide nanoparticles, the surface of ferrite nanoparticles is often modified by surfactants, silica, silicones or phosphoric acid derivatives to increase their stability in solution.

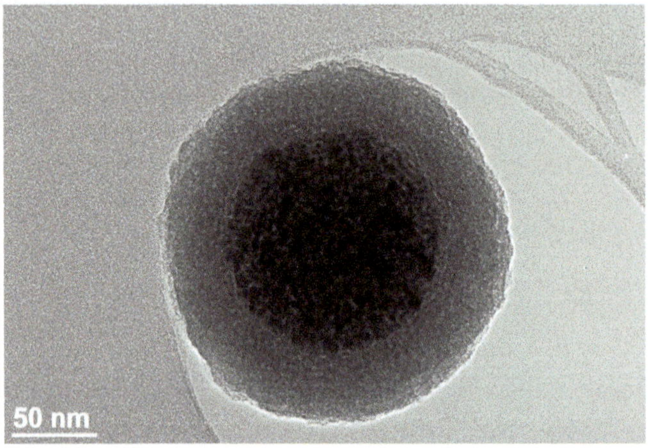

TEM image of a maghemite magnetic nanoparticle cluster with silica shell

Ferrites with a Shell

The surface of a maghemite or magnetite magnetic nanoparticle is relatively inert and does not usually allow strong covalent bonds with functionalization molecules. However, the reactivity of the magnetic nanoparticles can be improved by coating a layer of silica onto their surface. The silica shell can be easily modified with various surface functional groups via covalent bonds between organo-silane molecules and silica shell. In addition, some fluorescent dye molecules can be covalently bonded to the functionalized silica shell.

Ferrite nanoparticle clusters with narrow size distribution consisting of superparamagnetic oxide nanoparticles (~ 80 maghemite superparamagnetic nanoparticles per bead) coated with a silica shell have several advantages over metallic nanoparticles:

- Higher chemical stability (crucial for biomedical applications)
- Narrow size distribution (crucial for biomedical applications)
- Higher colloidal stability since they do not magnetically agglomerate
- Magnetic moment can be tuned with the nanoparticle cluster size
- Retained superparamagnetic properties (independent of the nanoparticle cluster size)
- Silica surface enables straightforward covalent functionalization

Metallic

Metallic nanoparticles may be beneficial for some technical applications due to their higher magnetic moment whereas oxides (maghemite, magnetite) would be beneficial for biomedical applications. This also implies that for the same moment, metallic nanoparticles can be made smaller than their oxide counterparts. On the other hand, metallic nanoparticles have the great disadvantage of being pyrophoric and reactive to oxidizing agents to various degrees. This makes their handling difficult and enables unwanted side reactions which makes them less appropriate for biomedical applications. Colloid formation for metallic particles is also much more challenging.

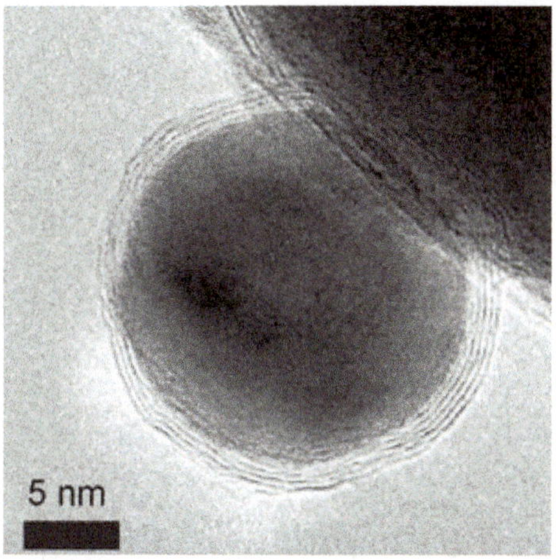

Cobalt nanoparticle with graphene shell (note: The individual graphene layers are visible)

Metallic with a Shell

The metallic core of magnetic nanoparticles may be passivated by gentle oxidation, surfactants, polymers and precious metals. In an oxygen environment, Co nanoparticles form an anti-ferromagnetic CoO layer on the surface of the Co nanoparticle. Recently, work has explored the synthesis and exchange bias effect in these Co core CoO shell nanoparticles with a gold outer shell. Nanoparticles with a magnetic core consisting either of elementary Iron or Cobalt with a nonreactive shell made of graphene have been synthesized recently. The advantages compared to ferrite or elemental nanoparticles are:

- Higher magnetization
- Higher stability in acidic and basic solution as well as organic solvents
- Chemistry on the graphene surface via methods already known for carbon nanotubes

Synthesis

Several methods exist for preparing magnetic nanoparticle.

Co-precipitation

Co-precipitation is a facile and convenient way to synthesize iron oxides (either Fe_3O_4 or γ-Fe_2O_3) from aqueous Fe^{2+}/Fe^{3+} salt solutions by the addition of a base under inert atmosphere at room temperature or at elevated temperature. The size, shape, and composition of the magnetic nanoparticles very much depends on the type of salts used (e.g.chlorides, sulfates, nitrates), the Fe^{2+}/Fe^{3+} ratio, the reaction temperature, the pH value and ionic strength of the media, and the mixing rate with the base solution used to provoke the precipitation. The co-precipitation approach has been used extensively to produce ferrite nanoparticles of controlled sizes and magnetic properties. A variety of experimental arrangements have been reported to facilitate continuous and large–scale co–precipitation of magnetic particles by rapid mixing. Recently, the growth rate of the magnetic nanoparticles was measured in real-time during the precipitation of magnetite nanoparticles by an integrated AC magnetic susceptometer within the mixing zone of the reactants.

Thermal Decomposition

Magnetic nanocrystals with smaller size can essentially be synthesized through the thermal decomposition of alkaline organometallic compounds in high-boiling organic solvents containing stabilizing surfactants.

Microemulsion

Using the microemulsion technique, metallic cobalt, cobalt/platinum alloys, and gold-coated cobalt/platinum nanoparticles have been synthesized in reverse micelles of cetyltrimethlyammonium bromide, using 1-butanol as the cosurfactant and octane as the oil phase.

Flame Spray Synthesis

Using flame spray pyrolysis and varying the reaction conditions, oxides, metal or carbon coated nanoparticles are produced at a rate of > 30 g/h.

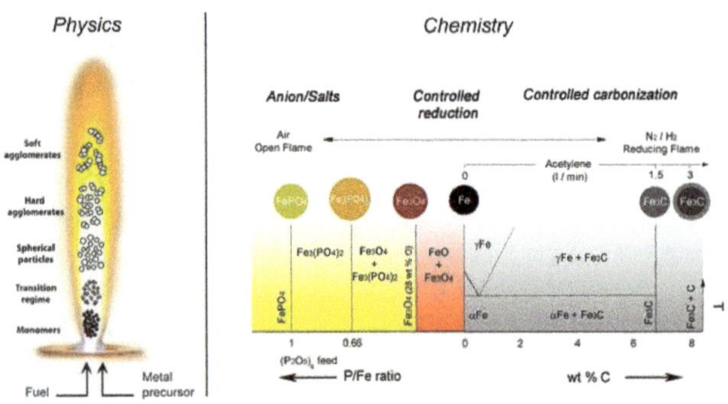

Various flame spray conditions and their impact on the resulting nanoparticles.

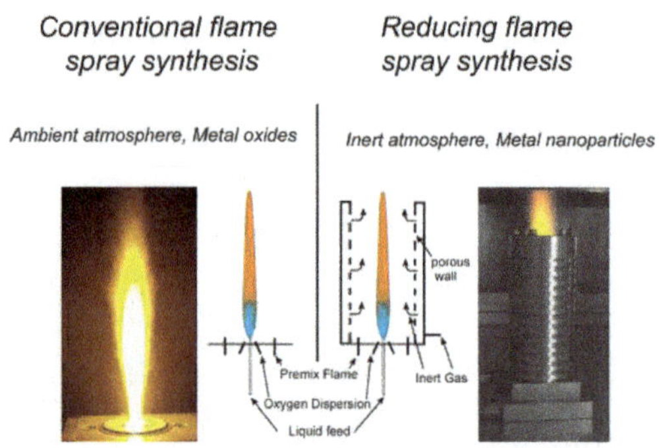

Operational layout differences between conventional and reducing flame spray synthesis.

Potential Applications

A wide variety of potential applications have been envisaged. Since magnetic nanoparticles are expensive to produce, there is interest in their recycling or for highly specialized applications. They are only used in scientific research. An industrial use has yet to be established.

The potential and versatility of magnetic chemistry arises from the fast and easy separation of the magnetic nanoparticles, eliminating tedious and costly separation processes usually applied in chemistry. Furthermore, the magnetic nanoparticles can be guided via a magnetic field to the desired location which could, for example, enable pinpoint precision in fighting cancer.

Medical Diagnostics and Treatments

Magnetic nanoparticles have been examined for use in an experimental cancer treatment called magnetic hyperthermia in which an alternating magnetic field (AMF) is used to heat the nanoparticles. To achieve sufficient magnetic nanoparticle heating, the AMF typically has a frequency between 100–500 kHz, although significant research has been done at lower frequencies as well as frequencies as high as 10 MHz, with the amplitude of the field usually between 8-16kAm^{-1}.

Affinity ligands such as epidermal growth factor (EGF), folic acid, aptamers, lectins etc. can be attached to the magnetic nanoparticle surface with the use of various chemistries. This enables targeting of magnetic nanoparticles to specific tissues or cells. This strategy is used in cancer research to target and treat tumors in combination with magnetic hyperthermia or nanoparticle-delivered cancer drugs. Despite research efforts, however, the accumulation of nanoparticles inside of cancer tumors of all types is sub-optimal, even with affinity ligands. Willhelm et al. conducted a broad analysis of nanoparticle delivery to tumors and concluded that the median amount of injected dose reaching a solid tumor is only 0.7%. The challenge of accumulating large amounts of nanoparticles inside of tumors is arguably the biggest obstacle facing nanomedicine in general. While direct injection is used in some cases, intravenous injection is most often preferred to obtain a good distribution of particles throughout the tumor. Magnetic nanoparticles have a distinct advantage in that they can accumulate in desired regions via magnetically guided delivery, although this technique still needs further development to achieve optimal delivery to solid tumors.

Another potential treatment of cancer includes attaching magnetic nanoparticles to free-floating cancer cells, allowing them to be captured and carried out of the body. The treatment has been tested in the laboratory on mice and will be looked at in survival studies.

Magnetic nanoparticles can be used for the detection of cancer. Blood can be inserted onto a microfluidic chip with magnetic nanoparticles in it. These magnetic nanoparticles are trapped inside due to an externally applied magnetic field as the blood is free to flow through. The magnetic nanoparticles are coated with antibodies targeting cancer cells or proteins. The magnetic nanoparticles can be recovered and the attached cancer-associated molecules can be assayed to test for their existence.

Magnetic nanoparticles can be conjugated with carbohydrates and used for detection of bacteria. Iron oxide particles have been used for the detection of Gram negative bacteria like Escherichia coli and for detection of Gram positive bacteria like Streptococcus suis.

Magnetic Immunoassay

Magnetic immunoassay (MIA) is a novel type of diagnostic immunoassay utilizing magnetic nanobeads as labels in lieu of conventional, enzymes, radioisotopes or fluorescent moieties. This assay involves the specific binding of an antibody to its antigen, where a magnetic label is conjugated to one element of the pair. The presence of magnetic nanobeads is then detected by a magnetic reader (magnetometer) which measures the magnetic field change induced by the beads. The signal measured by the magnetometer is proportional to the analyte (virus, toxin, bacteria, cardiac marker,etc.) quantity in the initial sample.

Waste Water Treatment

Thanks to the easy separation by applying a magnetic field and the very large surface to volume ratio, magnetic nanoparticles have a potential for treatment of contaminated water. In this method, attachment of EDTA-like chelators to carbon coated metal nanomagnets results in a magnetic reagent for the rapid removal of heavy metals from solutions or contaminated water by three orders of magnitude to concentrations as low as micrograms per Litre. Magnetic nanobeads or nanoparticle clusters composed of FDA-approved oxide superparamagnetic nanoparticles (e.g. maghemite, magnetite) hold much potential for waste water treatment since they express excellent biocompatibility which concerning the environmental impacts of the material is an advantage compared to metallic nanoparticles.

Supported Enzymes and Peptides

Enzymes, proteins, and other biologically and chemically active substances have been immobilized on magnetic nanoparticles. They are of interest as possible supports for solid phase synthesis.

This technology is potentially relevant to cellular labelling/cell separation, detoxification of biological fluids, tissue repair, drug delivery, magnetic resonance imaging, hyperthermia and magnetofection.

Catalyst Support

Magnetic nanoparticles are of potential use as a catalyst or catalyst supports. In chemistry, a catalyst support is the material, usually a solid with a high surface area, to which a catalyst is affixed. The reactivity of heterogeneous catalysts occurs at the surface atoms. Consequently, great effort is made to maximize the surface area of a catalyst by distributing it over the support. The support may be inert or participate in the catalytic reactions. Typical supports include various kinds of carbon, alumina, and silica. Immobilizing the catalytic center on top of nanoparticles with a large surface to volume ratio addresses this problem. In the case of magnetic nanoparticles it adds the property of facile a separation. An early example involved a rhodium catalysis attached to magnetic nanoparticles.

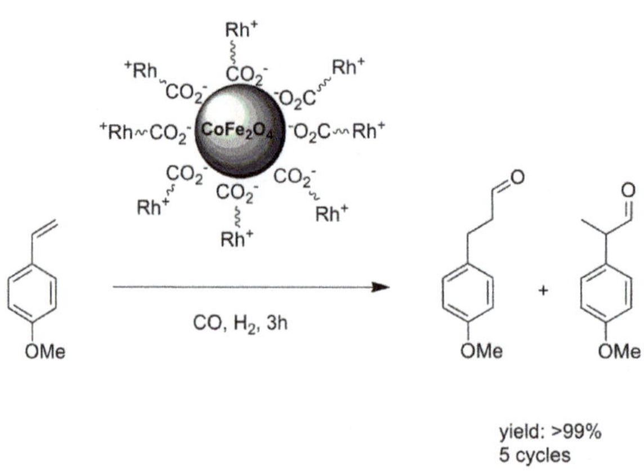

In another example, the stable radical TEMPO was attached to the graphene-coated cobalt nanoparticles via a diazonium reaction. The resulting catalyst was then used for the chemoselective oxidation of primary and secondary alcohols.

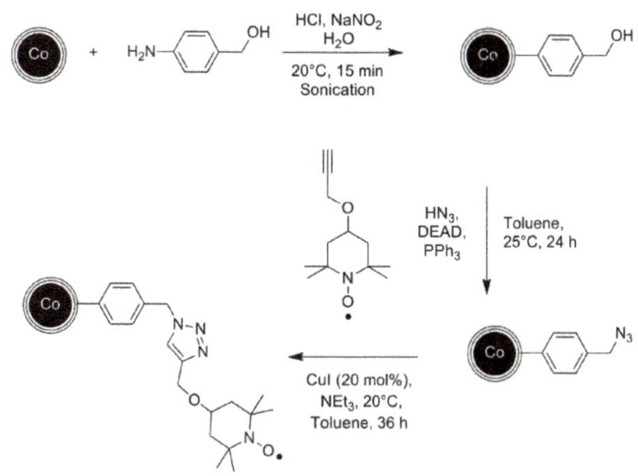

The catalytic reaction can be conducted in a continuous flow reactor instead of a batch reactor with no remains of the catalyst in the end product. Graphene coated cobalt nanoparticles have been used for that experiment since they exhibit a higher magnetization than Ferrite nanoparticles, which is essential for a fast and clean separation via external magnetic field.

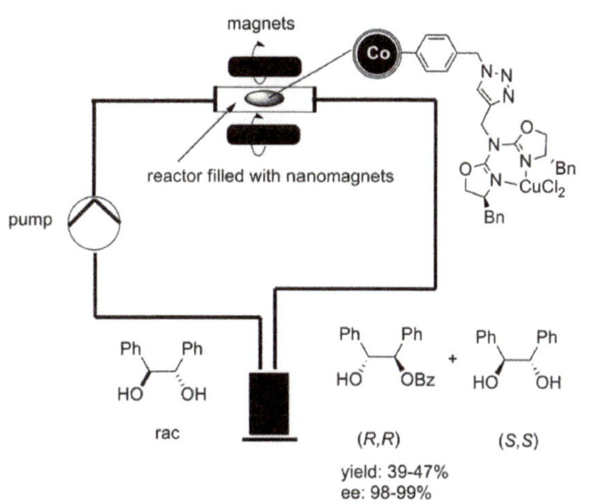

Biomedical Imaging

There are many applications for iron-oxide based nanoparticles in concert with magnetic resonance imaging. Magnetic CoPt nanoparticles are being used as an MRI contrast agent for transplanted neural stem cell detection.

Cancer Therapy

In magnetic fluid hyperthermia, nanoparticles of different types like Iron oxide, magnetite, maghemite or even gold are injected in tumor and then subjected under a high frequency magnetic field. These nanoparticles produce heat that typically increases tumor temperature to 40-46 °C, which can kill cancer cells. Another major potential of magnetic nanoparticles is the ability to combine heat (hyperthermia) and drug release for a cancer treatment. Numerous studies have shown particle constructs that can be loaded with a drug cargo and magnetic nanoparticles. The most prevalent construct is the "Magnetoliposome", which is a liposome with magnetic nanoparticles typically embedded in the lipid bilayer. Under an alternating magnetic field, the magnetic nanoparticles are heated, and this heat permeabilizes the membrane. This causes release of the loaded drug. This treatment option has a lot of potential as the combination of hyperthermia and drug release is likely to treat tumors better than either option alone, but it is still under development.

Information Storage

A promising candidate for high-density storage is the face-centered tetragonal phase FePt alloy. Grain sizes can be as small as 3 nanometers. If it's possible to modify the MNPs at this small scale, the information density that can be achieved with this media could easily surpass 1 Terabyte per square inch.

Genetic Engineering

Magnetic nanoparticles can be used for a variety of genetics applications. One application is the rapid isolation of mRNA. In one application, the magnetic bead is attached to a poly T tail. When mixed with mRNA, the poly A tail of the mRNA will attach to the bead's poly T tail and the isolation takes place simply by placing a magnet on the side of the tube and pouring out the liquid. Magnetic beads have also been used in plasmid assembly. Rapid genetic circuit construction has been achieved by the sequential addition of genes onto a growing genetic chain, using nanobeads as an anchor. This method has been shown to be much faster than previous methods, taking less than an hour to create functional multi-gene constructs in vitro.

Physical Modeling

There are a variety of mathematical models to describe the dynamics of the rotations of magnetic nanoparticles. Simple models include the Langevin function and the Stoner-Wohlfarth model which describe the magnetization of a nanoparticle at equilibrium. The Debye/Rosenszweig model can be used for low amplitude or high frequency oscillations of particles, which assumes linear response of the magnetization to an oscillating magnetic field. Non-equilibrium approaches include the Langevin equation formalism and the Fokker-Planck equation formalism, and these have been developed extensively to model applications such as magnetic nanoparticle hyperthermia, magnetic nanoparticle imaging (MPI), magnetic spectroscopy and biosensing etc.

Radioactive Nanoparticle

A radioactive nanoparticle is a nanoparticle that contains radioactive materials. Radioactive nanoparticles have applications in medical diagnostics, medical imaging, toxicokinetics, and environmental health, and are being investigated for applications in nuclear nanomedicine. Radioactive nanoparticles present special challenges in operational health physics and internal dosimetry that are not present for other substances, although existing radiation protection measures and hazard controls for nanoparticles generally apply.

Types and Applications

Engineered

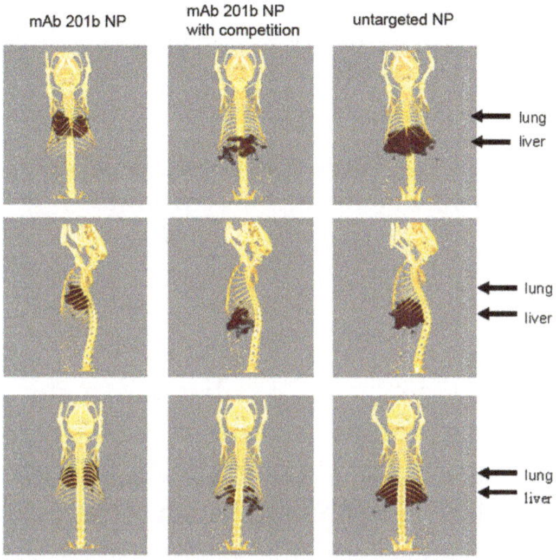

SPECT/CT images of injected gold-coated lanthanum/gadolinium phosphate nanoparticles containing the alpha-emitting radionuclide actinium-225 in a mouse. Depending on the surface functionalization, the particles migrate either to the lungs or the liver.

Engineered radioactive nanoparticles are used in medical imaging techniques such as positron emission tomography and single-photon emission computed tomography, and an aerosol of carbon nanoparticles containing technetium-99m are used in a commercially available procedure for ventilation/perfusion scintigraphy of the lungs. Engineered radioactive nanoparticles are also used as a radiolabel to detect the presence of the nanoparticles themselves in environmental health and toxicokinetics studies.

Engineered radioactive nanoparticles are being investigated for therapeutic use combining nuclear medicine with nanomedicine, especially for cancer. Neutron capture therapy is one such potential application. In addition, nanoparticles can help to sequester the toxic daughter nuclides of alpha emitters when used in radiotherapy.

Nuclear imaging is non-invasive and has high sensitivity, and nanoparticles are useful as a platform for combining multiple copies of targeting vectors and effectors in order to selectively deliver radioisotopes to a specific region of interest. Other benefits of nanoparticles for diagnostic and therapeutic use include increased blood and tumor retention time, as well as the possibility of using their unique physical and chemical properties in treatment. However, the nanoparticles must be engineered to avoid being recognized by the mononuclear phagocyte system and transported to the liver or spleen, often through manipulating their surface functionalization.

Targeting techniques include functionalizing radioactive nanoparticles with antibodies to target them to a specific tissue, and using magnetic nanoparticles that are attracted to a magnet placed over the tumor site. Technetium-99m, indium-111, and iodine-131 are common radioisotopes used for these purposes, with many others used as well. Radioactive nanoparticles can be produced by either synthesizing the nanoparticles directly from the radioactive materials, or by irradiating non-radioactive particles with neutrons or accelerated ions, sometimes *in situ*.

Natural and Incidental

As with all nanoparticles, radioactive nanoparticles can also be naturally occurring or incidentally produced as a byproduct of industrial processes. The main source of naturally occurring nanomaterials containing radionuclides is the decay of radon gas, whose immediate decay products are non-gaseous elements that precipitate into nanoscale particles along with atmospheric dust and vapors. Minor natural sources include primordial radionuclides present in the nanoscale portion of volcanic ash, and primordial and cosmogenic nuclides taken up by plants which are later burned. Radioactive nanoparticles may be incidentally produced by procedures in the nuclear industry such as nuclear reprocessing and the cutting of contaminated objects.

Health and Safety

Radioactive nanoparticles combine the hazards of radioactive materials with the hazards of nanomaterials. Inhalation exposure is the most common route of exposure to airborne particles in the workplace. Animal studies on some classes of nanoparticles indicate pulmonary effects including inflammation, granulomas, and pulmonary fibrosis, which were of similar or greater potency when compared with other known fibrogenic materials such as silica, asbestos, and ultrafine carbon black. Some studies in cells or animals have shown genotoxic or carcinogenic effects, or systemic cadiovascular effects from pulmonary exposure. The hazards of ionizing radiation depend on whether the exposure is acute or chronic, and includes effects like radiation-induced cancer and teratogenesis. In some cases, the inherent physicochemical toxicity of the nanoparticle itself may lead to lower exposure limits than those associated with the radioactivity alone, which is not the case with most radioactive materials.

Radioactive nanoparticles present special challenges in operational health physics and internal dosimetry that are not present for other substances, as the nanoparticles' toxicokinetics depend on their physical and chemical properties including size, shape, and surface chemistry. For example, inhaled nanoparticles will deposit in different locations in the lungs, and will be metabolized and transported through the body differently, than vapors or larger particles. There may also be hazards from associated processes such as strong magnetic fields and cryogens used in imaging equipment, and handling of lab animals in experimental studies. Effective risk assessment and communication is important, as both nanotechnology and radiation have unique considerations with public perception.

Hazard Controls

In general, most elements of a standard radiation protection program are applicable to radioactive nanomaterials, and many hazard controls for nanomaterials will be effective with the radioactive versions. The hierarchy of hazard controls encompasses a succession of five categories of control methods to reduce the risk of illness or injury. The two most effective are elimination and substitution, for example reducing dust exposure by eliminating a sonication process or substituting a nanomaterial slurry or suspension in a liquid solvent instead of a dry powder. Substitutions should consider both the radioactivity and physicochemical hazards of all the options, and also take into account that radioactive nanomaterials are easier to detect than non-radioactive substances.

A fume hood is an engineering control typically used to protect workers using nanoparticles

Engineering controls should be the primary form of protection, including local exhaust systems such as fume hoods, gloveboxes, biosafety cabinets, and vented balance enclosures; radiation shielding; and access control systems. The need for negative room pressure to prevent contamination of outside areas can conflict with the customary use of positive pressure when pharmaceuticals are being handled, although this can be overcome through use of a cascade pressure system, or by handling nanomaterials in enclosures.

Administrative controls include procedures to limit radiation doses, and contamination control procedures including encouraging good work practices and monitoring for contamination. Personal protective equipment is the least effective and should be used in conjunction with other hazard controls. In general, personal protective equipment intended for radioactive materials should be effective with radioactive nanomaterials, including impervious laboratory coats, goggles, safety gloves, and in some cases respirators, although the greater potential penetration through clothing and mobility in air of nanoparticles should be taken into account.

Nanosponges

Nanosponges are nano sized particles designed to look like red blood cell and protect the body. Nanoparticles have been used for unknown preventative measures, such as silverware and wound care products. Silverware contains small silver nanoparticles which contain an antimicrobial agent. Antimicrobial agents are able to prevent the spread of disease. When silverware was created that was not the intention of the silver product. In more recent years, many wound care products have been made with silver and other various nanoparticles. The universal use of silver in wound care products is mainly for the antimicrobial properties, which helps treat wounds of all severity. They are very minute particles used for a vast majority of things, one including healthcare. A specific type of nanoparticle seen recently in the healthcare field is a nanosponge. Nanosponges are nano sized particles designed to look like red blood cell and protect the body. They are also able to fight and prevent the spread of disease. Nanosponges target toxins, the toxins organize themselves to poke holes inside the red blood cell (nanosponge) membrane this method of poking holes is called pore-forming. After the toxins are attached to the nanosponge, it will lead to cell damage and cell death.

Nanosponges are introduced to the body by injection, and take the form of a red blood cell so that the bacteria or venom attacks it. Once it is attacked, it is trapped within the scaffolding of the nanosponge. After the nanosponge is full of toxins and cannot trap anymore it moves to the liver to filter out the toxins. Researchers are posed with is how to tackle all the different types of bacteria and venom, making a lot of different nanosponges for each specific bacteria and venom is nearly impossible. As of now they are focusing on toxins such as; E. Coli, MRSA, pneumonia, bee venom, snake venom and sea anemone venom. These venoms all use a pore-forming strategy, which is useful when dealing with nanosponges. A single nanosponge can capture many of the bacteria and venoms, instead of being tailored to each individually. Another obstacle researchers are faced with is the lifespan of the nanosponges. Once nanosponges are injected they can move rapidly through the blood system, and be found in the liver to be filtered out within hours. Researchers are working on a technique using hydrogel to coat the nanosponges to increase the life of the nanosponge and help them to remain stationary after injection. What essentially will happen is before the nanosponges are injected they will be coated in a hydrogel that with prolong life, and keep the nanosponges from moving around the body. Having a stationary location for all nanoparticles to reside can more effectively tackle specific regions affected with disease. It was found that 80 percent of nanosponges coated with a hydrogel lasted more than two days after injection. Only 20 percent of nanosponges not coated in the hydrogel lasted two hours after injection, and diffused to other locations in the body.

Nanosponges are known to move rapidly throughout the body, making them efficient in scanning and killing toxins, but irrelevant in tackling a stationary area of disease, or delivering drugs to a specific area. The hydrogel will help immensely in the longevity of the nanosponge, and will allow it to show its full potential in a serious situations. If the nanosponge follows healthcare into the future it can be a tremendous resource for preventative care. Imagine going to the doctor to receive a nanosponge shot to prevent you from acquiring some disease you may be prone to, or possibly a preventative measure before traveling out of the state or country. This technology could save many lives as being used as a preventative measure. Even when patients have to stay at the hospital for long periods of time, maybe due to a surgery or birth. These patients can be injected with nanosponges so they do not acquire Hospital Acquired Infections (HAI). This can reduce the cost to the patient but shortening their stay, and also the hospitals cost for less patients staying longer periods of time, due to a HAI.

Nanofluid

A nanofluid is a fluid containing nanometer-sized particles, called nanoparticles. These fluids are engineered colloidal suspensions of nanoparticles in a base fluid. The nanoparticles used in nanofluids are typically made of metals, oxides, carbides, or carbon nanotubes. Common base fluids include water, ethylene glycol and oil.

Nanofluids have novel properties that make them potentially useful in many applications in heat transfer, including microelectronics, fuel cells, pharmaceutical processes, and hybrid-powered engines, engine cooling/vehicle thermal management, domestic refrigerator, chiller, heat exchanger,in grinding, machining and in boiler flue gas temperature reduction. They exhibit enhanced thermal conductivity and the convective heat transfer coefficient compared to the base fluid. Knowledge of the rheological behaviour of nanofluids is found to be very critical in deciding their suitability for convective heat transfer applications Nanofluids also have special acoustical properties and in ultrasonic fields display additional shear-wave reconversion of an incident compressional wave; the effect becomes more pronounced as concentration increases.

In analysis such as computational fluid dynamics (CFD), nanofluids can be assumed to be single phase fluids. However, almost all of new academic paper use two-phase assumption. Classical theory of single phase fluids can be applied, where physical properties of nanofluid is taken as a function of properties of both constituents and their concentrations. An alternative approach simulates nanofluids using a two-component model.

The spreading of a nanofluid droplet is enhanced by the solid-like ordering structure of nanoparticles assembled near the contact line by diffusion, which gives rise to a structural disjoining pressure in the vicinity of the contact line. However, such enhancement is not observed for small droplets with diameter of nanometer scale, because the wetting time scale is much smaller than the diffusion time scale.

Synthesis

Nanofluids are produced by several techniques they are, 1.Direct Evaporation (1 step), 2.Gas condensation/dispersion (2 step), 3.Chemical vapour condensation (1 step), 4.Chemical precipitation (1 step). Several liquids including water, ethylene glycol, and oils have been used as base fluids. Although stabilization can be a challenge, on-going research indicates that it is possible. Nano-materials used so far in nanofluid synthesis include metallic particles, oxide particles, carbon nanotubes, graphene nano-flakes and ceramic particles.

Smart Cooling Nanofluids

Realizing the modest thermal conductivity enhancement in conventional nanofluids, a team of researchers at Indira Gandhi Centre for Atomic Research Centre, Kalpakkam developed a new class of magnetically polarizable nanofluids where the thermal conductivity enhancement up to 300% of basefluids is demonstrated. Fatty-acid-capped magnetite nanoparticles of different sizes (3-10 nm) have been synthesized for this purpose. It has been shown that both the thermal and rheological properties of such magnetic nanofluids are tunable by varying the magnetic field strength and orientation with respect to the direction of heat flow. Such response stimuli fluids are reversibly switchable and have applications in miniature devices such as micro- and nano-electromechanical systems. In 2013, Azizian et al. considered the effect of an external magnetic field on the convective heat transfer coefficient of water-based magnetite nanofluid experimentally under laminar flow regime. Up to 300% enhancement obtained at Re=745 and magnetic field gradient of 32.5 mT/mm. The effect of the magnetic field on the pressure drop was not as significant.

Nanoparticle Migration

In nanofluids, it is recognized that nanoparticles do not follow the fluid streamlines passively. In fact, there are some reasons that induce a slip velocity between the nanoparticles and the base fluid. Movements of nanoparticles has significant impact on rheological and thermophysical properties of the nanofluids. Therefore, investigating the nanoparticles motion is critical for evaluating the performance of nanoparticles inclusion to the base fluid as a heat transfer medium. Since the nanoparticles are very small (100 nm), Brownian and thermophoretic diffusivities are the main slip mechanisms in nanofluids, as Buongiorno declared. Brownian diffusion is due to random drifting of suspended nanoparticles in the base fluid which originates from continuous collisions among the nanoparticles and liquid molecules. Thermophoresis induces nanoparticle migration from warmer to colder region (in opposite direction of the temperature gradient), making a non-uniform nanoparticle volume fraction distribution.

In fact, theoretical models estimated that nanoparticles are non-homogeneously distributed. The level of non-uniformity is completely depend on thermal boundary conditions, the nanoparticle size, shape, and material. Rigorous readers encouraged to find more interesting results in open literature.

Response Stimuli Nanofluids for Sensing Applications

Researchers have invented a nanofluid-based ultrasensitive optical sensor that changes its colour on exposure to extremely low concentrations of toxic cations. The sensor is useful in detecting minute traces of cations in industrial and environmental samples. Existing techniques for monitoring cations levels in industrial and environmental samples are expensive, complex and time-consuming. The sensor is designed with a magnetic nanofluid that consists of nano-droplets with magnetic grains suspended in water. At a fixed magnetic field, a light source illuminates the nanofluid where the colour of the nanofluid changes depending on the cation concentration. This color change occurs within a second after exposure to cations, much faster than other existing cation sensing methods.

Such response stimulus nanofluids are also used to detect and image defects in ferromagnetic components. The photonic eye, as it has been called, is based on a magnetically polarizable nano-emulsion that changes colour when it comes into contact with a defective region in a sample. The device might be used to monitor structures such as rail tracks and pipelines.

Magnetically Responsive Photonic Crystals Nanofluids

Magnetic nanoparticle clusters or magnetic nanobeads with the size 80–150 nanometers form ordered structures along the direction of the external magnetic field with a regular interparticle spacing on the order of hundreds of nanometers resulting in strong diffraction of visible light in suspension.

Applications

Nanofluids are primarily used for their enhanced thermal properties as coolants in heat transfer equipment such as heat exchangers, electronic cooling system(such as flat plate) and radiators. Heat transfer over flat plate has been analyzed by many researchers. However, they are also useful for their controlled optical properties. Graphene based nanofluid has been found to enhance Polymerase chain reaction efficiency. Nanofluids in solar collectors is another application where nanofluids are employed for their tunable optical properties.

Magnetic-plasmonic Bifunctional Nanoparticles

Bifunctional plasmonic-magnetic nanoparticles (PM-NPs) consist of optical and magnetic components. They are useful for biomedical applications that require optical sensing/imaging/heating, magnetic stimulation/manipulation, or both of these functionalities. An example of these functionalities is that PM-NPs can attach to biological entities and separate them under an external magnetic field while simultaneously detecting their chemical nature via optical sensing. These functionalities are especially useful when studying tissues deep within an organism. Current available bifunctional PM-NPs are limited to quasi-zero-dimensional (0D) nanostructures. There are few reports of quasi-one-dimensional (1D) nanostructures, although 1D PM nanotubes provide advantages for biomedical applications that are not available with 0D nanostructures. A specific type of 1D PM nanotube has applications in targeted single-cell sensing.

SERS Characterization

A laser beam focused on two neighboring Ag nanoparticles can generate a high intensity electric field in the narrow junction due to LSP resonance as a result of coherent electron oscillation in the Ag nanoparticles. This junction is referred to as a "hotspot". If a molecule is in the vicinity of the hotspot, it's Raman scattering signals can be significantly amplified. This phenomenon is called surface enhanced Raman scattering (SERS). SERS EF is determined by the electric field intensity. Nanocapsules that have Ag nanoparticles on the inside and outside (double-layer nanotubes) exhibit 2x SERS intensity than those with a single layer of Ag nanoparticles.

Manufacturing SERS Nanocapsules and Arrays

Researchers at the University of Texas at Austin found a method for economically synthesizing SERS nanocapsules and creating arrays of the nanocapsules capable of ultrasensitive and position-predictable SERS detections. In doing so, the researchers also tackled two major problems previously associated with SERS sensing: 1. Lack of many hotspots with controlled gaps 2. Difficulty of assembling SERS probes at designated locations.

The nanocapsules created in this study comprised a 3-layer structure: the Ag/Ni/Ag nano core, a silica capsule, a uniformly distributed Ag nanoparticles on silica. Creating the nanocapsule works as follows: (1) The electrodeposition of a nanoporous anodized aluminum oxide and subsequent deposition of a Cu layer onto this membrane to create a three-electrode electrodeposition system; (2) building the nanowires via electrodeposition, starting at the bottom of the membrane's nanopores (the amount of electric charge passing through the circuit controls the length of the Ag/Ni/Ag nanowires; (3) dissolving the template membrane in NaOH and subsequently washing the nanowires via sonication and centrifuging in ethanol and deionized water; (4) coating nanocore in silica via hydrolysis of tetraethyl orthosilicate; (5) isolating Ag ions and then incubating with the silica-coated nanocapsule to reduce ionic Ag into metallic Ag nanoparticles on the silica.

In terms of functionality, the nanocore of Ag and Ni allows for manipulating the nanocapsule with electric tweezers. Ag/Ni/Ag nanocores prove necessary for steering/orientation via the AC field. The silica capsule serves to separate the nanocore from the Ag nanoparticles on the outside, preventing the plasmonic quenching effect. The Ag nanodots act as a hotspot layer (with an enhanced electric field) and perform the sensing task.

The researchers arranged the nanocapsules into arrays using magnets of Cr/Ni/Ag. The Cr anchored the magnet in substrate; the Ni worked in concert with the Ni in the nanocapsule core by providing an attractive magnetic field; the Au layer tuned the magnetic interaction force between nanomagnet and nanocapsule. The researchers used a quadruple electrode chip to create an array of nanocapsules. The AC and DC fields could be oriented parallel or perpendicular to each other in order to create the electric tweezers (the mechanism could transport capsules parallel or perpendicular to their orientations depending on whether or not AC and DC fields were directed parallel or perpendicular to one another). The Ag/Ni/Ag nanocores proved necessary for steering/orientation via the AC field (when removed no net torque resulted). The electric tweezers then moved the nanocapsules to their nanomagnet anchors, utilizing the attraction between Ni in capsule and magnet.

These nanocapsule arrays prove consistent, with the variation of Raman intensity between sensors +/-9%. This low error margin comes from the controlled sizing and spacing of Ag nanoparticles and the large number of such hotspots on the nanocapsule. Previously there existed only randomly placed hotspots, which produced more variation between nanocapsule sensitivity and therefore increased the error margin. The nanocapsules can detect Raman spectra at small concentrations (for example, 1,2-bi-(4-pyridyl) ethylene (BPE) at 10-12 M) and as concentration increases, the detection sensitivity increases logarithmically.

Plasmonic Simulation

SERS enhancement can be attributed to two factors: electric field (E-field) enhancement due to the plasmonic resonance of nanoparticles and chemical enhancement due to charge transfer between the molecules and metal particles. E-field enhancement is a major contributor to SERS. A group of metallic particles on the same surface can provide stronger E-field enhancement than that of single particles or dimers. A group of nanoparticles with controlled distributions further enhance the E field. Narrower junctions between nanoparticles exhibit a strong E field. For smaller nanoparticles, the normalized electric field increases as size increases. However, for nanoparticles of diameters >40 nm, E-field enhancement reduces. This is because the plasmonic resonance shifts to a longer wavelength. When a double-layer nanotube is placed on top of a glass substrate and excited by a surface normal beam, the hot-spot is at the bottom of the nanotube. This is because the bottom hot-spot is surrounded by silica, while the top is surrounded by air. In addition, the inter-particle coupling through the nanoparticle chain at the outer surface enhances the electric field at the bottom of the nanotube. Not only does the inner layer add more hot-spots for SERS sensing, but it also increases the intensity of the hot spots in the outer layer nanoparticles. However, when the core is filled with platinum, the opposite is observed. The hot spots at the top of the nanotube are enhanced by 2x, while those at the bottom are reduced. This is because at the top of the nanotube, light is reflected to enhance the hot spot, while no light is able to pass through the platinum, which reduces the hot spots on the bottom. A similar phenomenon occurs when the tube is filled with nickel. Thickness of the silica also affects enhancement of SERS. Near-field enhancement can be observed on nanotubes with a silica coating ranging from 70 to 150 nm. However, if the thickness is increased, near-field enhancement will disappear.

Magnetic Characterization

These 1D PM nanotubes also offer tunable magnetic properties for controlled manipulation. When the Ni segment is relatively large, the anisotropy direction is along the long axis. When the Ni segment is relatively low, magnetic anisotropy is transverse to the length of the nanotubes with zero remanence. This is a way of controlling the magnetic anisotropy, and can be used to affect the way the tubes align with a magnetic field.

Single-Cell Bioanalysis

Because of their various functionalities, 1D PM nanotubes can be used for various applications, including analysis of membrane composition of a single animal cell among many. Because of the bifunctionality of the nanotubes, it is possible to precisely transport the nanotube to a specific living cell, and analyze various features of that cell.

Nanoshell

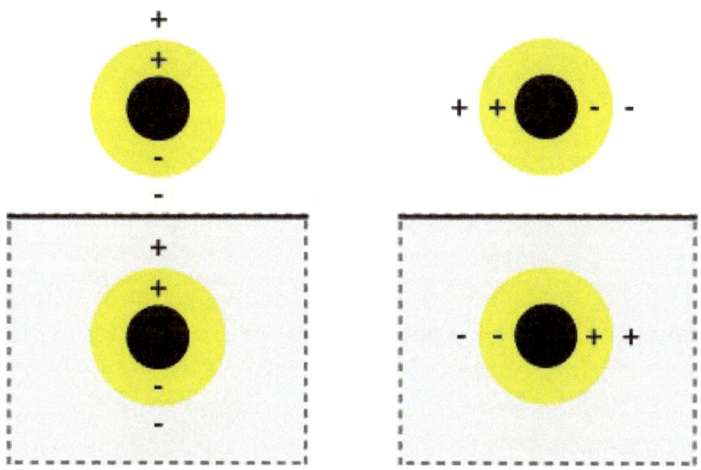

A nanoshell, or rather a nanoshell plasmon, is a type of spherical nanoparticle consisting of a dielectric core which is covered by a thin metallic shell (usually gold). These nanoshells involve a quasiparticle called a plasmon which is a collective excitation or quantum plasma oscillation where the electrons simultaneously oscillate with respect to all the ions.

The simultaneous oscillation can be called plasmon hybridization where the tunability of the oscillation is associated with mixture of the inner and outer shell where they hybridize to give a lower energy or higher energy. This lower energy couples strongly to incident light, whereas the higher energy is an anti-bonding and weakly combines to incident light. The hybridization interaction is stronger for thinner shell layers, hence, the thickness of the shell and overall particle radius determines which wavelength of light it couples with. Nanoshells can be varied across a broad range of the light spectrum that spans the visible and near infrared regions. The interaction of light and nanoparticles affects the placement of charges which affects the coupling strength. Incident light polarized parallel to the substrate gives a s-polarization, hence the charges are further from the substrate surface which gives a stronger interaction between the shell and core. Otherwise, a p-polarization is formed which gives a more strongly shifted plasmon energy causing a weaker interaction and coupling.

Discovery

The discovery of the nanoshell was made by Professor Naomi J. Halas and her team at Rice University in 2003. When she and her team discovered nanoshells, they weren't initially sure what potential such nanoshells held. "We said, 'Gee, what could it be good for?'" Halas told CNN. After many suggestions, cancer therapy came out of ongoing collaborations with bioengineers looking for different types of biomedical applications. "One of our visions", Halas stated, "no less than single visit diagnosis and treatment of cancer". In 2003 Halas was awarded for Best Discovery of 2003 by *Nanotechnology Now*.

Production

A state of the art method for synthesizing gold nanoshells is the use of the Microfluidic Composite Foams. This method has the potential to replace the standard lithographic method of synthesizing plasmonic nanoshells. The production process described below was an experiment performed by Suhanya Duraiswamy and Saif A. Khan of the Department of Chemical and Biomolecular Engineering in Singapore. Although this method was an experiment, it represents the future of nanoshells synthesis.

The materials required for the production of the nanoshells are the following; Tetraethyl orthosilicate, ammonium hydroxide, hydroxylamine hydrochloride, 3-aminopropyl tris, hydrogentetrachloroaurate(III) trihydrate, tetrakis(hydroxymethyl) phosphonium chloride, sodium hydroxide, potassium carbonate, ethanol, Ultrapure water and glassware washed in aqua regia and rinsed thoroughly in water.

The first step in synthesizing nanoshells in this method is by creating the device for the reaction to take place within. Microfluidic device patterns were fabricated onto silicon wafers by standard photolithography using negative photoresist SU-8 2050. Devices were subsequently molded in poly(dimethyl siloxane) (PDMS) using the soft lithography technique.(40) Briefly, PDMS was molded onto the SU-8 masters at 70 °C for 4 h, peeled, cut, and cleaned. Inlet and outlet holes (1/16-in. o.d.) were punched into the device. The microchannels were irreversibly bonded to a glass slide precoated with a thin layer of PDMS after a brief 35 s air plasma treatment. The microchannels have rectangular cross-section and are 300 μm wide, 155 μm deep, and 0.45 m long.

The actual production of the nanoparticles involves pumping "silicone oil, a mixture of gold-seeded silica particles and gold-plating solution and reducing agent solution to the microfluidic device while nitrogen gas was delivered from a cylinder." The plating solution was then left to age, in a controlled environment, for longer than 24 hours. After the aging process, the fluid is collected from the Microfluidic Device and placed in a centrifuge. The resulting liquid has a layer of oil on the surface with a solution below that contains the nanoshells.

The reason this method is revolutionary is that the size and relative thickness of the gold nanoshell can be controlled by changing the amount of time the reaction is allowed to take place as well as the concentration of the plating solution. Thereby allowing researchers to tailor the particles to suit their given needs. Albeit for optics or cancer treatment.

Cancer Treatment

Gold-shelled nanoparticles, which are spherical nanoparticles with silica cores and gold shells, are used in cancer therapy and bio-imaging enhancement. Theranostic probes – capable of detection and treatment of cancer in a single treatment - are nanoparticles that have binding sites on their shell that allow them to attach to a desired location (typically cancerous cells) then can be imaged through dual modality imagery (an imaging strategy that uses x-rays and radionuclide imaging) and through near-infrared fluorescence. The reason gold nanoparticles are used is due to their vivid optical properties which are controlled by their size, geometry, and their surface plasmons. Gold nanoparticles (such as AuNPs) have the benefit of being biocompatible and the flexibility to have multiple different molecules, and fundamental materials, attached to their shell (almost anything that can normally be attached to gold can be attached to the gold nano-shell, which can be used in helping identifying and treating cancer). The treatment of cancer is possible only because of the scattering and absorption that occurs for plasmonics. Under scattering, the gold-plated nano-particles become visible to imaging processes that are tuned to the correct wavelength which is dependent upon the size and geometry of the particles. Under absorption, photothermal ablation occurs, which heats the nanoparticles and their immediate surroundings to temperatures capable of killing the cancer cells. This is accomplished with minimal damage to cells in the body due to the utilization of the "water window" (the spectral range between 800 and 1300 nm). As the human body is mostly water, this optimizes the light used versus the effects rendered.

These gold nanoshells are shuttled into tumors by the use of phagocytosis, where phagocytes engulf the nanoshells through the cell membrane to form an internal phagosome, or macrophage. After this it is shuttled into a cell and enzymes are usually used to metabolize it and shuttle it back out of the cell. These nanoshells are not metabolized so for them to be effective they just need to be within the tumor cells and photo-induced cell death (as described above) is used to terminate the tumor cells.

Nanoparticle-based therapeutics have been successfully delivered into tumors by exploiting the enhanced permeability and retention effect, a property that permits nanoscale structures to be taken up passively into tumors without the assistance of antibodies. Delivery of nanoshells into the important regions of tumors can be very difficult. This is where most nanoshells try to exploit the tumor's natural recruitment of monocytes for delivery as seen in the figure. This delivery system is called a "Trojan Horse".

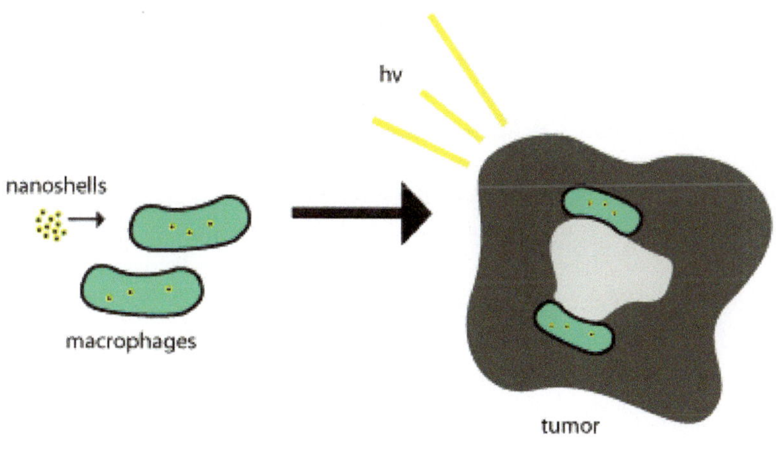

This process works so well since tumors are about ¾ macrophages and once monocytes are brought into the tumor, it differentiates into macrophages which would also be need to maintain the cargo nanoparticles. Once the nanoshells are at the necrotic center, near-infrared illumination is used to destroy the tumor associated macrophages.

Additionally, these nanoparticles can be made to release antisense DNA oligonucleotides when under photo-activation. These oligonucleotides are used in conjunction with the photo-thermal ablation treatments to perform gene-therapy. This is accomplished because nanoparticle complexes are delivered inside of cells then undergo light induced release of DNA from their surface. This will allow for the internal manipulation of a cell and provide a means for monitoring a group cells return to equilibrium.

Another example of nanoshell plasmonics in cancer treatment involves placing drugs inside of the nanoparticle and using it as a vehicle to deliver toxic drugs to cancerous sites only. This is accomplished by coating the outside of a nanoparticle with iron oxide (allowing for easy tracking with an MRI machine), then once the area of the tumor is coated with the drug-filled nanoparticles, the nanoparticles can be activated using resonant light waves to release the drug.

Nanosized Aluminium Oxide

Nanosized aluminium oxide (nanosized alumina) occurs in the form of spherical or nearly spherical nanoparticles, and in the form of oriented or undirected fibers.

Properties

Properties, of the final material, defined as the set of properties of the solid Aluminium oxide and specific properties of nanostructures.

Properties of Nanoscale Colloidal Alumina Particles

- Small diameter of the particles/fibers (2-10 nm)
- High specific surface area (>100 m^2/g)
- High defectiveness of the material surface and specific structure of the nanoparticles (the volume and size of pores, degree of crystallinity, phase composition, structure, and composition of the surface — modification possibility)

Properties of the Nanoscale Fibers of Aluminium Oxide

- The ratio of length-diameter about 20,000,000:1
- A high degree of orientation of the fibers
- The weak interaction of the fibers among themselves
- Absence of surface pores
- High surface concentration of hydroxyl groups

Production

Methods of obtaining Powders of Aluminium Oxide Nanometer Scale

Industrially obtained alumina nano fibers of Nafen brand

1. Grinding powder alumina particles of a nanometer level (for example, 10-50 nm). For example, using a planetary mill using grinding bodies of size less than 0.1 μm.

2. The decomposition of fresh chemically-synthesized AlOOH or $Al(OH)_3$ to aluminium oxide in the rapid achievement of the temperature of decomposition 175 °C and use for it the pressure of 5 bars within thirty minutes. The sooner of the temperature of decomposition of the hydroxo-compounds of aluminium is achieved, the smaller the resulting particles nano oxide in size.

Alumina Nano Fibers

The oxidation of the surface of some liquid metal alloys leads to the formation of loose or porous 3D nanostructures. For the first time this effect was observed in the system aluminium-mercury and published more than 100 years ago. The fibers of such kind do not occur in nature and only grown by artificial means. Depending on the method of synthesis can be produced various nanostructures, such as aerogel from oxyhydroxide aluminium (AlOOH or $Al_2O_3*nH_2O$, where n=1-4, are easily turned into aluminium oxide) or nano-fibers of aluminium oxide (Al_2O_3).

At this moment, the main ways of production are:

1. The method for selective oxidation of aluminium on the surface of the molten Ga-AI in a humid atmosphere at a temperature of 20 to 70 °C (Method of IPCE RAS).

2. The liquid metal technology of synthesis of nanostructural aerogel AlOOH from molten Ga-Bi and Al-Al (Institute of RF IPPE named after A. I. Leipunsky, Obninsk city).

3. Growing fiber nano oxide of aluminium on the surface of the aluminum melt (a Method of industrial synthesis, developed and patented by the ANF Technology).

Application

- Adsorbent (to capture hydrocarbon impurities from the air; for extracting fluorine from a variety of media (the ability of aluminum oxide to chemosensitivity fluorine ions used for the purification of water with increased fluorine content; for vapor recovery of hydrogen fluoride from gases of super phosphate and electrolysis) for brightening solutions in sugar production; to trap solvents; adsorption purification of oils (first transformer); adsorbent for gas and liquid adsorption chromatography (adsorption); for ion-exchange and sediment-sorption chromatography in aqueous solution (ion exchange and precipitation); as an inert carrier during liquid-distribution chromatography)
- Desiccant (for drying of gases (deep dehydration to dew points of -60 °C and below); the preservation of instruments and equipment, and also for such systems as the respiratory valves, tanks, transformers, etc.; to create protective atmospheres during long-term storage of food and pharmaceutical products)
- Sorbent of the ions of metals from solutions of their salts, for example, $CsNO_3$, $AgNO_3$, $Ba(NO_3)_2$, $Sr(NO_3)_2$, $Pb(NO_3)_2$, etc., with the possibility of obtaining of metal oxides on the surface of the fibers during annealing
- Sorbent of radionuclides from wastewaters of nuclear power plants
- Inert (reinforcing) filler
- Ceramics and composites (including composite metals) — high toughness, fire resistance and anti-friction properties, insulating properties. Known use in several products such as burner discharge lamp, the substrate of integrated circuits, shut-off elements ceramic pipeline valves, prostheses, etc.
- Abrasive (composed of means for ultra-fine polishing)
- Refractory (high-temperature component for heat insulation)

In addition to these areas, used as a catalyst and carrier of catalysts. Nanoscale oxide due to the small diameter of the particles/fibers, high specific surface area and activity associated with the defects, and the specific structure of the nanoparticles (the volume and size of pores, degree of crystallinity, phase composition, structure, and composition of the surface) strongly enhances the catalytic properties, and increases the range of massive aluminium oxide as a catalyst.

Nanoparticle–biomolecule Conjugate

A nanoparticle–biomolecule conjugate is a nanoparticle with biomolecules attached to its surface. Nanoparticles are minuscule particles, typically measured in nanometers (nm), that are used in nanobiotechnology to explore the functions of biomolecules. Properties of the ultrafine particles are characterized by the components on their surfaces more so than larger structures, such as cells, due to large surface area-to-volume ratios. Large surface area-to-volume-ratios of nanoparticles optimize the potential for interactions with biomolecules.

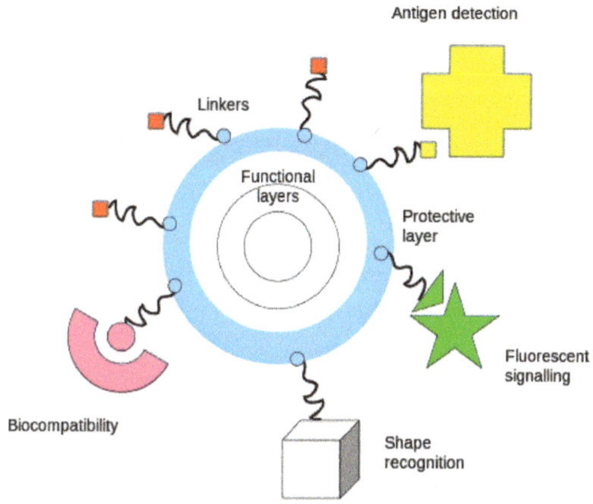

Attachments on nanoparticles make them more biocompatible.

Characterization

Major characteristics of nanoparticles include volume, structure, and visual properties that make them valuable in nanobiotechnology. Depending on specific properties of size, structure, and luminescence, nanoparticles can be used for different applications. Imaging techniques are used to identify such properties and give more information about the tested sample. Techniques used to characterize nanoparticles are also useful in studying how nanoparticles interact with biomolecules, such as amino acids or DNA, and include magnetic resonance imaging (MRI), denoted by the solubility of the nanoparticles in water and fluorescent. MRI can be applied in the medical field to visualize structures; atomic force microscopy (AFM) that gives a topographic view of the sample on a substrate; transmission electron microscopy (TEM) that gives a top view, but with a different technique then that of atomic force microscopy; Raman spectroscopy or surface enhanced Raman spectroscopy (SERS) gives information about wavelengths and energy in the sample. ultraviolet-visible spectroscopy (UV-Vis) measures the wavelengths where light is absorbed; X-Ray diffraction (XRD) generally gives an idea of the chemical composition of the sample.

Chemistry

Physical

Nanomolecules can be created from virtually any element, but the majority produced in today's industry use carbon as the basis upon which the molecules are built around. Carbon can bond with nearly any element, allowing many possibilities when it comes to creating a specific molecule. Scientists can create thousands upon thousands of individual nanomolecules from a simple carbon basis. Some of the most famous nanomolecules currently in existence are solely carbon; these include carbon nanotubes and buckminsterfullerenes. In contrast with nanomolecules, the chemical components of nanoparticles usually consist of metals, such as iron, gold, silver, and platinum.

Interactions between nanoparticles and molecules change depending on the nanoparticle's core. Nanoparticle properties depend not only on the composition of the core material, but also on varying thicknesses of material used. Magnetic properties are particularly useful in molecule manipulation, and thus metals are often used as core material. Metals contain inherent magnetic properties that allow for manipulation of molecular assembly. As nanoparticles interact with molecules via ligand properties, molecular assembly can be controlled by external magnetic fields interacting with magnetic properties in the nanoparticles. Significant problems with producing nanoparticles initially arise once these nanoparticles are generated in solution. Without the use of a stabilizing agent, nanoparticles tend to stick together once the stirring is stopped. In order to counteract this, a certain collidial stabilizer is generally added. These stabilizers bind to the nanoparticles in a way that prevents other particles from bonding with them. Some effective stabilizers found so far include citrate, cellulose, and sodium borohydride.

Application Chemistry

Nanoparticles are desirable in today's industry for their high surface area-to-volume ratio in comparison with larger particles of the same elements. Because chemical reactions occur at a rate directly proportional to the available surface area of reactant compounds, nanoparticles can generate reactions at a much faster rate than larger particles of equal mass. Nanoparticles therefore are among the most efficient means of producing reactions and are inherently valuable in the chemical industry. The same property makes them valuable in interactions with molecules.

Applications with Biomolecules and Biological Processes

Nanoparticles have the potential to greatly influence biological processes. The potency of a nanoparticle increases as its surface area-to-volume-ratio does. Attachments of ligands to the surface of nanoparticles allow them to interact with biomolecules.

Identification of Biomolecules

Nanoparticles are valuable tools in identification of biomolecules, through the use of bio-tagging or labeling. Attachments of ligands or molecular coatings to the surface of a nanoparticle facilitate nanoparticle-molecule interaction, and make them biocompatible. Conjugation can be achieved through intermolecular attractions between the nanoparticle and biomolecule such as covalent bonding, chemisorption, and noncovalent interactions.

To enhance visualization, nanoparticles can also be made to fluoresce by controlling the size and shape of a nanoparticle probe. Fluorescence increases luminescence by increasing the range of wavelengths the emitted light can reach, allowing for biomarkers with a variety of colors. This technique is used to track the efficacy of protein transfer both in vivo and in vitro in terms of genetic alternations.

Biological Process Control

Biological processes can be controlled through transcription regulation, gene regulation, and enzyme inhibition processes that can be regulated using nanoparticles. Nanoparticles can play a part in gene regulation through ionic bonding between positively charged cationic ligands on the surfaces of nanoparticles and negatively charged anionic nucleic acids present in DNA. In an experiment, a nanoparticle-DNA complex inhibited transcription by T7 RNA polymerase, signifying strong bonding in the complex. A high affinity of the nanoparticle-DNA complex indicates strong bonding and a favorable use of nanoparticles. Attaching ionic ligands to nanoparticles allows control over enzyme activity. An example of enzyme inhibition is given by binding of a-chymotrypsin (ChT), an enzyme with a largely cationic active site. When a-chymotrypsin is incubated with anionic (negatively charged) nanoparticles, ChT activity is inhibited as anionic nanoparticles bind to the active site. Enzyme activity can be restored by the addition of cationic surfactants. Alkyl surfactants form a bilayer around ChT, whereas thiol and alcohol surfactants alter the surface of ChT such that interactions with nanoparticles are interrupted. Though formation of a protein-nanoparticle complex can inhibit enzyme activity, studies show that it can also stabilize protein structure, and significantly protect the protein from denaturization. Attachments of ligands to segments of nanoparticles selected for functionalization of metallic properties can be used to generate a magnetic nanowire, which generates a magnetic field that allows for the manipulation of cellular assemblies.

Genetic Alteration

Nanoparticles can also be used in conjunction with DNA to perform genetic alterations. These are frequently monitored through the use of fluorescent materials, allowing scientists to judge if these tagged proteins have successfully been transmitted—for example green fluorescent protein, or GFP. Nanoparticles are significantly less cytotoxic than currently used organic methods, providing a more efficient method of monitoring genetic alternations. They also do not degrade or bleach with time, as organic dyes do. Suspensions of nanoparticles with the same size and shapes (mono-dispersed) with functional groups attached to their surfaces can also electrostatically bind to DNA, protecting them from several types of degradation. Because the fluorescence of these nanoparticles does not degrade, cellular localization can be tracked without the use of additional tagging, with GFPs or other methods. The 'unpacking' of the DNA can be detected in live cells using luminescence resonance energy transfer (LRET) technology.

Medical Implications

Small molecules *in vivo* have a short retention time, but the use of larger nanoparticles does not. These nanoparticles can be used to avoid immune response, which aids in the treatment of chronic diseases. It has been investigated as a potential cancer therapy and also has the potential to affect the understanding of genetic disorders. Nanoparticles also have the potential to aid in site-specific drug delivery by improving the amount of unmodified drug that is circulated within the system, which also decreases the necessary dosage frequency. The targeted nature of nanoparticles also means that non-targeted organs are much less likely to experience side effects of drugs intended for other areas.

Another example is the delivery of nucleic acids to alter key genes associated with cancer progression. For example, Conde et al have shown that by twisting RNA strands into a triple helix and embedding them in a biocompatible gel, they can not only deliver the strands efficiently but also use them to shrink aggressive tumors in mice. Using this technique, the researchers dramatically improved cancer survival rates by simultaneously turning on a tumor-suppressing microRNA and de-activating one that causes cancer. They believe their approach could also be used for delivering other types of RNA, as well as DNA and other therapeutic molecules.

Studying Cell Interactions

Cellular interactions occur at a microscopic level and cannot be easily observed even with the advanced microscopes available today. Due to difficulties observing reactions at the molecular level, indirect methods are used which greatly limits the scope of the understanding that can be gained by studying these processes essential to life. Advances in the material industry has evolved a new field known as nanobiotechnology, that uses nanoparticles to study interactions at the biomolecular level.

One area of research featuring nanobiotechnology is the extracellular matrices of cells (ECM). The ECM is primarily composed of interwoven fibers of collagen and elastin that have diameters ranging from 10–300 nm. In addition to holding the cell in place, the ECM has a variety of other functions including providing a point of attachment for the ECM of other cells and transmembrane receptors that are essential for life. Until recently it has been nearly impossible to study the physical forces that help cells maintain their functionality, but nanobiotechnology has given us the ability to learn more about these interactions. Using the unique properties of nanoparticles, it is possible to control how the nanoparticles adhere to certain patterns present in the ECM, and as a result can understand how changes in the ECM's shape can affect cell functionality.

Using nanobiotechnology to study the ECM allows scientists to investigate the binding interactions that occur between the ECM and its supporting environment. Investigators were able to study these interactions by utilizing tools such as optical tweezers, which have the ability to trap nano-scale objects with focused light. The tweezers can affect the binding of a substrate to the ECM by attempting to draw the substrate away from it. The light emitted from the tweezers was used to restrain ECM-coated microbeads, and the changes in the force exerted by the ECM onto the substrate were studied by modulating the effect of the optical tweezers. Experiments showed that the force exerted by the ECM on the substrate positively correlated with the force of the tweezers, which led to the subsequent discovery that the ECM and the transmembrane proteins are able to sense external forces, and can adapt to overcome these forces.

Nanotechnology Crossing the Blood-brain Barrier

The blood-brain barrier (BBB) is composed of a system of capillaries that has an especially dense lining of endothelial cells which protects the central nervous system (CNS) against the diffusion of foreign substances into the cerebrospinal fluid. These harmful objects include microscopic bacteria, large hydrophobic molecules, certain hormones and neurotransmitters, and low-lipid-soluble molecules. The BBB prevents these harmful particles from entering the brain via tight junctions between endothelial cells and metabolic barriers. The thoroughness with which the BBB does its job makes it difficult to treat diseases of the brain such as cancer, Alzheimer's, and autism, because it is very difficult to transport drugs across the BBB. Currently, in order to deliver therapeutic molecules into the brain, doctors must use highly invasive techniques such as drilling directly into the brain, or sabotaging the integrity of the BBB through biochemical means. Due to their small size and large surface area, nanoparticles offer a promising solution for neurotherapeutics.

Nanotechnology is helpful in delivering drugs and other molecules across the blood-brain barrier (BBB). Nanoparticles allow drugs, or other foreign molecules, to efficiently cross the BBB by camouflaging themselves and tricking the brain into providing them with the ability to cross the BBB in a process called the Trojan Horse Method. Using nanotechnology is advantageous because only the engineered complex is necessary whereas in ordinary applications the active compound must carry out the reaction. This allows for maximum efficacy of the active drug. Also, the use of nanoparticles results in the attraction of proteins to the surfaces of cells, giving cell membranes a biological identity. They also use endogenous active transport where transferrin, an iron binding protein, is linked to rod-shaped semiconductor nanocrystals, in order to move across the BBB into the brain. This discovery is a promising development towards designing an efficient nanoparticle-based drug delivery system.

Self-assembly of Nanoparticles

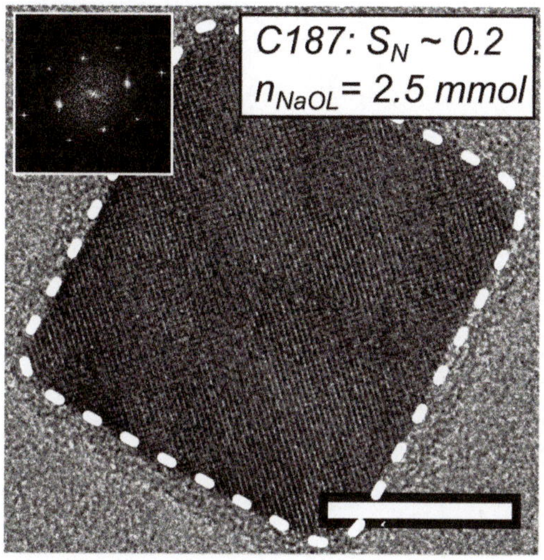

Transmission electron microscopy image of an iron oxide nanoparticle. Regularly arranged dots within the dashed border are columns of Fe atoms. Left inset is the corresponding electron diffraction pattern. Scale bar: 10 nm.

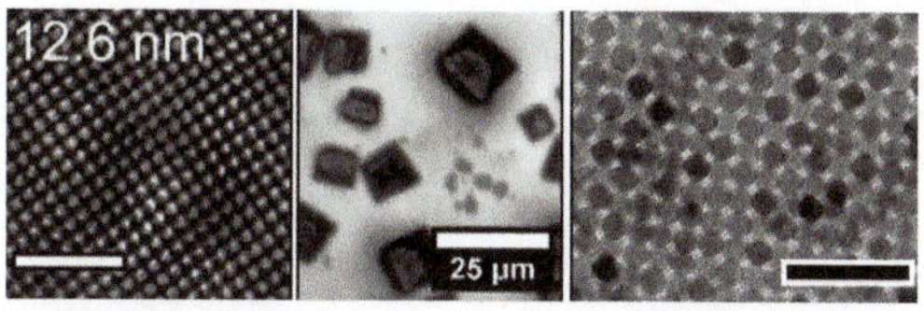

Iron oxide nanoparticles can be dispersed in an organic solvent (toluene). Upon its evaporation, they may self-assemble (left and right panels) into micron-sized mesocrystals (center) or multilayers (right). Each dot in the left image is a traditional "atomic" crystal shown in the image above. Scale bars: 100 nm (left), 25 μm (center), 50 nm (right).

Self-assembly is a phenomenon where the components of a system assemble themselves spontaneously via an interaction to form a larger functional unit. This spontaneous organization can be due to direct specific interaction and/or indirectly through their environment. Due to the increasing technological advancements, the study of materials in the nanometre scale is becoming more important. The spatial arrangements of these self-assembled nanoparticles can be potentially used to build increasingly complex structures leading to a wide variety of materials that can be used for different purposes.

At the molecular level, intermolecular force hold the spontaneous gathering of molecules into a well-defined and stable structure together. In chemical solutions, self-assembly is an outcome of random motion of molecules and the affinity of their binding sites for one another. In the area of nanotechnology, developing a simple, efficient method to organize molecules and molecular clusters into precise, pre-determined structure is crucial.

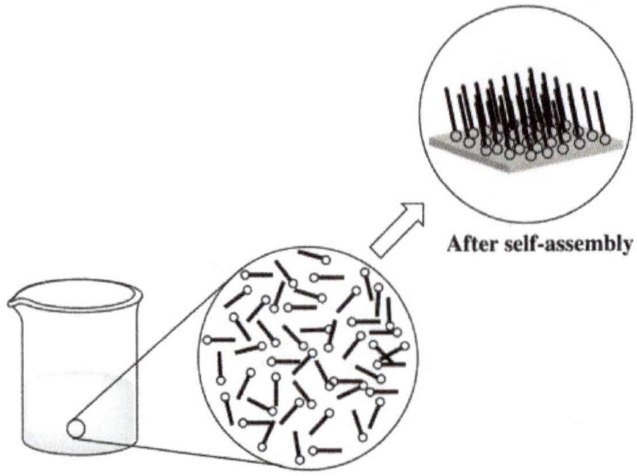

An example of self-assembly of nanoparticles in a solution. In this diagram, it can be seen that a disordered system formed an organized structure which can be due to specific interactions among the particles.

History

The study of self-assembly of nanoparticles is a new area in nanotechnology research. Scientists who are focusing their study on nanotechnology recognized that some properties of atoms and molecules enable them to arrange themselves into patterns. There are a variety of applications where the self-assembly of nanoparticles can be useful. For example, building sensors to detect chemical and biological molecules. In addition, it can also be used on creating computer chips with smaller component sizes, which can then allow more computing power to be stored on a chip. Recently, researchers were able to watch nanoparticles self-assemble for the very first time in real-time. This research was done by scientists at the U.S. Department of Energy's (DOE) Argonne National Laboratory. They used a transmission electron microscope (TEM) located at the Center for Nanoscale Materials to capture the quick movement of the nanoparticles into a self-assembled system.

Thermodynamics

Self-assembly is an equilibrium process where the assembled components are in equilibrium with the individual components. Self-assembly is driven by the minimization of Gibbs free energy. The minimization of Gibbs free energy is attained by the minimization of repulsive and the maximization of attractive molecular interactions. In addition, the lower free energy is usually a result of a weaker intermolecular force between self-assembled moieties and is essentially enthalpic in nature.

The thermodynamics of the self-assembly process can be represented by a simple Gibbs free energy equation:

$$\Delta G_{SA} = \Delta H_{SA} - T\Delta S_{SA}$$

where if ΔG_{SA} is negative, self-assembly is a spontaneous process. ΔH_{SA} is the enthalpy change of the process and is largely determined by the potential energy/intermolecular forces between the assembling entities. ΔS_{SA} is the change in entropy associated with the formation of the ordered arrangement. In general, the organization is accompanied by a decrease in entropy and in order for the assembly to be spontaneous the enthalpy term must be negative and in excess of the entropy term. This equation shows that as the value of $T\Delta S_{SA}$ approaches the value of ΔH_{SA} and above a critical temperature, the self-assembly process will become progressively less likely to occur and spontaneous self-assembly will not happen.

The self-assembly is governed by the normal processes of nucleation and growth. Small assemblies are formed because of their increased lifetime as the attractive interactions between the components lower the Gibbs free energy. As the assembly grows, the Gibbs free energy continues to decrease until the assembly becomes stable enough to last for a long period of time. The necessity of the self-assembly to be an equilibrium process is defined by the organization of the structure which requires non-ideal arrangements to be formed before the lowest energy configuration is found.

Defects

In a self-assembled structure a reasonable concentration of defects is present. Defects can be defined thermodynamically. In most cases, the thermodynamic driving force for self-assembly is provided by weak intermolecular interactions and is usually of the same order of magnitude as the entropy term. In order for a self-assembling system to reach the minimum free energy configuration, there has to be enough thermal energy to allow the mass transport of the self-assembling molecules. For defect formation, the free energy of single defect formation is given by:

$$\Delta G_{DF} = \Delta H_{DF} - T\Delta S_{DF}$$

The enthalpy term, ΔH_{DF} does not necessarily reflect the intermolecular forces between the molecules, it is the energy cost associated with disrupting the pattern and may be thought of as a region where optimum arrangement does not occur and the reduction of enthalpy associated with ideal self-assembly did not occur. An example of this can be seen in a system of hexagonally packed cylinders where defect regions of lamellar structure exist.

If ΔG_{DF} is negative, there will be a finite number of defects in the system and the concentration will be given by:

$$\frac{N}{N_0} = exp(\frac{-\Delta E_{act}}{RT})$$

N is the number of defects in a matrix of N_0 self-assembled particles or features and ΔE_{act} is the activation energy of defect formation. The activation energy, ΔE_{act}, should not be confused with ΔH_{DF}. The activation energy represents the energy difference between the initial ideally arranges state and a transition state towards the defective structure. At low defect concentrations, defect formation is entropy driven until a critical concentration of defects allows the activation energy term to compensate for entropy. There is usually an equilibrium defect density indicated at the minimum free energy. The activation energy for defect formation increases this equilibrium defect density.

Particle Interaction

Intermolecular forces govern the particle interaction in self-assembled systems. The forces tend to be intermolecular in type rather than ionic or covalent because ionic or covalent bonds will "lock" the assembly into non-equilibrium structures. The types intermolecular forces seen in self-assembly processes are van der Waals, hydrogen bonds, and weak polar forces, just to name a few. In self-assembly, regular structural arrangements are frequently observed, therefore there must be a balance of attractive and repulsive between molecules otherwise an equilibrium distance will not exist between the particles. The repulsive forces can be electron cloud-electron cloud overlap or electrostatic repulsion.

Processing

The processes by which nanoparticles self-assemble are widespread and important. Understanding why and how self-assembly occurs is key in reproducing and optimizing results. Typically, nanoparticles will self-assemble for one or both of two reasons: molecular interactions and external direction. Recently, researchers are more interested in the latter of these two reasons, and are able to apply external directing to a large variety of systems.

Self-assembly by Molecular Interactions

Nanoparticles have the ability to assemble chemically through covalent or noncovalent interactions with their capping ligand. The terminal functional group(s) on the particle are known as capping ligands. As these ligands tend to be complex and sophisticated, self-assembly can provide a simpler pathway for nanoparticle organization by synthesizing efficient functional groups. For instance, DNA oligomers have been a key ligand for nanoparticle building blocks to be self-assembling via sequence-based specific organization.However, to deliver precise and scalable (programmable) assembly for a desired structure, a careful positioning of ligand molecules onto the nanoparticle counterpart should be required at the building block (precursor) level, such as direction, geometry, morphology, affinity, etc.The successful design of ligand-building block units can play an essential role in manufacturing a wide-range of new nano systems, such as nanosensor systems, nanomachines/nanobots, nanocomputers, and many more uncharted systems.

Intermolecular Forces

Nanoparticles can self-assemble as a result of their intermolecular forces. As systems look to minimize their free energy, self-assembly is one option for the system to achieve its lowest free energy thermodynamically. Nanoparticles can be programmed to self-assemble by changing the functionality of their side groups, taking advantage of weak and specific intermolecular forces to spontaneously order the particles. These direct interparticle interactions can be typical intermolecular forces such as hydrogen bonding or Van der Waals forces, but can also be internal characteristics, such as hydrophobicity or hydrophilicity. For example, lipophilic nanoparticles have the tendency to self-assemble and form crystals as solvents are evaporated. While these aggregations are based on intermolecular forces, external factors such as temperature and pH also play a role in spontaneous self-assembly.

Hamaker Interaction

As nanoparticle interactions take place on a nanoscale, the particle interactions must be scaled similarly. Hamaker interactions take into account the polarization characteristics of a large number of nearby particles and the effects they have on each other. Hamaker interactions sum all of the forces between all particles and the solvent(s) involved in the system. While Hamaker theory generally describes a macroscopic system, the vast number of nanoparticles in a self-assembling system allows the term to be applicable. Hamaker constants for nanoparticles are calculated using Lifshitz theory, and can often be found in literature.

Hamaker Constants for Nanoparticles in Water	
Material	A_{131}
Fe_3O_4	22
Υ-Fe_2O_3	26
α-Fe_2O_3	29
Ag	33
Au	45
All values reported in zJ	

Externally Directed Self-assembly

The natural ability of nanoparticles to self-assemble can be replicated in systems that do not intrinsically self-assemble. Directed self-assembly (DSA) attempts to mimic the chemical properties of self-assembling systems, while simultaneously controlling the thermodynamic system to maximize self-assembly.

Electric and Magnetic Fields

External fields are the most common directors of self-assembly. Electric and magnetic fields allow induced interactions to align the particles. The fields take advantage of the polarizability of the nanoparticle and its functional groups. When these field-induced interactions overcome random Brownian motion, particles join to form chains and then assemble. At more modest field strengths, ordered crystal structures are established due to the induced dipole interactions. Electric and magnetic field direction requires a constant balance between thermal energy and interaction energies.

Flow Fields

Macroscopic Viscous Flow

Macroscopic viscous flow fields can direct self-assembly of a random solution of particles into ordered crystals. However, the assembled particles tend to disassemble when the flow is stopped or removed. Shear flows are useful for jammed suspensions or random close packing. As these systems begin in nonequilibrium, flow fields are useful in that they help the system relax towards ordered equilibrium. Flow fields are also useful when dealing with complex matrices that themselves have rheological behavior. Flow can induce anisotropic viseoelastic stresses, which helps to overcome the matrix and cause self-assembly.

Large Amplitude Oscillatory Shear (LAOS)

Large amplitude oscillatory shear (LAOS) is most effective for particles that are 100 nm-1 μm in size. Hard and soft shears can order in steady shear. However, this assembly strongly relies on particle volume fraction, particle interaction potentials, polydisterity, and shear rate and strain. The large amount of directing factors can cause complications in directing self-assembly by LAOS. Diblock copolymer micells have recently been studied in regard to structuring nanoparticles in bulk.

Combination of Fields

The most effective self-assembly director is a combination of fields. If the fields and conditions are optimized, self-assembly can be permanent and complete. When a field combination is used with nanoparticles that are tailored to be intrinsically responsive, the most complete assembly is observed. Combinations of fields allow the benefits of self-assembly, such as scalability and simplicity, to be maintained while being able to control orientation and structure formation. Field combinations possess the greatest potential for future directed self-assembly work.

Interfaces

Solid Interfaces

Nano-particles can self-assemble on solid surfaces after applying external forces (like magnetic, electric, and flow). Templates made of microstructures like carbon nanotubes or block polymers can also be used to assist in self-assembly; they cause directed self-assembly (DSA) in which active sites are embedded to selectively induce nanoparticle deposition. Such templates are considered as any object onto which different particles can be arranged into a structure with a morphology similar to that of the template. Carbon nanotubes (microstructures), single molecules, or block copolymers are common templates. Nanoparticles are often shown to self-assemble within distances of nanometers and micrometers, but block copolymer templates can be used to form well-defined self-assemblies over macroscopic distances. By incorporating active sites to the surfaces of nanotubes and polymers, the functionalization of these templates can be transformed to favor self-assembly of specified nanoparticles.

Liquid Interfaces

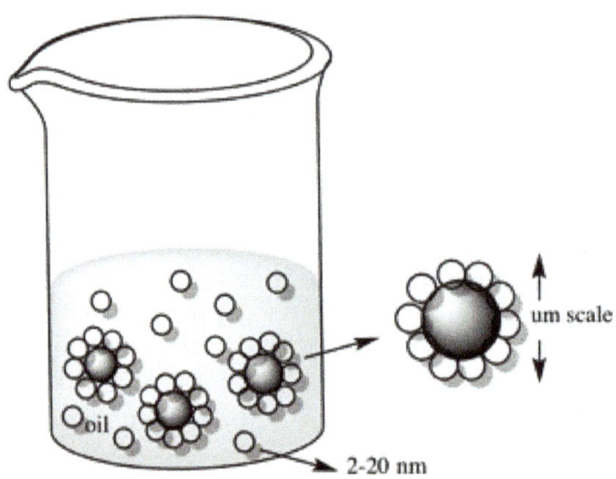

Self-assembly of solid nanoparticles at oil-water interface

Pickering and Ramsden explained the idea of pickering emulsions when experimenting with paraffin-water emulsions with solid particles like iron oxide and silicon dioxide. They observed that the micron-sized colloids generated a resistant film at the interface between the two immiscible phases, inhibiting the coalescence of the emulsion drops. These Pickering emulsions, as shown in the figure below, are formed from the self-assembly of colloidal particles in two-part liquid systems, such as oil-water systems. The desorption energy, which is directly related to the stability of emulsions depends on the particle size, particle-particle interaction and, of course, particle-water and particle-oil interactions.

A decrease in total free energy was observed to be a result from the assembly of nanoparticles at an oil/water (O/W) interface. This is shown in the following equation in which a particle with radius r at an interface between oil (O) and water (W) results in a decrease of the initial interfacial energy E_o to E_1; this difference in energy is ΔE_1.

$$E_0 - E_1 = -\frac{\pi r^2}{\gamma_{O/W}}[\gamma_{O/W} - (\gamma_{P/W} - \gamma_{P/O})]^2$$

At a fixed $\gamma_{P/O}$, $\gamma_{P/W}$, and $\gamma_{O/W}$, the equation shows that the stability of a particle assembly is determined by the radius square. When moving to the interface, particles reduce the unfavorable contact between the immiscible fluids and decrease the interfacial energy. The decrease in total free energy for microscopic particles is much larger than that of thermal energy; this results in an effective confinement of large colloids to the interface. They are irreversibly bound to the interface. Nanoscopic particles are restricted to the interface by an energy reduction comparable to thermal energy. Thus, nanoparticles are easily displaced from the interface. A constant particle exchange then occurs at the interface; the rate of this exchange depends on particle size. The thermally activated escape of small particles occurs more often than larger particles. For the equilibrium state of assembly, the total gain in free energy is smaller for smaller particles. Thus, large nanoparticles assemblies are more stable. This size dependence allows nanoparticles to self-assemble at the interface to attain its equilibrium structure. Micrometer- size colloids, on the other hand, may be confined in a non-equilibrium state.

The interfacial tension and the wettability of a particle surface affect the desorption energy. The contact angle θ between the solid and the oil/water interface determines its wettability. As shown in the figure below, a contact angle θ greater than 90° favors a water-in-oil emersion while a contact angle θ less than 90° favors oil-in-water emulsion. These contact angles affect the stability of the emulsion.

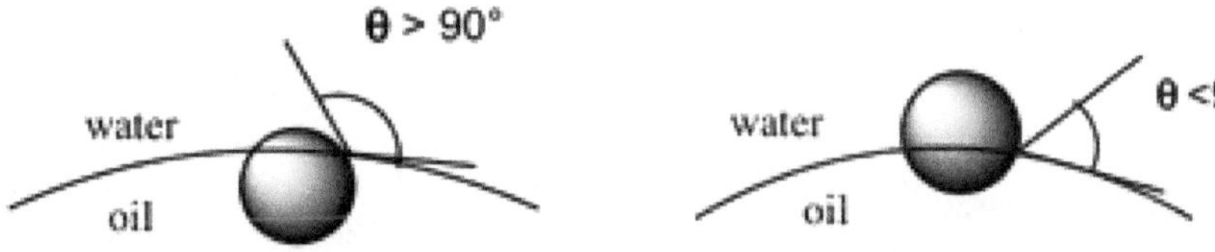

Effects of contact angle on wettability

A maximum desorption energy peak is observed at a contact angle of 90°. When the contact angle is greater than or less than this point, the desorption energy gradually decreases; thus, the stability of the emulsion decreases as well.

Applications

Material that consists with nano-particles is called "nanostructured material". The phase of "nanostructured material" implies two important ideas:

1. that at least some of the heterogeneity in materials is determined by the size range of nanostructures (about 1–100 nm), and
2. these nanostructures might be synthesized and distributed (or organized), at least in part. The study of self-assembly nanoparticles is important to understand the interaction between single particle in terms of applying them into different applications.

Electronics

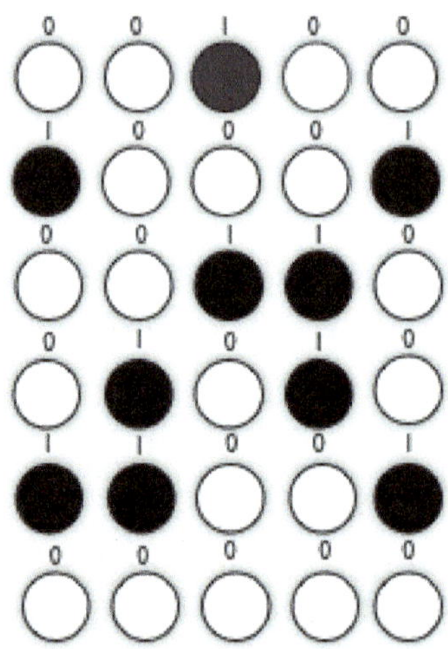

Model of multidimensional array of nano-particles. A particle could have two spins, spin up or down. Based on the spin directions, nano-particles will be able to store 0 and 1. Therefore, nanostructural material has a great potential for future use in electronic devices.

Self-assembly of nanoscale structures from functional nanoparticles has provided a powerful path to developing small and powerful electronic components. In the recent decade, it has been confirmed that organic compounds can be conductors or semiconductors. In addition to insulator, one of the opportunities in nanomaterials science is to use organic synthesis and molecular design to make electronically usefully structures starting with organic molecules. The difficulty of applied nanostructure material is nanoscale objects have always been difficult to work with because they cannot be characterized by molecular techniques and they are too small to see optically. 2D self-assembly monodisperse particle colloids has a strong potential in dense magnetic storage media. Each colloid particle has the ability to store information as known as binary number 0 and 1 after applying it to a strong magnetic field. In the meantime, it requires a nanoscale sensor or detector in order to selectively choose the colloid particle.

Biological Applications

Drug Delivery

Block copolymers offer the ability to self-assemble into uniform, nanosized micelles and accumulate in tumors via the enhanced permeability and retention effect. Polymer composition can be chosen to control the micelle size and compatibility with the drug of choice. The challenges of this application are the difficulty of reproducing or controlling the size of self-assembly nano micelle, preparing predictable size-distribution, and the stability of the micelle with high drug load content.

Magnetic Drug Delivery

Magnetic nanochains are a class of new magnetoresponsive and superparamagnetic nanostructures with highly anisotropic shapes (chain-like) which can be manipulated using magnetic field and magnetic field gradient. The magnetic nanochains possess attractive properties which are significant added value for many potential uses including magneto-mechanical actuation-associated nanomedicines in low and super-low frequency alternating magnetic field and magnetic drug delivery.

Self-assembled RNA triple-helix Nanoparticles Doped on Hydrogels for Cancer Therapy

In 2015, Conde et al. developed a self-assembled dual-color RNA triple-helix structure comprising two miRNAs, a miR mimic (tumor suppressor miRNA) and an antagomiR (oncomiR inhibitor) provides outstanding capability to synergistically abrogate tumors. The authors hypothesized that efficacious delivery of the RNA triple-helix hydrogel nanoconjugates would be achieved by coating the breast tumor with the adhesive hydrogel scaffold that they have previously shown able to enhance the stability of embedded nanoparticles used for local gene and drug delivery using smart gold nanobeacons.

Conjugation of RNA triple-helices to dendrimers allows the formation of stable triplex nanoparticles, which form an RNA-triple-helix adhesive scaffold on interaction with dextran aldehyde, the latter able to chemically interact and adhere to natural tissue amines in the tumor. The authors also show that the self-assembled RNA-triple-helix conjugates remain functional *in vitro* and *in vivo*, and that they lead to nearly 90% levels of tumor shrinkage two weeks post-gel implantation in a triple-negative breast-cancer mouse model. These findings suggest that the RNA-triple-helix hydrogels can be used as an efficient anticancer platform to locally modulate the expression of endogenous miRs in cancer.

The authors reported on a novel strategy for concomitant oncomiR inhibition and tumor suppressor miR replacement therapy using a RNA triple-helix hydrogel scaffold that affords highly efficacious local anticancer therapy. Self-assembled RNA triple-helix conjugates remain functional *in vitro* with high selective uptake and control over miR expression compared to their respective single-stranded or double-stranded forms. Hence, cancer gene delivery systems should provide potent, selective and specific treatment to tumor cells only, unlike the standard delivery of most conventional chemotherapeutic drugs. This approach can be implemented to design self-assembled triplex structures from any other miR combination, or from other genetic materials including antisense DNA or siRNA to treat a range of diseases.

Nanofabrics

Nanofabrics are textiles engineered with small particles that give ordinary materials advantageous properties such as superhydrophobicity, odor and moisture elimination, increased elasticity and strength, and bacterial resistance. Depending on the desired property, a nanofabric is either constructed from nanoscopic fibers called nanofibers, or is formed by applying a solution containing nanoparticles to a regular fabric. Nanofabrics research is an interdisciplinary effort involving bioengineering, molecular chemistry, physics, electrical engineering, computer science, and systems engineering. Applications of nanofabrics have the potential to revolutionize textile manufacturing and areas of medicine such as drug delivery and tissue engineering.

Introduction

A fiber that has a width of less than 1000 nanometers (1000 nm or 1 μm) is generally defined as a nanofiber. A nanoparticle is defined as a small group of atoms or molecules with a radius of less than 100 nanometers (100 nm). Particles on the nanoscale have a very high surface area to volume ratio, whereas this ratio is much lower for objects on the macroscopic scale. A high relative surface area means that a large proportion of a particle's mass exists on its surface, so nanofibers and nanoparticles show a greater level of interaction with other materials. The high surface area to volume ratio observed in very small particles is what makes it possible to create many special properties exhibited by nanofabrics.

The use of nanoparticles and nanofibers to produce specialized nanofabrics became a subject of interest after the sol-gel and electrospinning techniques were fully developed in the 1980s. Since 2000, dramatic increases in global funding have accelerated research efforts in nanotechnology, including nanofabrics research.

Sol-gel

The sol-gel process is used to create gel-like solutions which can be applied to textiles as a liquid finish to create nanofabrics with novel properties. The process begins with dissolving nanoparticles in a liquid solvent (often an alcohol). Once dissolved, several chemical reactions take place that cause the nanoparticles to grow and establish a network throughout the liquid. The network transforms the solution into a colloid (a suspension of solid particles in a liquid) with a gelatinous texture. Finally, the colloid must go through a drying process to remove excess solvent from the mixture before it can be used to treat fabrics. The sol-gel process is used in a similar fashion to make polymer nanofibers, which are long, ultra-thin chains of proteins bonded together.

Electrospinning

Electrospinning extracts nanofibers from polymer solutions (synthesized by the sol-gel process) and collects them to form nonwoven nanofabrics. A strong electric field is applied to the solution to charge the polymer strands. The solution is put into a syringe and aimed at an oppositely charged collector plate. When the force of attraction between the polymer nanofibers and the collector plate exceed the surface tension of the solution, the nanofibers are released from the solution and deposit onto the collector plate. The deposited fibers form a porous nanofabric that can aid in drug delivery and tissue engineering depending on the type of polymer used.

Applications

Textile Manufacturing

When nanoengineered coatings are applied to fabrics, the nanoparticles readily form bonds with the fibers of the material. The high surface area relative to the volume of particles increases their chemical reactivity, allowing them to stick to materials more permanently. Fabrics treated with nanoparticle coatings during manufacturing produce materials that kill bacteria, eliminate moisture and odor, and prevent static electricity. Polymer nanofiber coatings applied to textiles bond to the material at one end of the polymer, forming a surface of tiny, hair-like structures. The polymer "hairs" create a thin layer that prevents liquids from making contact with the actual fabric. Nanofabrics with dirt-proof, stain-proof, and superhydrophobic properties are possible as a result of the layer formed by polymer nanofibers.

Development of nanofabrics for use in the clothing and textiles industry is still in its early stages. Some applications such as bacteria-resistant clothing are not yet practical from an economic standpoint. For example, a Cornell University student's prototype for a bactericidal jacket cost $10,000 alone, so it may be a long time before nanofabric clothing is on the market.

Drug Delivery

Nanofabrics used in medicine can deliver antibiotics, anticancer drugs, proteins, and DNA in precise quantities. Electrospinning creates porous nanofabrics that can be loaded with the desired drug which are then applied to the tissue of the targeted area. The drug passes through the tissue by diffusion, a process in which substances move through a membrane from high to low concentration. The rate at which the drug is administered can be changed by altering the composition of the nanofabric.

Tissue Engineering

Nonwoven fabrics made by electrospinning have the potential to assist in the growth of organ tissue, bone, neurons, tendons, and ligaments. Polymer nanofabrics can act either as a scaffold to support damaged tissue or as a synthetic substitute for actual tissue. Depending on the function, the nanofabric can be made of natural or synthetic polymers, or a combination of both.

Environmental Implications

As nanotechnology advances, many studies have been conducted to determine the effects nanoengineered materials can have on the environment. Most textiles can lose up to 20% of their mass during their lifetime, so nanoparticles used in production of nanofabrics are at risk of being released into the air and waterways.

Nano-silver is expected to have as much as 49.5% of its global production taken by the nanotextiles industry due to its antibacterial properties. It is predicted that 20% of the nano-silver used in the nanofabrics industry will be released into waterways which could cause harm to microorgansims. However, more than 90% of nano-silver is removed during treatment at wastewater facilities, so it is likely that the environmental impact will be minimal. A study on aluminum oxide nanoparticles showed that inhalation caused inflammation in rat lungs. Aluminum oxide nanoparticles are not used in large quantity, so its health risks are negligible. Other studies conducted for nanoparticles suggest that their environmental impact should be low as the nanotextiles industry continues to grow.

Nanoremediation

Nanoremediation is the use of nanoparticles for environmental remediation. It is being explored to treat ground water, wastewater, soil, sediment, or other contaminated environmental materials. Nanoremediation is an emerging industry; by 2009, nanoremediation technologies had been documented in at least 44 cleanup sites around the world, predominantly in the United States. In Europe, nanoremediation is being investigated by the EC funded NanoRem Project. A report produced by the NanoRem consortium has identified around 70 nanoremediation projects worldwide at pilot or full scale. During nanoremediation, a nanoparticle agent must be brought into contact with the target contaminant under conditions that allow a detoxifying or immobilizing reaction. This process typically involves a pump-and-treat process or *in situ* application.

Some nanoremediation methods, particularly the use of nano zero-valent iron for groundwater cleanup, have been deployed at full-scale cleanup sites. Other methods remain in research phases.

Applications

Nanoremediation has been most widely used for groundwater treatment, with additional extensive research in wastewater treatment. Nanoremediation has also been tested for soil and sediment cleanup. Even more preliminary research is exploring the use of nanoparticles to remove toxic materials from gases.

Groundwater Remediation

Currently, groundwater remediation is the most common commercial application of nanoremediation technologies. Using nanomaterials, especially zero-valent metals (ZVMs), for groundwater remediation is an emerging approach that is promising due to the availability and effectiveness of many nanomaterials for degrading or sequestering contaminants.

Nanotechnology offers the potential to effectively treat contaminants *in situ*, avoiding excavation or the need to pump contaminated water out of the ground. The process begins with nanoparticles being injected into a contaminated aquifer via an injection well. The nanoparticles are then transported by groundwater flow to the source of contamination. Upon contact, nanoparticles can sequester contaminants (via adsorption or complexation), immobilizing them, or they can degrade the contaminants to less harmful compounds. Contaminant transformations are typically redox reactions. When the nanoparticle is the oxidant or reductant, it is considered reactive.

The ability to inject nanoparticles to the subsurface and transport them to the contaminant source is imperative for successful treatment. Reactive nanoparticles can be injected into a well where they will then be transported down gradient to the contaminated area. Drilling and packing a well is quite expensive. Direct push wells cost less than drilled wells and are the most often used delivery tool for remediation with nanoiron. A nanoparticle slurry can be injected along the vertical range of the probe to provide treatment to specific aquifer regions.

Surface Water Treatment

The use of various nanomaterials, including carbon nanotubes and TiO_2, shows promise for treatment of surface water, including for purification, disinfection, and desalination. Target contaminants in surface waters include heavy metals, organic contaminants, and pathogens. In this context, nanoparticles may be used as sorbents, as reactive agents (photocatalysts or redox agents), or in membranes used for nanofiltration.

Trace Contaminant Detection

Nanoparticles may assist in detecting trace levels of contaminants in field settings, contributing to effective remediation. Instruments that can operate outside of a laboratory often are not sensitive enough to detect trace contaminants. Rapid, portable, and cost-effective measurement systems for trace contaminants in groundwater and other environmental media would thus enhance contaminant detection and cleanup. One potential method is to separate the analyte from the sample and concentrate them to a smaller volume, easing detection and measurement. When small quantities of solid sorbents are used to absorb the target for concentration, this method is referred to as solid-phase microextraction.

With their high reactivity and large surface area, nanoparticles may be effective sorbents to help concentrate target contaminants for solid-phase microextraction, particularly in the form of self-assembled monolayers on mesoporous supports. The mesoporous silica structure, made through a surfactant templated sol-gel process, gives these self-assembled monolayers high surface area and a rigid open pore structure. This material may be an effective sorbent for many targets, including heavy metals such as mercury, lead, and cadmium, chromate and arsenate, and radionuclides such as ^{99}Tc, ^{137}CS, uranium, and the actinides.

Mechanism

The small size of nanoparticles leads to several characteristics that may enhance remediation. Nanomaterials are highly reactive because of their high surface area per unit mass. Their small particle size also allows nanoparticles to enter small pores in soil or sediment that larger particles might not penetrate, granting them access to contaminants sorbed to soil and increasing the likelihood of contact with the target contaminant.

Because nanomaterials are so tiny, their movement is largely governed by Brownian motion as compared to gravity. Thus, the flow of groundwater can be sufficient to transport the particles. Nanoparticles then can remain suspended in solution longer to establish an *in situ* treatment zone.

Once a nanoparticle contacts the contaminant, it may degrade the contaminant, typically through a redox reaction, or adsorb to the contaminant to immobilize it. In some cases, such as with magnetic nano-iron, adsorbed complexes may be separated from the treated substrate, removing the contaminant. Target contaminants include organic molecules such as pesticides or organic solvents and metals such as arsenic or lead. Some research is also exploring the use of nanoparticles to remove excessive nutrients such as nitrogen and phosphorus.

Materials

A variety of compounds, including some that are used as macro-sized particles for remediation, are being studied for use in nanoremediation. These materials include zero-valent metals like zero-valent iron, calcium carbonate, carbon-based compounds such as graphene or carbon nanotubes, and metal oxides such as titanium dioxide and iron oxide.

Nano Zero-valent Iron

As of 2012, nano zero-valent iron (nZVI) was the nanoscale material most commonly used in bench and field remediation tests. nZVI may be mixed or coated with another metal, such as palladium, silver, or copper, that acts as a catalyst in what is called a bimetallic nanoparticle. nZVI may also be emulsified with a surfactant and an oil, creating a membrane that enhances the nanoparticle's ability to interact with hydrophobic liquids and protects it against reactions with materials dissolved in water. Commercial nZVI particle sizes may sometimes exceed true "nano" dimensions (100 nm or less in diameter).

nZVI appears to be useful for degrading organic contaminants, including chlorinated organic compounds such as polychlorinated biphenyls (PCBs) and trichloroethene (TCE), as well as immobilizing or removing metals. nZVI and other nanoparticles that do not require light can be injected belowground into the contaminated zone for *in situ* groundwater remediation and, potentially, soil remediation.

nZVI nanoparticles can be prepared by using sodium borohydride as the key reductant. $NaBH_4$ (0.2 M) is added into $FeCl_3 \cdot 6H_2$ (0.05 M) solution (~1:1 volume ratio). Ferric iron is reduced via the following reaction:

$$4Fe^{3+} + 3BH_4^- + 9H_2O \rightarrow 4Fe^0 + 3H_2BO_3^- + 12H^+ + 6H_2$$

Palladized Fe particles are prepared by soaking the nanoscale iron particles with an ethanol solution of 1wt% of palladium acetate ($[Pd(C_2H_3O_2)2]_3$). This causes the reduction and deposition of Pd on the Fe surface:

$$Pd^{2+} + Fe^o \rightarrow Pd^o + Fe^{2+}$$

Similar methods may be used to prepared Fe/Pt, Fe/Ag, Fe/Ni, Fe/Co, and Fe/Cu bimetallic particles. With the above methods, nanoparticles of diameter 50-70 nm may be produced. The average specific surface area of Pd/Fe particles is about 35 m^2/g. Ferrous iron salt has also been successfully used as the precursor.

Titanium Dioxide

Titanium dioxide (TiO_2) is also a leading candidate for nanoremediation and wastewater treatment, although as of 2010 it is reported to have not yet been expanded to full-scale commercialization. When exposed to ultraviolet light, such as in sunlight, titanium dioxide produces hydroxyl radicals, which are highly reactive and can oxidize contaminants. Hydroxyl radicals are used for water treatment in methods generally termed advanced oxidation processes. Because light is required for this reaction, TiO_2 is not appropriate for underground *in situ* remediation, but it may be used for wastewater treatment or pump-and-treat groundwater remediation.

TiO_2 is inexpensive, chemically stable, and insoluble in water. TiO_2 has a wide band gap energy (3.2 eV) that requires the use of UV light, as opposed to visible light only, for photocatalytic activation.

To enhance the efficiency of its photocatalysis, research has investigated modifications to TiO_2 or alternative photocatalysts that might use a greater portion of photons in the visible light spectrum. Potential modifications include doping TiO_2 with metals, nitrogen, or carbon.

Challenges

When using *in situ* remediation the reactive products must be considered for two reasons. One reason is that a reactive product might be more harmful or mobile than the parent compound. Another reason is that the products can affect the effectiveness and/or cost of remediation. TCE (trichloroethylene), under reducing conditions by nanoiron, may sequentially dechlorinate to DCE (dichloroethene) and VC (vinyl chloride). VC is known to be more harmful than TCE, meaning this process would be undesirable.

Nanoparticles also react with non-target compounds. Bare nanoparticles tend to clump together and also react rapidly with soil, sediment, or other material in ground water. For *in situ* remediation, this action inhibits the particles from dispersing in the contaminated area, reducing their effectiveness for remediation. Coatings or other treatment may allow nanoparticles to disperse farther and potentially reach a greater portion of the contaminated zone. Coatings for nZVI include surfactants, polyelectrolyte coatings, emulsification layers, and protective shells made from silica or carbon.

Such designs may also affect the nanoparticles' ability to react with contaminants, their uptake by organisms, and their toxicity. A continuing area of research involves the potential for nanoparticles used for remediation to disperse widely and harm wildlife, plants, or people.

In some cases, bioremediation may be used deliberately at the same site or with the same material as nanoremediation. Ongoing research is investigating how nanoparticles may interact with simultaneous biological remediation.

References

- Reeves, Daniel B.; Weaver, John B. (15 December 2012). "Simulations of magnetic nanoparticle Brownian motion". Journal of Applied Physics. 112 (12): 124311. ISSN 0021-8979. doi:10.1063/1.477032
- Prasad, Paras N. (2012-05-11). Introduction to Nanomedicine and Nanobioengineering. John Wiley & Sons. pp. 121–124. ISBN 9781118351079
- "Radiation Safety Aspects of Nanotechnology". National Council on Radiation Protection and Measurements. 2017-03-02. Retrieved 2017-07-07
- Forrester, D. M.; et al. (2016). "Experimental verification of nanofluid shear-wave reconversion in ultrasonic fields". Nanoscale. doi:10.1039/C5NR07396K
- Buongiorno, J. (March 2006). "Convective Transport in Nanofluids". Journal of Heat Transfer. American Society Of Mechanical Engineers. 128 (3): 240. doi:10.1115/1.2150834. Retrieved 27 March 2010
- Rotello, Vincent (2004). Nanoparticles Building Blocks for Nanotechnology. New York: Springer Science + Business Media Inc. ISBN 0306482878
- "Current Intelligence Bulletin 63: Occupational Exposure to Titanium Dioxide". U.S. National Institute for Occupational Safety and Health. April 2011. pp. v–vii, 73–78. Retrieved 2017-04-27
- Mahendran, V. (2012). "Nanofluid based opticalsensor for rapid visual inspection of defects in ferromagnetic materials". Appl. Phys. Lett. 100: 073104. doi:10.1063/1.3684969

- Wang, "Transmission Electron Microscopy of Shape-Controlled Nanocrystals and Their Assemblies", Journal of Physical Chemistry B. 104. (2000): 1153-1175. DOI 10.1021/jp993593c
- Whitesides, G.M.; Kriebel, J.K.; Mayers, B.T. (2005). "Self-Assembly and Nanostructured Materials" (PDF). Nanoscale Assembly – Chemical Techniques. Springer. ISBN 978-0-387-23608-7. doi:10.1007/b106927
- Das, Sarit K.; Stephen U. S. Choi; Wenhua Yu; T. Pradeep (2007). Nanofluids: Science and Technology. Wiley-Interscience. p. 397. Retrieved 27 March 2010
- Bardan, R; Lal, S; Joshi, A; Halas, Nj (May 2011). "Theranostic Nanoshells: From Probe Design to Imaging and Treatment of Cancer". Accounts of Chemical Research. 44 (10): 936–946. doi:10.1021/ar200023x
- Li, Fei; et al. (2008). "Structural and luminescent properties of ZnO nanorods and ZnO/ZnS nanocomposites". Journal of Alloys and Compounds. 474: 531–535. doi:10.1016/j.jallcom.2008.06.149
- Martín, Isabel García; Frigell, Jens; Llop, Jordi; Marradi, Marco (2016-03-22). "Radiolabelling of NPs Using Radiometals: ^{99m}Tc, ^{68}Ga, ^{67}Ga, ^{89}Zr, and ^{64}Cu". In Llop, Jordi; Gomez-Vallejo, Vanessa; Gibson, Peter Neil. Isotopes in Nanoparticles. Pan Stanford. pp. 183–229. ISBN 9789814669085. doi:10.1201/b19950-9
- Salata, OV (30 April 2004). "Applications of Nanoparticles in Biology and Medicine". Journal of Nanobiotechnology. 2 (1): 3. PMC 419715. PMID 15119954. doi:10.1186/1477-3155-2-3. Retrieved 29 April 2011
- Grzelczak, S.; Vermant, J; Furst, E.M.; Liz-Marzan, L.M. (2010). "Directed Self-Assembly of Nanoparticles". ACS Nano. 4 (7): 3591–3605. doi:10.1021/nn100869j
- Klein, L.C.; G.J. Garvey (1980). "Kinetics of the Sol-Gel Transition". Journal of Non-Crystalline Solids. 38: 45–50. doi:10.1016/0022-3093(80)90392-0
- Reese, Torsten; Gómez-Vallejo, Vanessa; Ferreira, Paola; Llop, Jordi (2016-03-22). "Health and Safety Considerations for Radiolabelled Nanoparticles". In Llop, Jordi; Gomez-Vallejo, Vanessa; Gibson, Peter Neil. Isotopes in Nanoparticles. Pan Stanford. pp. 493–512. ISBN 9789814669085. doi:10.1201/b19950-19
- Lubick, Naomi (2008-03-01). "Risks of Nanotechnology Remain Uncertain". Environmental Science & Technology. 42 (6): 1821–1824. ISSN 0013-936X. doi:10.1021/es087058e. Retrieved 2014-07-25
- Kuznetsov, A.V.; Nield, D.A. "Natural convective boundary-layer flow of a nanofluid past a vertical plate". International Journal of Thermal Sciences. 49 (2): 243–247. doi:10.1016/j.ijthermalsci.2009.07.015

2

Classification of Nanoparticles by Composition

Copper nanoparticles are particles that are formed by natural processes or by chemical synthesis. Silver nanoparticle, platinum nanoparticles, colloidal gold, cobalt oxide nanoparticles, gold nanobeacon, nanoscale iron particles, nanofiber, iron oxide nanoparticles, iron-platinum nanoparticles and nanoceramic are other classifications of nanoparticles by composition. This chapter has been carefully written to provide an easy understanding of the varied classifications of nanoparticles by composition.

Copper Nanoparticles

Copper nanoparticles are copper based particles 1 to 100 nm in size. Like many other forms of nanoparticles, copper nanoparticles can be formed by natural processes or through chemical synthesis. These nanoparticles are of particular interest due to their historical application as coloring agents and their modern-day biomedical ones.

Historical Uses

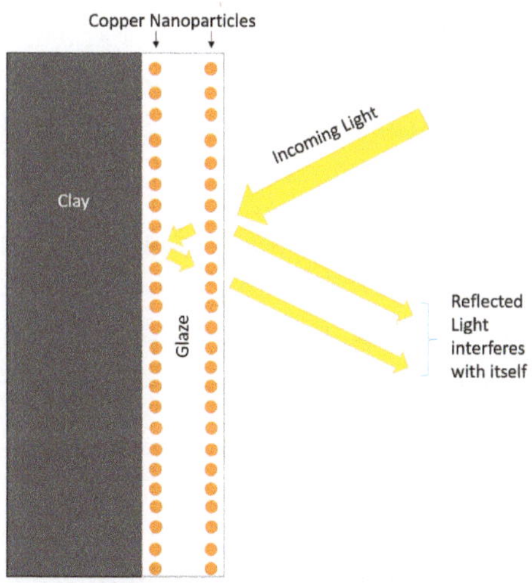

The luster effect is caused by interference effects of light reflecting off two layers of copper nanoparticles in the pottery's glaze.

One of the earliest uses of copper nanoparticles was to color glass and ceramics during the ninth century in Mesopotamia. This was done by creating a glaze with copper and silver salts and applying it to clay pottery. When the pottery was baked at high temperatures in reducing conditions, the metal ions migrated to the outer part of the glaze and were reduced to metals. The end result was a double layer of metal nanoparticles with a small amount of glaze in between them. When the finished pottery was exposed to light, the light would penetrate and reflect off the first layer. The light penetrating the first layer would reflect off the second layer of nanoparticles and cause interference effects with light reflecting off the first layer, creating a luster effect that results from both constructive and destructive interference.

Synthesis

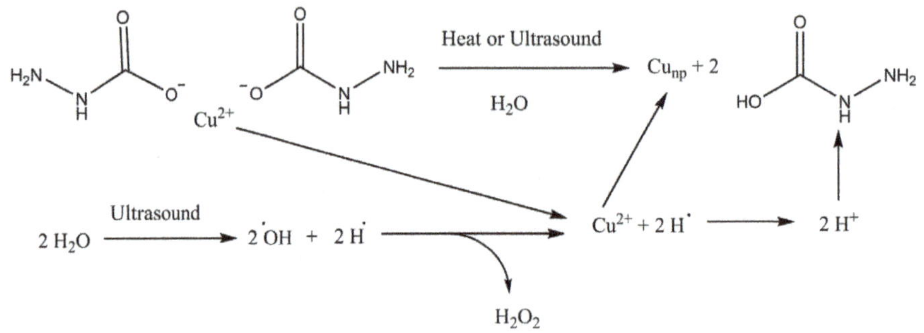

One method of synthesizing copper nanoparticles involves the copper (II) hydrazine carboxylate salt which undergoes a radical reaction with radical hydrogen produced by ultrasounds to form nanoparticles, hydrogen peroxide, and hydrazine carboxylic acid.

Various methods have been described to chemically synthesize copper nanoparticles. An older method involves the reduction of copper hydrazine carboxylate in an aqueous solution using reflux or by heating through ultrasound under an inert argon atmosphere. This results in a combination of copper oxide and pure copper nanoparticle clusters, depending on the method used. A more modern synthesis utilizes copper chloride in a room temperature reaction with sodium citrate or myristic acid in an aqueous solution containing sodium formaldehyde sulfoxylate (SFS) to obtain a pure copper nanoparticle powder. While these syntheses generate fairly consistent copper nanoparticles, the possibility of controlling the sizes and shapes of copper nanoparticles has also been reported. The reduction of copper(II) acetylacetonate in organic solvent with oleyl amine and oleic acid causes the formation of rod and cube-shaped nanoparticles while variations in reaction temperature affect the size of the synthesized particles.

Another method of synthesis involves using copper (II) hydrazine carboxylate salt with ultrasound or heat in water to generate a radical reaction, as shown in the figure to the above. Copper nanoparticles can also be synthesized using green chemistry to reduce the environmental impact of the reaction. Copper chloride can be reduced using only L-ascorbic acid in a heated aqueous solution to produce stable copper nanoparticles.

Characteristics

Copper nanoparticles display unique characteristics including catalytic and antifungal/antibacterial activities that are not observed in commercial copper. First of all, copper nanoparticles demonstrate a very strong catalytic activity, a property that can be attributed to their large catalytic surface area. With the small size and great porosity, the nanoparticles are able to achieve a higher reaction yield and a shorter reaction time when utilized as reagents in organic and organometallic synthesis. In fact, copper nanoparticles that are used in a condensation reaction of iodobenzene attained about 88% conversion to biphenyl, while the commercial copper exhibited only a conversion of 43%.

Copper nanoparticles that are extremely small and have a high surface to volume ratio can also serve as antifungal/antibacterial agents. The antimicrobial activity is induced by their close interaction with microbial membranes and their metal ions released in solutions. As the nanoparticles oxidize slowly in solutions, cupric ions are released from them and they can create toxic hydroxyl free radicals when the lipid membrane is nearby. Then, the free radicals disassemble lipids in cell membranes through oxidation to degenerate the membranes. As a result, the intracellular substances seep out of cells through the destructed membranes; the cells are no longer able to sustain fundamental biochemical processes. In the end, all these alterations inside of the cell caused by the free radicals lead to cell death.

Applications

Copper nanoparticles with great catalytic activities can be applied to biosensors and electrochemical sensors. Redox reactions utilized in those sensors are generally irreversible and also require high overpotentials (more energy) to run. In fact, the nanoparticles have the ability to make the redox reactions reversible and to lower the overpotentials when applied to the sensors.

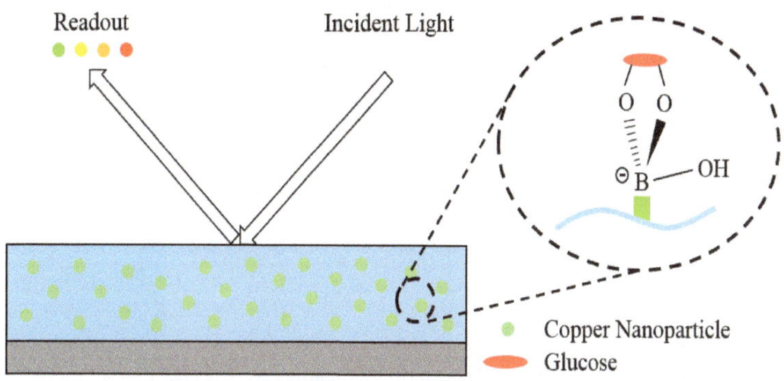

A polyacrylamide hydrogel with copper nanoparticles inside is able to determine glucose levels in a sample added to the gel. As phenylboronic acid groups on the hydrogel polymers bind the glucose molecules, the gel becomes swollen. As a result, the copper nanoparticles move apart, changing how incident light is diffracted by the gel. As the glucose levels decrease, the color of gel changes from red to orange to yellow to green.

One of the examples is a glucose sensor. With the use of copper nanoparticles, the sensor does not require any enzyme and therefore has no need to deal with enzyme degradation and denaturation. As described in the figure, depending on the level of glucose, the nanoparticles in the sensor diffract the incident light at a different angle. Consequently, the resulting diffracted light gives a different color based on the level of glucose. In fact, the nanoparticles enable the sensor to be more stable at high temperatures and varying pH, and more resistant to toxic chemicals. Moreover, using nanoparticles, native amino acids can be detected. A copper nanoparticle-plated screen-printed carbon electrode functions as a stable and effective sensing system for all 20 amino acid detection.

Colloidal Gold

Colloidal gold is a sol or colloidal suspension of nanoparticles of gold in a fluid, usually water. The liquid is usually either an intense red colour (for particles less than 100 nm) or blue/purple (for larger particles). Due to the unique optical, electronic, and molecular-recognition properties of gold nanoparticles, they are the subject of substantial research, with applications in a wide variety of areas, including electron microscopy, electronics, nanotechnology, and materials science.

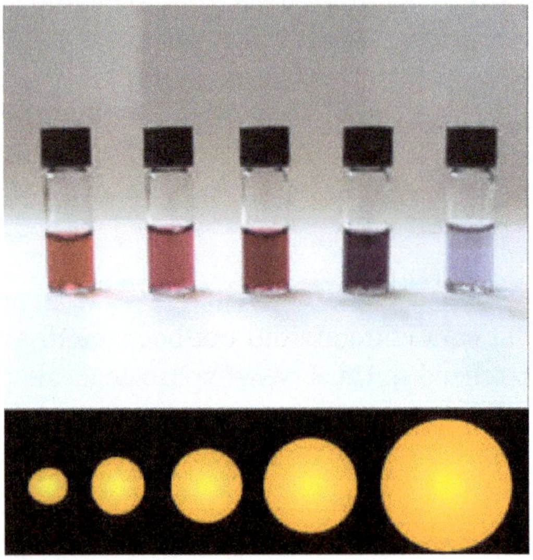

Suspensions of gold nanoparticles of various sizes. The size difference causes the difference in colors

The properties of colloidal gold nanoparticles, and thus their applications, depend strongly upon their size and shape. For example, rodlike particles have both transverse and longitudinal absorption peak, and anisotropy of the shape affects their self-assembly.

History

This cranberry glass bowl was made by adding a gold salt (probably gold chloride) to molten glass

Used since ancient times, the synthesis of colloidal gold was crucial to the 4th-century Lycurgus Cup, which changes color depending on the location of light source. Later it was used as a method of staining glass.

During the Middle Ages, soluble gold, a solution containing gold salt, had a reputation for its curative property for various diseases. In 1618, Francis Anthony, a philosopher and member of the medical profession, published a book called *Panacea Aurea, sive tractatus duo de ipsius Auro*

Potabili (Latin: gold potion, or two treatments of potable gold). The book introduces information on the formation of colloidal gold and its medical uses. About half a century later, English botanist Nicholas Culpepper published book in 1656, *Treatise of Aurum Potabile*, solely discussing the medical uses of colloidal gold.

In 1676, Johann Kunckel, a German chemist, published a book on the manufacture of stained glass. In his book *Valuable Observations or Remarks About the Fixed and Volatile Salts-Auro and Argento Potabile, Spiritu Mundi and the Like*, Kunckel assumed that the pink color of Aurum Potabile came from small particles of metallic gold, not visible to human eyes. In 1842, John Herschel invented a photographic process called chrysotype (from the Greek meaning "gold") that used colloidal gold to record images on paper.

Modern scientific evaluation of colloidal gold did not begin until Michael Faraday's work in the 1850s. In 1856, in a basement laboratory of Royal Institution, Faraday accidentally created a ruby red solution while mounting pieces of gold leaf onto microscope slides. Since he was already interested in the properties of light and matter, Faraday further investigated the optical properties of the colloidal gold. He prepared the first pure sample of colloidal gold, which he called 'activated gold', in 1857. He used phosphorus to reduce a solution of gold chloride. The colloidal gold Faraday made 150 years ago is still optically active. For a long time, the composition of the 'ruby' gold was unclear. Several chemists suspected it to be a gold tin compound, due to its preparation. Faraday recognized that the color was actually due to the miniature size of the gold particles. He noted the light scattering properties of suspended gold microparticles, which is now called Faraday-Tyndall effect.

In 1898, Richard Adolf Zsigmondy prepared the first colloidal gold in diluted solution. Apart from Zsigmondy, Theodor Svedberg, who invented ultracentrifugation, and Gustav Mie, who provided the theory for scattering and absorption by spherical particles, were also interested in the synthesis and properties of colloidal gold.

With advances in various analytical technologies in the 20th century, studies on gold nanoparticles has accelerated. Advanced microscopy methods, such as atomic force microscopy and electron microscopy, have contributed the most to nanoparticle research. Due to their comparably easy synthesis and high stability, various gold particles have been studied for their practical uses. Different types of gold nanoparticle are already used in many industries, such as medicine and electronics. For example, several FDA-approved nanoparticles are currently used in drug delivery.

Physical Properties

Optical

Colloidal gold has been used by artists for centuries because of the nanoparticle's interactions with visible light. Gold nanoparticles absorb and scatter light with incredible efficiency. Ranging from vibrant reds to blues to black and finally to clear and colorless, colloidal gold has the ability to exhibit a wide range of colors depending on particle size, shape, local refractive index, and aggregation state. These colors occur because of a phenomenon called Localized Surface Plasmon Resonance (LSPR), in which conduction electrons on the surface of the nanoparticle oscillate in resonance with incident light.

Effect of Size

As a general rule, the wavelength of light absorbed increases as a function of increasing nano particle size. For example, pseudo-spherical gold nanoparticles with diameters ~ 30 nm have a peak LSPR absorption at ~530 nm.

Effect of Local Refractive Index

Changes in the apparent color of a gold nanoparticle solution can also be caused by the environment in which the colloidal gold is suspended The optical properties of gold nanoparticles depends on the refractive index near the nanoparticle surface, therefore both the molecules directly attached to the nanoparticle surface (i.e. nanoparticle ligands) and/or the nanoparticle solvent both may influence observed optical features. As the refractive index near the gold surface increases, the NP LSPR will shift to longer wavelengths In addition to solvent environment, the extinction peak can be tuned by coating the nanoparticles with non-conducting shells such as silica, bio molecules, or aluminium oxide.

Effect of Aggregation

When gold nano particles aggregate, the optical properties of the particle change, because the effective particle size, shape, and dielectric environment all change. Gold nanoparticles aggregate particular crown ether.

Applications

Electron Microscopy

Colloidal gold and various derivatives have long been among the most widely used labels for antigens in biological electron microscopy. Colloidal gold particles can be attached to many traditional biological probes such as antibodies, lectins, superantigens, glycans, nucleic acids, and receptors. Particles of different sizes are easily distinguishable in electron micrographs, allowing simultaneous multiple-labelling experiments.

In addition to biological probes, gold nanoparticles can be transferred to various mineral substrates, such as mica, single crystal silicon, and atomically flat gold(III), to be observed under atomic force microscopy (AFM).

Medical Research

Drug Delivery System

Gold nanoparticles can be used to optimize the biodistribution of drugs to diseased organs, tissues or cells, in order to improve and target drug delivery. It is important to realize that the nanoparticle-mediated drug delivery is feasible only if the drug distribution is otherwise inadequate. These cases include drug targeting of difficult, unstable molecules (proteins, siRNA, DNA), delivery to the difficult sites (brain, retina, tumors, intracellular organelles) and drugs with serious side effects (e.g. anti-cancer agents). The performance of the nanoparticles depends on the size and surface functionalities in the particles. Also, the drug release and particle disintegration can vary depending on the system (e.g. biodegradable polymers sensitive to pH). An optimal nanodrug delivery system ensures that the active drug is available at the site of action for the correct time and duration, and their concentration should be above the minimal effective concentration (MEC) and below the minimal toxic concentration (MTC).

Gold nanoparticles are being investigated as carriers for drugs such as Paclitaxel. The administration of hydrophobic drugs require molecular encapsulation and it is found that nanosized particles are particularly efficient in evading the reticuloendothelial system.

Gold nanoparticles are also used to circumvent multidrug resistance (MDR) mechanisms. Mechanisms of MDR include decreased uptake of drugs, reduced intracellular drug concentration by activation of the efflux transporters, modifications in cellular pathways by altering cell cycle checkpoints, increased metabolism of drugs, induced emergency response genes to impair apoptotic pathways and altered DNA repair mechanisms.

Tumor Detection

In cancer research, colloidal gold can be used to target tumors and provide detection using SERS (surface enhanced Raman spectroscopy) *in vivo*. These gold nanoparticles are surrounded with Raman reporters, which provide light emission that is over 200 times brighter than quantum dots. It was found that the Raman reporters were stabilized when the nanoparticles were encapsulated with a thiol-modified polyethylene glycol coat. This allows for compatibility and circulation *in vivo*. To specifically target tumor cells, the polyethylenegylated gold particles are conjugated with an antibody (or an antibody fragment such as scFv), against, e.g. epidermal growth factor receptor, which is sometimes overexpressed in cells of certain cancer types. Using SERS, these pegylated gold nanoparticles can then detect the location of the tumor.

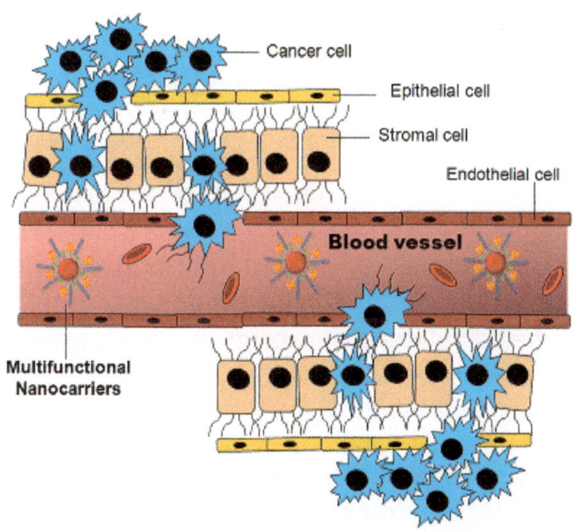

Tumor targeting via multifunctional nanocarriers. Cancer cells reduce adhesion to neighboring cells and migrate into the vasculature-rich stroma. Once at the vasculature, cells can freely enter the bloodstream. Once the tumor is directly connected to the main blood circulation system, multifunctional nanocarriers can interact directly with cancer cells and effectively target tumors.

Gold nanoparticles accumulate in tumors, due to the leakiness of tumor vasculature, and can be used as contrast agents for enhanced imaging in a time-resolved optical tomography system using short-pulse lasers for skin cancer detection in mouse model. It is found that intravenously administrated spherical gold nanoparticles broadened the temporal profile of reflected optical signals and enhanced the contrast between surrounding normal tissue and tumors.

Therefore, gold nanoparticles have the potential to join numerous therapeutic functions into a single platform, by targeting specific tumor cells, tissues and organs. Actually, Conde et al. reported the evaluation of the inflammatory response and therapeutic siRNA silencing via RGD-nanoparticles in a lung cancer mouse model. This study reported the use of siRNA/RGD gold nanoparticles capable of targeting tumor cells in two lung cancer xenograft mouse models, resulting in successful and significant *c-Myc*oncogene downregulation followed by tumor growth inhibition and prolonged survival of the animals. This delivery system can achieve translocation of siRNA duplexes directly into the tumour cell cytoplasm and accomplish successful silencing of an oncogene expression. Actually, RGD/siRNA-AuNPs can target preferentially and be taken up by tumor cells via integrin $\alpha v\beta 3$-receptor-mediated endocytosis with no cytotoxicity, showing that can accumulate in tumor tissues overexpressing $\alpha v\beta 3$ integrins and selectively delivered *c-Myc* siRNA to suppress tumor growth and angiogenesis.

Gene Therapy

Gene therapy is receiving increasing attention and, in particular, small-interference RNA (siRNA) shows importance in novel molecular approaches in the knockdown of specific gene expression in cancerous cells. The major obstacle to clinical application is the uncertainty about how to deliver therapeutic siRNAs with maximal therapeutic impact. Gold nanoparticles have shown potential as intracellular delivery vehicles for siRNA oligonucleotides with maximal therapeutic impact.

Recently, Conde et al. provided evidence of *in vitro* and *in vivo* RNAi triggering via the synthesis of a library of novel multifunctional gold nanoparticles, using a hierarchical approach including three biological systems of increasing complexity: *in vitro* cultured human cells, *in vivo* freshwater polyp (*Hydra vulgaris*), and *in vivo* mice models. The authors developed effective conjugation strategies to combine, in a highly controlled way, specific biomolecules to the surface of gold nanoparticles such as: (a) biofunctional spacers: Poly(ethylene glycol) (PEG) spacers used to increase solubility and biocompatibility; (b) cell penetrating peptides such as TAT and RGD peptides: A novel class of membrane translocating agents named cell penetrating peptides (CPPs) that exploit more than one mechanism of endocytosis to overcome the lipophilic barrier of the cellular membranes and deliver large molecules and even small particles inside the cell for their biological actions; and (c) siRNA complementary to a master regulator gene, the protooncogene c-myc, were bond covalently (thiol-siRNA) and ionically (naked/unmodified siRNA) to gold nanoparticles.

Gold nanoparticles have also shown potential as intracellular delivery vehicles for antisense oligonucleotides (ssDNA,dsDNA) by providing protection against intracellular nucleases and ease of functionalization for selective targeting. Recently, Conde et al. developed a new theranostic system capable of intersecting all RNA pathways: from gene specific downregulation to silencing the silencers, i.e. siRNA and miRNA pathways. The authors reported the development gold nanoparticles functionalized with a fluorophore labeled hairpin-DNA, i.e. gold nanobeacons, capable of efficiently silencing single gene expression, exogenous siRNA and endogenous miRNAs while yielding a quantifiable fluorescence signal directly proportional to the level of silencing. This method describes a gold nanoparticle-based nanobeacon as an innovative theranostic approach for detection and inhibition of sequence-specific DNA and RNA for *in vitro* and *ex vivo* applications. Under hairpin configuration, proximity to gold nanoparticles leads to fluorescence quenching; hybridization to a complementary target restores fluorescence emission due to the gold nanobeacons' conformational reorganization that causes the fluorophore and the gold nanoparticle to part from each other. This concept can easily be extended and adapted to assist the in vitro evaluation of silencing potential of a given sequence to be later used for *ex vivo* gene silencing and RNAi approaches, with the ability to monitor real-time gene delivery action.

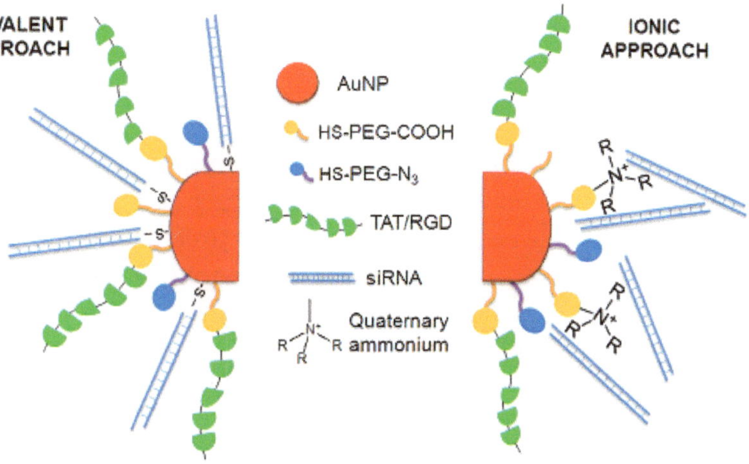

Multifunctional siRNA-gold nanoparticles with several biomolecules: PEG, cell penetration and cell adhesion peptides and siRNA. Two different approaches were employed to conjugate the siRNA to the gold nanoparticle: (1) Covalent approach: use of thiolated siRNA for gold-thiol binding to the nanoparticle; (2) Ionic approach: interaction of the negatively charged siRNA to the modified surface of the AuNP through ionic interactions.

Photothermal Agents

Gold nanorods are being investigated as photothermal agents for in-vivo applications. Gold nanorods are rod-shaped gold nanoparticles whose aspect ratios tune the surface plasmon resonance (SPR) band from the visible to near-infrared wavelength. The total extinction of light at the SPR is made up of both absorption and scattering. For the smaller axial diameter nanorods (~10 nm), absorption dominates, whereas for the larger axial diameter nanorods (>35 nm) scattering can dominate. As a consequence, for in-vivo applications, small diameter gold nanorods are being used as photothermal converters of near-infrared light due to their high absorption cross-sections. Since near-infrared light transmits readily through human skin and tissue, these nanorods can be used as ablation components for cancer, and other targets. When coated with polymers, gold nanorods have been observed to circulate in-vivo with half-lives longer than 6 hours, bodily residence times around 72 hours, and little to no uptake in any internal organs except the liver. Apart from rod-like gold nanoparticles, also spherical colloidal gold nanoparticles are recently used as markers in combination with photothermal single particle microscopy.

Radiotherapy Dose Enhancer

Following work by Hainfield et al. there has been considerable interest in the use of gold and other heavy-atom containing nanoparticles to enhance the dose delivered to tumors. Since the gold nanoparticles are taken up by the tumors more than the nearby healthy tissue, the dose is selectively enhanced. The biological effectiveness of this type of therapy seems to be due to the local deposition of the radiation dose near the nanoparticles. This mechanism is the same as occurs in heavy ion therapy.

Detection of Toxic Gas

Researchers have developed simple inexpensive methods for on-site detection of hydrogen sulfide H_2S present in air based on the antiaggregation of gold nanoparticles (AuNPs). Dissolving H_2S into a weak alkaline buff solution leads to the formation of HS-, which can stabilize AuNPs and ensure they maintain their red color allowing for visual detection of toxic levels of H_2S.

Gold Nanoparticle based Biosensor

Gold nanoparticles are incorporated into biosensors to enhance its stability, sensitivity, and selectivity. Nanoparticle properties such as small size, high surface-to-volume ratio, and high surface energy allow immobilization of large range of biomolecules. Gold nanoparticle, in particular, could also act as "electron wire" to transport electrons and its amplification effect on electromagnetic light allows it to function as signal amplifiers. Main types of gold nanoparticle based biosensors are optical and electrochemical biosensor.

Optical Biosensor

Gold nanoparticles improve the sensitivity of optical sensor by response to the change in local refractive index. The angle of the incidence light for surface plasmon resonance, an interaction between light wave and conducting electrons in metal, changes when other substances are bounded to the metal surface. Because gold is very sensitive to its surroundings' dielectric constant, binding of an analyte would significantly shift gold nanoparticle's SPR and therefore allow more sensitive detection. Gold nanoparticle could also amplify the SPR signal. When the plasmon wave pass through the gold nanoparticle, the charge density in the wave and the electron I the gold interacted and resulted in higher energy response, so called electron coupling. Since the analyte and bio-receptor now bind to the gold, it increases the apparent mass of the analyte and therefore amplified the signal. These properties had been used to build DNA sensor with 1000-fold sensitive than without the Au NP. Humidity senor was also built by altering the atom interspacing between molecules with humidity change, the interspacing change would also result in a change of the Au NP's LSPR.

Electrochemical Biosensor

Electrochemical sensor covert biological information into electrical signals that could be detected. The conductivity and biocompatibility of Au NP allow it to act as "electron wire". It transfers electron between the electrode and the active site of the enzyme. It could be accomplished in two ways: attach the Au NP to either the enzyme or the electrode. GNP-glucose oxidase monolayer electrode was constructed use these two methods. The Au NP allowed more freedom in the enzyme's orientation and therefore more sensitive and stable detection. Au NP also acts as immobilization platform for the enzyme. Most biomolecules denatures or lose its activity when interacted with the electrode. The biocompatibility and high surface energy of Au allow it to bind to a large amount of protein without altering its activity and results in a more sensitive sensor. Moreover, Au NP also catalyzes biological reactions. Gold nanoparticle under 2 nm has shown catalytic activity to the oxidation of styrene.

Thin Films

Gold nanoparticles capped with organic ligands, such as alkanethiol molecules, can self-assemble into large monolayers ($> cm^2$). The particles are first prepared in organic solvent, such as chloroform or toluene, and are then spread into monolayers either on a liquid surface or on a solid substrate.

The mechanical properties of nanoparticle monolayers have been studied extensively. For 5 nm spheres capped with dodecanethiol, the Young's modulus of the monolayer is on the order of GPa. The mechanics of the membranes are guided by strong interactions between ligand shells on adjacent particles. Upon fracture, the films crack perpendicular to the direction of strain at a fracture stress of 11 ± 2.6 MPa, comparable to that of cross-linked polymer films. Free-standing nanoparticle membranes exhibit bending rigidity on the order of 10^5 eV, higher than what is predicted in theory for continuum plates of the same thickness, due to nonlocal microstructural constraints such as nonlocal coupling of particle rotational degrees of freedom. On the other hand, resistance to bending is found to be greatly reduced in nanoparticle monolayers that are supported at the air/water interface, possibly due to screening of ligand interactions in a wet environment.

Surface Chemistry

In many different types of colloidal gold syntheses, the interface of the nanoparticles can display widely different character – ranging from an interface similar to a self-assembled monolayer to a disordered boundary with no repeating patterns. Beyond the Au-Ligand interface, conjugation of the interfacial ligands with various functional moieties (from small organic molecules to polymers to DNA to RNA) afford colloidal gold much of its vast functionality.

Ligand Exchange/Functionalization

After initial nanoparticle synthesis, colloidal gold ligands are often exchanged with new ligands designed for specific applications. For example, Au NPs produced via the Turkevich-style (or Citrate Reduction) method are readily reacted via ligand exchange reactions, due to the relatively weak binding between the carboxyl groups and the surfaces of the NPs. This ligand exchange can produce conjugation with a number of biomolecules from DNA to RNA to proteins to polymers (such as PEG) to increase biocompatibility and functionality. For example, ligands have been shown to enhance catalytic activity by mediating interactions between adsorbates and the active gold surfaces for specific oxygenation reactions. Ligand exchange can also be used to promote phase transfer of the colloidal particles. Ligand exchange is also possible with alkane thiol-arrested NPs produced from the Brust-type synthesis method, although higher temperatures are needed to promote the rate of the ligand detachment. An alternative method for further functionalization is achieved through the conjugation of the ligands with other molecules, though this method can cause the colloidal stability of the Au NPs to breakdown.

Ligand Removal

In many cases, as in various high-temperature catalytic applications of Au, the removal of the capping ligands produces more desirable physicochemical properties. The removal of ligands from colloidal gold while maintaining a relatively constant number of Au atoms per Au NP can be difficult due to the tendency for these bare clusters to aggregate. The removal of ligands is partially achievable by simply washing away all excess capping ligands, though this method is ineffective in removing all capping ligand. More often ligand removal achieved under high temperature or light ablation followed by washing. Alternatively, the ligands can be electrochemically etched off.

Surface Structure and Chemical Environment

The precise structure of the ligands on the surface of colloidal gold NPs impact the properties of the colloidal gold particles. Binding conformations and surface packing of the capping ligands at the surface of the colloidal gold NPs tend to differ greatly from bulk surface model adsorption, largely due to the high curvature observed at the nanoparticle surfaces. Thiolate-gold interfaces at the nanoscale have been well-studied and the thiolate ligands are observed to pull Au atoms off of the surface of the particles to for "staple" motifs that have significant Thiyl-Au(0) character. The citrate-gold surface, on the other hand, is relatively less-studied due to the vast number of binding conformations of the citrate to the curved gold surfaces. A study performed in 2014 identified that the most-preferred binding of the citrate involves two carboxylic acids and the hydroxyl group of the citrate binds three surface metal atoms.

Health and Safety

As gold nanoparticles (AuNPs) are further investigated for targeted drug delivery in humans, their toxicity needs to be considered. For the most part, it is suggested that AuNPs are biocompatible, but it is important to ask at what concentration they would be toxic, and if that concentration falls within the range of used concentrations. Toxicity can be tested *in vitro* and *in vivo*. *In vitro* toxicity results can vary depending on the type of the cellular growth media with different protein compositions, the method used to determine cellular toxicity (cell health, cell stress, how many cells are taken into a cell), and the capping ligands in solution.*In vivo* assessments can determine the general health of an organism (abnormal behavior, weight loss, average life span) as well as tissue specific toxicology (kidney, liver, blood) and inflammation and oxidative responses.*In vitro* experiments are more popular than *in vivo*experiments because *in vitro* experiments are more simplistic to perform than *in vivo* experiments.

Toxicity and Hazards in Synthesis

While AuNPs themselves appear to have low or negligible toxicity, and the literature shows that the toxicity has much more to do with the ligands rather than the particles themselves, the synthesis of them involves chemicals that are hazardous. Sodium borohydride, a harsh reagent, is used to reduce the gold ions to gold metal. The gold ions usually come from chloroauric acid, a potent acid. Because of the high toxicity and hazard of reagents used to synthesize AuNPs, the need for more "green" methods of synthesis arose.

Toxicity due to Capping Ligands

Some of the capping ligands associated with AuNPs can be toxic while others are nontoxic. In gold nanorods (AuNRs), it has been shown that a strong cytotoxicity was associated with CTAB-stabilized AuNRs at low concentration, but it is thought that free CTAB was the culprit in toxicity. Modifications that overcoat these AuNRs reduces this toxicity in human colon cancer cells (HT-29) by preventing CTAB molecules from desorbing from the AuNRs back into the solution. Ligand toxicity can also be seen in AuNPs. Compared to the 90% toxicity of HAuCl4 at the same concentration, AuNPs with carboxylate termini were shown to be non-toxic. Large AuNPs conjugated with biotin, cysteine, citrate, and glucose were not toxic in human leukemia cells (K562) for concentrations up to 0.25 M. Also, citrate-capped gold nanospheres (AuNSs) have been proven to be compatible with human blood and did not cause platelet aggregation or an immune response. However, citrate-capped gold nanoparticles sizes 8-37 nm were found to be lethally toxic for mice, causing shorter lifespans, severe sickness, loss of appetite and weight, hair discoloration, and damage to the liver, spleen, and lungs; gold nanoparticles accumulated in the spleen and liver after traveling a section of the immune system. There are mixed-views for polyethylene glycol (PEG)-modified AuNPs. These AuNPs were found to be toxic in mouse liver by injection, causing cell death and minor inflammation. However, AuNPs conjugated with PEG copolymers showed negligible toxicity towards human colon cells (Caco-2). AuNP toxicity also depends on the overall charge of the ligands. In certain doses, AuNSs that have positively-charged ligands are toxic in monkey kidney cells (Cos-1), human red blood cells, and E. coli because of the AuNSs interaction with the negatively-charged cell membrane; AuNSs with negatively-charged ligands have been found to be nontoxic in these species. In addition to the previously mentioned "in vivo" and "in vitro " experiments, other similar experiments have been performed. Alkylthiolate-AuNPs with trimethlyammonium ligand termini mediate the translocation of DNA across mammalian cell membranes "in vitro" at a high level, which is detrimental to these cells. Corneal haze in rabbits have been healed "in vivo" by using polyethylemnimine-capped gold nanoparticles that were transfected with a gene that promotes wound healing and inhibits corneal fibrosis.

Toxicity due to Size of Nanoparticles

Toxicity in certain systems can also be dependent on the size of the nanoparticle. AuNSs size 1.4 nm were found to be toxic in human skin cancer cells (SK-Mel-28), human cervical cancer cells (HeLa), mouse fibroblast cells (L929), and mouse macrophages (J774A.1), while 0.8, 1.2, and 1.8 nm sized AuNSs were less toxic by a six-fold amount and 15 nm AuNSs were nontoxic. There is some evidence for AuNP buildup after injection in "in vivo" studies, but this is very size dependent. 1.8 nm AuNPs were found to be almost totally trapped in the lungs of rats. Different sized AuNPs were found to buildup in the blood, brain, stomach, pancreas, kidneys, liver, and spleen.

Synthesis

Generally, gold nanoparticles are produced in a liquid ("liquid chemical methods") by reduction of chloroauric acid ($H[AuCl_4]$). After dissolving $H[AuCl_4]$, the solution is rapidly stirred while a reducing agent is added. This causes Au^{3+} ions to be reduced to Au^+ ions. Then a disproportionation reaction occurs whereby 3 Au^+ ions give rise to Au^{3+} and 2 Au^0 atoms. The Au^0 atoms act as center of nucleation around which further Au^+ ions gets reduced. To prevent the particles from aggregating, some sort of stabilizing agent that sticks to the nanoparticle surface is usually added. In the Turkevich method of Au NP synthesis, citrate initially acts as the reducing agent and finally as the capping agent which stabilizes the Au NP through electrostatic interactions between the lone pair of electrons on the oxygen and the metal surface. Also, gold colloids can be synthesised without stabilizers by laser ablation in liquids.

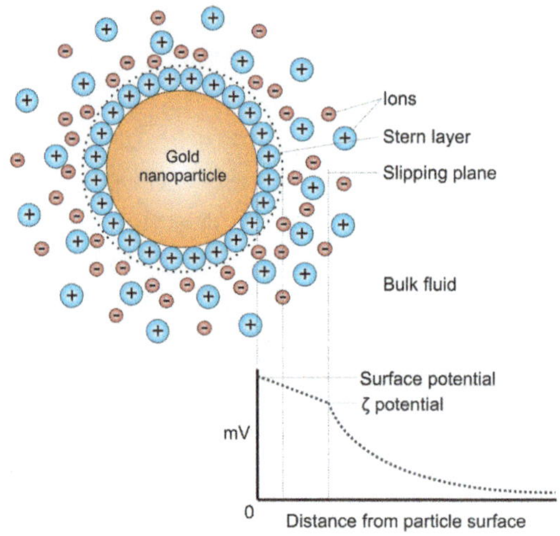

Potential difference as a function of distance from particle surface

They can be functionalized with various organic ligands to create organic-inorganic hybrids with advanced functionality.

Turkevich Method

The method pioneered by J. Turkevich et al. in 1951 and refined by G. Frens in the 1970s, is the simplest one available. In general, it is used to produce modestly monodisperse spherical gold nanoparticles suspended in water of around 10–20 nm in diameter. Larger particles can be produced, but this comes at the cost of monodispersity and shape. It involves the reaction of small amounts of hot chloroauric acid with small amounts of sodium citrate solution. The colloidal gold will form because the citrate ions act as both a reducing agent and a capping agent. A capping agent is used in nanoparticle synthesis to stop particle growth and aggregation. A good capping agent has a high affinity for the new nuclei so it will bind to surface atoms which stabilizes the surface energy of the new nuclei and makes so that they cannot bind to other nuclei.

Recently, the evolution of the spherical gold nanoparticles in the Turkevich reaction has been elucidated. It is interesting to note that extensive networks of gold nanowires are formed as a transient intermediate. These gold nanowires are responsible for the dark appearance of the reaction solution before it turns ruby-red.

To produce larger particles, less sodium citrate should be added (possibly down to 0.05%, after which there simply would not be enough to reduce all the gold). The reduction in the amount of sodium citrate will reduce the amount of the citrate ions available for stabilizing the particles, and this will cause the small particles to aggregate into bigger ones (until the total surface area of all particles becomes small enough to be covered by the existing citrate ions).

Brust-schiffrin Method

This method was discovered by Brust and Schiffrin in the early 1990s, and can be used to produce gold nanoparticles in organic liquids that are normally not miscible with water (like toluene). It involves the reaction of a chlorauric acid solution with tetraoctylammonium bromide (TOAB) solution in toluene and sodium borohydride as an anti-coagulant and a reducing agent, respectively.

Here, the gold nanoparticles will be around 5–6 nm. $NaBH_4$ is the reducing agent, and TOAB is both the phase transfer catalyst and the stabilizing agent.

It is important to note that TOAB does not bind to the gold nanoparticles particularly strongly, so the solution will aggregate gradually over the course of approximately two weeks. To prevent this, one can add a stronger binding agent, like a thiol (in particular, alkanethiols), which will bind to gold, producing a near-permanent solution. Alkanethiol protected gold nanoparticles can be precipitated and then redissolved. Thiols are better binding agents because there is a strong affinity for the gold-sulfur bonds that form when the two substances react with each other. Tetra-dodecanthiol is a commonly used strong binding agent to synthesize smaller particles. Some of the phase transfer agent may remain bound to the purified nanoparticles, this may affect physical properties such as solubility. In order to remove as much of this agent as possible, the nanoparticles must be further purified by soxhlet extraction.

Perrault Method

This approach, discovered by Perrault and Chan in 2009, uses hydroquinone to reduce $HAuCl_4$ in an aqueous solution that contains 15 nm gold nanoparticle seeds. This seed-based method of synthesis is similar to that used in photographic film development, in which silver grains within the film grow through addition of reduced silver onto their surface. Likewise, gold nanoparticles can act in conjunction with hydroquinone to catalyze reduction of ionic gold onto their surface. The presence of a stabilizer such as citrate results in controlled deposition of gold atoms onto the particles, and growth. Typically, the nanoparticle seeds are produced using the citrate method. The hydroquinone method complements that of Frens, as it extends the range of monodispersed spherical particle sizes that can be produced. Whereas the Frens method is ideal for particles of 12–20 nm, the hydroquinone method can produce particles of at least 30–300 nm.

Martin Method

This simple method, discovered by Martin and Eah in 2010, generates nearly monodisperse "naked" gold nanoparticles in water. Precisely controlling the reduction stoichiometry by adjusting the ratio of $NaBH_4$-NaOH ions to $HAuCl_4$-HCl ions within the "sweet zone," along with heating, enables reproducible diameter tuning between 3–6 nm. The aqueous particles are colloidally stable due to their high charge from the excess ions in solution. These particles can be coated with various hydrophilic functionalities, or mixed with hydrophobic molecules for applications in non-polar solvents. In non-polar solvents the nanoparticles remain highly charged, and self-assemble on liquid droplets to form 2D monolayer films of monodisperse nanoparticles.

Nanotech Applications

Bacillus licheniformis can be used in synthesis of gold nanocubes with sizes between 10 and 100 nanometres. Gold nanoparticles are usually synthesized at high temperatures in organic solvents or using toxic reagents. The bacteria produce them in much milder conditions.

Navarro et al. Method

The precise control of particle size with a low polydispersity of spherical gold nanoparticles remains difficult for particles larger than 30 nm. In order to provide maximum control on the NP structure, Navarro and co-workers used a modified Turkevitch-Frens procedure using sodium acetylacetonate (Na(acac) as the reducing agent and sodium citrate as the stabilizer.

Sonolysis

Another method for the experimental generation of gold particles is by sonolysis. The first method of this type was invented by Baigent and Müller. This work pioneered the use of ultrasound to provide the energy for the processes involved and allowed the creation of gold particles with a diameter of under 10 nm. In another method using ultrasound, the reaction of an aqueous solution of $HAuCl_4$ with glucose, the reducing agents are hydroxyl radicals and sugar pyrolysisradicals (forming at the interfacial region between the collapsing cavities and the bulk water) and the morphology obtained is that of nanoribbons with width 30–50 nm and length of several micrometers. These ribbons are very flexible and can bend with angles larger than 90°. When glucose is replaced by cyclodextrin (a glucose oligomer), only spherical gold particles are obtained, suggesting that glucose is essential in directing the morphology toward a ribbon.

Block Copolymer-mediated Method

An economical, environmentally benign and fast synthesis methodology for gold nanoparticles using block copolymer has been developed by Sakai et al. In this synthesis methodology, block copolymer plays the dual role of a reducing agent as well as a stabilizing agent. The formation of gold nanoparticles comprises three main steps: reduction of gold salt ion by block copolymers in the solution and formation of gold clusters, adsorption of block copolymers on gold clusters and further reduction of gold salt ions on the surfaces of these gold clusters for the growth of gold particles in steps, and finally its stabilization by block copolymers. But this method usually has a limited-yield (nanoparticle concentration), which does not increase with the increase in the gold salt concentration. Recently, Ray et al. demonstrated that the presence of an additional reductant (trisodium citrate) in 1:1 mole ratio with gold salt enhances the yield by manyfold at ambient conditions and room temperature.

"Green Chemistry" based Methods

Methods Employing Phytochemicals

Phytochemicals found in various plant sources have been utilized as a means of developing a more economical and environmentally friendly synthetic pathway in the formation of gold nanoparticles. In accordance to the principles of "green chemistry," these methods employ the use of non-toxic chemicals, marginal energy consumption, renewable materials, and environmentally benign solvents to minimize the use, disposal, and health repercussions of hazardous chemicals. Additionally, these methods provide a more efficient synthetic pathway through a one-step process without the use of supplementary surfactants or polymers, capping agents, or templates to restrict agglomeration of the gold nanoparticles. This method is effective in producing well-defined gold nanoparticles since the phytochemicals perform a dual role as both a reducing agent of gold and as a stabilizer in the formation of a sturdy coating on the nanoparticles. One "green" method that has been employed in the formation of gold nanoparticles utilizes the phytochemicals and polyphenols in Darjeeling black tea leaves, with water acting as a benign solvent at room temperature.

The phytochemicals in the black tea reduce $HAuCl_4$ and stabilize the aggregation of the gold atoms as the nanoparticle is formed. In addition to the "green" benefits of using black tea, the size of the nanoparticle is influenced by the concentration of the tea, and the absorbance and size of the gold nanoparticles formed can be easily determined using UV-Vis spectrometry abiding that an increase in λmax correlates to an increase in the size of the nanoparticle.

Other "green" methods that have been studied and employed include the use of *Elettaria cardamomum* (cardamom) and cinnamon in the synthesis of gold nanoparticles, as well as *Syzygium aromaticum* (cloves) in the formation of copper nanoparticles and table sugar in the formation of silver nanoparticles.

Gold Nanoparticles as a Benign Starting Material to Access Gold Sponges

Besides using phytochemicals from plant sources to act as reducing agents and stabilizing agents, several other approaches have been taken to achieve more "green" approaches to gold nanoparticle syntheses. One such approach employs thiolated poly(ethylene glycol) (PEG Thiol) to destabilize gold nanoparticles prepared by citrate reduction so that they self-assemble into mesoporous gold sponges. Mesoporous gold sponges are attractive materials for molecular sensing by Surface-Enhanced Raman Spectroscopy (SERS), for catalysis, and for fuel cell construction. The following approach is "green" because PEG Thiol is biocompatible, and because it requires relatively little energy; PEG Thiol-triggered self-assembly of mesoporous gold sponges occurs at room temperature. By contrast, the most popular method of generating mesoporous gold sponges, dealloying Au-Ag alloys, employs electrochemical corrosion.

Gold Nanoparticle Synthesis in Flow

Another green approach is a modification to the Turkevich citrate reduction making use of flow chemistry, reported by Bayazit et al. Flow chemistry is an appealing replacement for many heated batch reactions. By exposing more surface area of the reaction to the heating element, flow reactors heat a reaction faster and more evenly than a batch reactor can, promoting rapid nucleation and smaller particle sizes with higher monodispersity.

Self-heating Synthesis of Gold Nanoparticles and Gold Nanowires

Another green approach is a self-heating synthesis method reported by Yin et al. This method utilize the rapid heating effect from neutralization between primary amines and carboxylic acids to reduce gold precursor, and doesn't require external heating apparatus. The synthesis of gold nanoparticles or gold nanowires can be achieved within minutes by this method, which is suitable for large scale continuous production of colloidal gold.

Gold Nanobeacon

Gold nanobeacons are gold nanoparticles functionalized with a fluorophore-labelled hairpin-DNA.

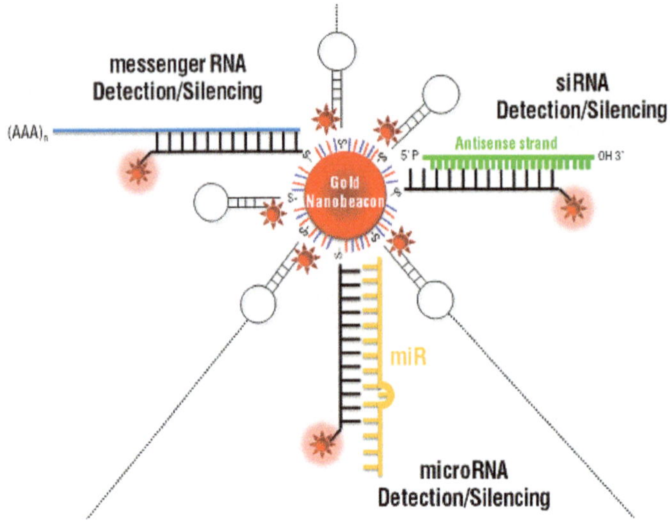

Universal gene silencing tool based on gold nanoparticles functionalized with a fluorophore labeled hairpin-DNA – Gold nanobeacon (Au-nanobeacon) – that effectively detects and silences the specific target (siRNA, miRNA or gene specific mRNA) while simultaneously signalling its action via fluorescence emission in cancer cells.

Theranostics

Recently, Conde et al. developed a new theranostic system capable of intersecting all RNA pathways: from gene specific downregulation to silencing the silencers, i.e. siRNA and miRNA pathways. The authors reported the development of new nanomaterials, i.e. gold nanoparticles functionalized with a fluorophore labeled hairpin-DNA.

Bioresponsive Antisense DNA Gold Nanobeacons for the Inhibition of Cancer Cells and Metastasis

Gold nanobeacons can be used as a tool for cancer theranostics. Recently, Conde et al. developed a nanomaterial platform based on gold nanobeacons to detect, target and inhibit the expression of a mutant Kras gene in an in vivo murine gastric cancer model. The conjugation of fluorescently-labeled antisense DNA hairpin oligonucleotides to the surface of gold nanoparticles enables using their localized surface plasmon resonance properties to directly track the delivery to the primary gastric tumor and to lung metastatic sites. The fluorescently labeled nanobeacons reports on the interaction with the target as the fluorescent Cy3 signal is quenched by the gold nanoparticle and only emit light following conjugation to the Kras target owing to reorganization and opening of the nanobeacons, thus increasing the distance between the dye and the quencher. The systemic administration of the anti-Kras nanobeacons resulted in approximately 60% tumor size reduction and a 90% reduction in tumor vascularization. More important, the inhibition of the Kras gene expression in gastric tumors prevents the occurrence of metastasis to lung (80% reduction), increasing mice survival in more than 85%. The developed platform can be easily adjusted to hybridize with any specific target and provide facile diagnosis and treatment for neoplastic diseases.

2-pair FRET/NSET Nanoswitch to Sense, Inhibit and Deliver Drugs

Now with one single local application using hydrogel scaffolds embedded with a 2-pair FRET/ NSET (Fluorescence Resonance Energy Transfer/NanoSurface Energy Transfer) gold nanobeacons (also known as dark-gold nanobeacons) we are able to overcome drug resistance by detecting and silencing a multidrug resistance protein (MRP1), before chemotherapeutic drug delivery in vivo. Our platform contains hydrogel embedded with dark-gold nanoparticles modified with 5-fluorouracil (5-FU)-intercalated nanobeacons that serve as an ON/OFF molecular nanoswitch triggered by the increased MRP1 expression within the tumour tissue microenvironment. This nanoswitch can sense and overcome multidrug resistance (MDR) prior to local drug release. These nanoprobes comprise a thiol-DNA-hairpin labelled with a NIR dye and a thiol-DNA oligo labelled with a dark quencher (BHQ2), polyethylene glycol (PEG), and intercalated drug- all of which are conjugated to a gold nanoparticle core. These dark-gold nanobeacons are then loaded with 5-FU that intercalates in the beacon stem (dsDNA part) of the DNA-hairpin oligo. Under hairpin configuration, the proximity of the NIR dye to the dark quencher leads to fluorescence quenching. Hybridization of the DNA hairpin to a complementary mRNA target restores fluorescence emission due to the gold nanobeacons' conformational reorganization that causes the fluorophore and the quencher to part from each other, yielding a quantitative response. On the other hand, the release of the 5-FU drug can only occur when DNA hairpin hybridizes with the complementary target and can be measured once the distance from the 5-FU and the gold core increases, escalating the drug emission. To evaluate the efficiency of the dark-gold nanobeacon probes in sensing and in overcoming MDR in vivo, an orthotopic breast cancer mouse model was developed by injecting 5-FU resistant MDAMB-231 cells to the mammary fat pad of female SCID hairless congenic mice. Efficacious and local delivery of the dark-gold nanobeacon probes is achieved by the implantation of a hydrogel disk on top of the triple-negative breast tumours using a polyamidoamine (PAMAM G5) dendrimer cross-linked with dextran aldehyde, which provides enhanced stability of the embedded nanoparticles. Despite the cross-resistance to 5-FU, more than 90% tumour reduction is achieved in vivo, following 80% MRP1 silencing compared with the continuous tumour growth following only drug or nonsense nanobeacon administration.

Silver Nanoparticle

Silver nanoparticles are nanoparticles of silver of between 1 nm and 100 nm in size. While frequently described as being 'silver' some are composed of a large percentage of silver oxide due to their large ratio of surface-to-bulk silver atoms. Numerous shapes of nanoparticles can be constructed depending on the application at hand. Commonly used are spherical silver nanoparticles but diamond, octagonal and thin sheets are also popular.

Their extremely large surface area permits the coordination of a vast number of ligands. The properties of silver nanoparticles applicable to human treatments are under investigation in laboratory and animal studies, assessing potential efficacy, toxicity, and costs.

Synthetic Methods

Wet Chemistry

The most common methods for nanoparticle synthesis fall under the category of wet chemistry, or the nucleation of particles within a solution. This nucleation occurs when a silver ion complex, usually $AgNO_3$ or $AgClO_4$, is reduced to colloidal silver in the presence of a reducing agent. When the concentration increases enough, dissolved metallic silver ions bind together to form a stable surface. The surface is energetically unfavorable when the cluster is small, because the energy gained by decreasing the concentration of dissolved particles is not as high as the energy lost from creating a new surface. When the cluster reaches a certain size, known as the critical radius, it becomes energetically favorable, and thus stable enough to continue to grow. This nucleus then remains in the system and grows as more silver atoms diffuse through the solution and attach to the surface When the dissolved concentration of atomic silver decreases enough, it is no longer possible for enough atoms to bind together to form a stable nucleus. At this nucleation threshold, new nanoparticles stop being formed, and the remaining dissolved silver is absorbed by diffusion into the growing nanoparticles in the solution.

As the particles grow, other molecules in the solution diffuse and attach to the surface. This process stabilizes the surface energy of the particle and blocks new silver ions from reaching the surface. The attachment of these capping/stabilizing agents slows and eventually stops the growth of the particle. The most common capping ligands are trisodium citrate and polyvinylpyrrolidone (PVP), but many others are also used in varying conditions to synthesize particles with particular sizes, shapes, and surface properties.

There are many different wet synthesis methods, including the use of reducing sugars, citrate reduction, reduction via sodium borohydride, the silver mirror reaction, the polyol process, seed-mediated growth, and light-mediated growth. Each of these methods, or a combination of methods, will offer differing degrees of control over the size distribution as well as distributions of geometric arrangements of the nanoparticle.

A new, very promising wet-chemical technique was found by Elsupikhe et al. (2015). They have developed a green ultrasonically-assisted synthesis. Under ultrasound treatment, silver nanoparticles (AgNP) are synthesized with κ-carrageenan as a natural stabilizer. The reaction is performed at ambient temperature and produces silver nanoparticles with fcc crystal structure without impurities. The concentration of κ-carrageenan is used to influence particle size distribution of the AgNPs.

Monosaccharide Reduction

There are many ways silver nanoparticles can be synthesized; one method is through monosaccharides. This includes glucose, fructose, maltose, maltodextrin, etc. but not sucrose. It is also a simple method to reduce silver ions back to silver nanoparticles as it usually involves a one-step process,. There have been methods that indicated that these reducing sugars are essential to the formation of silver nanoparticles. Many studies indicated that this method of green synthesis, specifically using Cacumen platycladi extract, enabled the reduction of silver. Additionally, the size of the nanoparticle could be controlled depending on the concentration of the extract. The studies indicate that the higher concentrations correlated to an increased number of nanoparticles. Smaller nanoparticles were formed at high pH levels due to the concentration of the monosaccharides.

Another method of silver nanoparticle synthesis includes the use of reducing sugars with alkali starch and silver nitrate. The reducing sugars have free aldehyde and ketone groups, which enable them to be oxidized into gluconate. The monosaccharide must have a free ketone group because in order to act as a reducing agent it first undergoes tautomerization. In addition, if the aldehydes are bound, it will be stuck in cyclic form and cannot act as a reducing agent. For example, glucose has an aldehyde functional group that is able to reduce silver cations to silver atoms and is then oxidized to gluconic acid. The reaction for the sugars to be oxidized occurs in aqueous solutions. The capping agent is also not present when heated.

Citrate Reduction

An early, and very common, method for synthesizing silver nanoparticles is citrate reduction. This method was first recorded by M. C. Lea, who successfully produced a citrate-stabilized silver colloid in 1889. Citrate reduction involves the reduction of a silver source particle, usually $AgNO_3$ or $AgClO_4$, to colloidal silver using trisodium citrate, $Na_3C_6H_5O_7$. The synthesis is usually performed at an elevated temperature (~100 °C) to maximize the monodispersity (uniformity in both size and shape) of the particle. In this method, the citrate ion traditionally acts as both the reducing agent and the capping ligand, making it a useful process for AgNP production due to its relative ease and short reaction time. However, the silver particles formed may exhibit broad size distributions and form several different particle geometries simultaneously. The addition of stronger reducing agents to the reaction is often used to synthesize particles of a more uniform size and shape.

Reduction Via Sodium Borohydride

The synthesis of silver nanoparticles by sodium borohydride ($NaBH_4$) reduction occurs by the following reaction:

$$Ag^{+} + BH_4^{-} + 3H_2O \rightarrow Ag^{0} + B(OH)_3 + 3.5H_2$$

The reduced metal atoms will form nanoparticle nuclei. Overall, this process is similar to the above reduction method using citrate. The benefit of using sodium borohydride is increased monodispersity of the final particle population. The reason for the increased monodispersity when using $NaBH_4$ is that it is a stronger reducing agent than citrate. The impact of reducing agent strength can be seen by inspecting a LaMer diagram which describes the nucleation and growth of nanoparticles.

When silver nitrate ($AgNO_3$) is reduced by a weak reducing agent like citrate, the reduction rate is lower which means that new nuclei are forming and old nuclei are growing concurrently. This is the reason that the citrate reaction has low monodispersity. Because $NaBH_4$ is a much stronger reducing agent, the concentration of silver nitrate is reduced rapidly which shortens the time during which new nuclei form and grow concurrently yielding a monodispersed population of silver nanoparticles.

Particles formed by reduction must have their surfaces stabilized to prevent undesirable particle agglomeration (when multiple particles bond together), growth, or coarsening. The driving force for these phenomena is the minimization of surface energy (nanoparticles have a large surface to volume ratio). This tendency to reduce surface energy in the system can be counteracted by adding species which will adsorb to the surface of the nanoparticles and lowers the activity of the particle surface thus preventing particle agglomeration according to the DLVO theory and preventing growth by occupying attachment sites for metal atoms. Chemical species that adsorb to the surface of nanoparticles are called ligands. Some of these surface stabilizing species are: $NaBH_4$ in large amounts, poly(vinyl pyrrolidone) (PVP), sodium dodecyl sulfate (SDS), and/or dodecane thiol.

Once the particles have been formed in solution they must be separated and collected. There are several general methods to remove nanoparticles from solution, including evaporating the solvent phase or the addition of chemicals to the solution that lower the solubility of the nanoparticles in the solution. Both methods force the precipitation of the nanoparticles.

Polyol Process

The polyol process is a particularly useful method because it yields a high degree of control over both the size and geometry of the resulting nanoparticles. In general, the polyol synthesis begins with the heating of a polyol compound such as ethylene glycol, 1,5-pentanediol, or 1,2-propylene glycol7. An Ag^+ species and a capping agent are added (although the polyol itself is also often the capping agent). The Ag^+ species is then reduced by the polyol to colloidal nanoparticles. The polyol process is highly sensitive to reaction conditions such as temperature, chemical environment, and concentration of substrates. Therefore, by changing these variables, various sizes and geometries can be selected for such as quasi-spheres, pyramids, spheres, and wires. Further study has examined the mechanism for this process as well as resulting geometries under various reaction conditions in greater detail.

Seed-mediated Growth

Seed-mediated growth is a synthetic method in which small, stable nuclei are grown in a separate chemical environment to a desired size and shape. Seed-mediated methods consist of two different stages: nucleation and growth. Variation of certain factors in the synthesis (e.g. ligand, nucleation time, reducing agent, etc.), can control the final size and shape of nanoparticles, making seed-mediated growth a popular synthetic approach to controlling morphology of nanoparticles.

The nucleation stage of seed-mediated growth consists of the reduction of metal ions in a precursor to metal atoms. In order to control the size distribution of the seeds, the period of nucleation should be made short for monodispersity. The LaMer model illustrates this concept. Seeds typically consist small nanoparticles, stabilized by a ligand. Ligands are small, usually organic molecules that bind to the surface of particles, preventing seeds from further growth. Ligands are necessary as they increase the energy barrier of coagulation, preventing agglomeration. The balance between attractive and repulsive forces within colloidal solutions can be modeled by DLVO theory. Ligand binding affinity, and selectivity can be used to control shape and growth. For seed synthesis, a ligand with medium to low binding affinity should be chosen as to allow for exchange during growth phase.

The growth of nanoseeds involves placing the seeds into a growth solution. The growth solution requires a low concentration of a metal precursor, ligands that will readily exchange with preexisting seed ligands, and a weak or very low concentration of reducing agent. The reducing agent must not be strong enough to reduce metal precursor in the growth solution in the absence of seeds. Otherwise, the growth solution will form new nucleation sites instead of growing on preexisting ones (seeds). Growth is the result of the competition between surface energy (which increases unfavorably with growth) and bulk energy (which decreases favorably with growth). The balance between the energetics of growth and dissolution is the reason for uniform growth only on preexisting seeds (and no new nucleation). Growth occurs by the addition of metal atoms from the growth solution to the seeds, and ligand exchange between the growth ligands (which have a higher bonding affinity) and the seed ligands.

Range and direction of growth can be controlled by nanoseed, concentration of metal precursor, ligand, and reaction conditions (heat, pressure, etc.). Controlling stoichiometric conditions of growth solution controls ultimate size of particle. For example, a low concentration of metal seeds to metal precursor in the growth solution will produce larger particles. Capping agent has been shown to control direction of growth and thereby shape. Ligands can have varying affinities for binding across a particle. Differential binding within a particle can result in dissimilar growth across particle. This produces anisotropic particles with nonspherical shapes including prisms, cubes, and rods.

Light-mediated Growth

Light-mediated syntheses have also been explored where light can promote formation of various silver nanoparticle morphologies.

Silver Mirror Reaction

The silver mirror reaction involves the conversion of silver nitrate to Ag(NH3)OH. Ag(NH3)OH is subsequently reduced into colloidal silver using an aldehyde containing molecule such as a sugar. The silver mirror reaction is as follows:

$$2\left(Ag\left(NH_3\right)_2\right)^+ + RCHO + 2OH^- \rightarrow RCOOH + 2Ag + 4NH_3.$$

The size and shape of the nanoparticles produced are difficult to control and often have wide distributions. However, this method is often used to apply thin coatings of silver particles onto surfaces and further study into producing more uniformly sized nanoparticles is being done.

Ion Implantation

Ion implantation has been used to create silver nanoparticles embedded in glass, polyurethane, silicone, polyethylene, and poly(methyl methacrylate). Particles are embedded in the substrate by means of bombardment at high accelerating voltages. At a fixed current density of the ion beam up to a certain value, the size of the embedded silver nanoparticles has been found to be monodisperse within the population, after which only an increase in the ion concentration is observed. A further increase in the ion beam dose has been found to reduce both the nanoparticle size and density in the target substrate, whereas an ion beam operating at a high accelerating voltage with a gradually increasing current density has been found to result in a gradual increase in the nanoparticle size. There are a few competing mechanisms which may result in the decrease in nanoparticle size; destruction of NPs upon collision, sputtering of the sample surface, particle fusion upon heating and dissociation.

The formation of embedded nanoparticles is complex, and all of the controlling parameters and factors have not yet been investigated. Computer simulation is still difficult as it involves processes of diffusion and clustering, however it can be broken down into a few different sub-processes such as implantation, diffusion, and growth. Upon implantation, silver ions will reach different depths within the substrate which approaches a Gaussian distribution with the mean centered at X depth. High temperature conditions during the initial stages of implantation will increase the impurity diffusion in the substrate and as a result limit the impinging ion saturation, which is required for nanoparticle nucleation. Both the implant temperature and ion beam current density are crucial to control in order to obtain a monodisperse nanoparticle size and depth distribution. A low current density may be used to counter the thermal agitation from the ion beam and a buildup of surface charge. After implantation on the surface, the beam currents may be raised as the surface conductivity will increase. The rate at which impurities diffuse drops quickly after the formation of the nanoparticles, which act as a mobile ion trap. This suggests that the beginning of the implantation process is critical for control of the spacing and depth of the resulting nanoparticles, as well as control of the substrate temperature and ion beam density. The presence and nature of these particles can be analyzed using numerous spectroscopy and microscopy instruments. Nanoparticles synthesized in the substrate exhibit surface plasmon resonances as evidenced by characteristic absorption bands; these features undergo spectral shifts depending on the nanoparticle size and surface asperities, however the optical properties also strongly depend on the substrate material of the composite.

Biological Synthesis

The biological synthesis of nanoparticles has provided a means for improved techniques compared to the traditional methods that call for the use of harmful reducing agents like sodium borohydride. Many of these methods could improve their environmental footprint by replacing these relatively strong reducing agents. The problems with the chemical production of silver nanoparticles is usually involves high cost and the longevity of the particles is short lived due to aggregation. The harshness of standard chemical methods has sparked the use of using biological organisms to reduce silver ions in solution into colloidal nanoparticles.

In addition, precise control over shape and size is vital during nanoparticle synthesis since the NPs therapeutic properties are intimately dependent on such factors. Hence, the primary focus of research in biogenic synthesis is in developing methods that consistently reproduce NPs with precise properties.

Fungi and Bacteria

Bacterial and fungal synthesis of nanoparticles is practical because bacteria and fungi are easy to handle and can be modified genetically with ease. This provides a means to develop biomolecules that can synthesize AgNPs of varying shapes and sizes in high yield, which is at the forefront of current challenges in nanoparticle synthesis. Fungal strains such as Verticillium and bacterial strains such as K. pneumoniae can be used in the synthesis of silver nanoparticles. When the fungus/bacteria is added to solution, protein biomass is released into the solution.Electron donating residues such as tryptophan and tyrosine reduce silver ions in solution contributed by silver nitrate. These methods have been found to effectively create stable monodisperse nanoparticles without the use of harmful reducing agents.

A method has been found of reducing silver ions by the introduction of the fungus *Fusarium ox-*

ysporum. The nanoparticles formed in this method have a size range between 5 and 15 nm and consist of silver hydrosol. The reduction of the silver nanoparticles is thought to come from an enzymatic process and silver nanoparticles produced are extremely stable due to interactions with proteins that are excreted by the fungi.

Bacterium found in silver mines, Pseudomonas stutzeri AG259, were able to construct silver particles in the shapes of triangles and hexagons. The size of these nanoparticles had a large range in size and some of them reached sizes larger than the usual nanoscale with a size of 200 nm. The silver nanoparticles were found in the organic matrix of the bacteria.

Lactic acid producing bacteria have been used to produce silver nanoparticles. The bacteria *Lactobacillus* spp., *Pediococcus pentosaceus, Enteroccus faeciumI*, and *Lactococcus garvieae* have been found to be able to reduce silver ions into silver nanoparticles. The production of the nanoparticles takes place in the cell from the interactions between the silver ions and the organic compounds of the cell. It was found that the bacterium *Lactobacillus fermentum* created the smallest silver nanoparticles with an average size of 11.2 nm. It was also found that this bacterium produced the nanoparticles with the smallest size distribution and the nanoparticles were found mostly on the outside of the cells. It was also found that there was an increase in the pH increased the rate of which the nanoparticles were produced and the amount of particles produced.

Plants

The reduction of silver ions into silver nanoparticles has also been achieved using geranium leaves. It has been found that adding geranium leaf extract to silver nitrate solutions causes their silver ions to be quickly reduced and that the nanoparticles produced are particularly stable. The silver nanoparticles produced in solution had a size range between 16 and 40 nm.

In another study different plant leaf extracts were used to reduce silver ions. It was found that out of Camellia sinensis (green tea), pine, persimmon, ginko, magnolia, and platanus that the magnolia leaf extract was the best at creating silver nanoparticles. This method created particles with a disperse size range of 15 to 500 nm, but it was also found that the particle size could be controlled by varying the reaction temperature. The speed at which the ions were reduced by the magnolia leaf extract was comparable to those of using chemicals to reduce.

The use of plants, microbes, and fungi in the production of silver nanoparticles is leading the way to more environmentally sound production of silver nanoparticles.

A green method is available for synthesizing silver nanoparticles using Amaranthus gangeticus Linn leaf extract.

Products and Functionalization

Synthetic protocols for silver nanoparticle production can be modified to produce silver nanoparticles with non-spherical geometries and also to functionalize nanoparticles with different materials, such as silica. Creating silver nanoparticles of different shapes and surface coatings allows for greater control over their size-specific properties.

Anisotropic Structures

Silver nanoparticles can be synthesized in a variety of non-spherical (anisotropic) shapes. Because silver, like other noble metals, exhibits a size and shape dependent optical effect known as localized surface plasmon resonance (LSPR) at the nanoscale, the ability to synthesize Ag nanoparticles in different shapes vastly increases the ability to tune their optical behavior. For example, the wavelength at which LSPR occurs for a nanoparticle of one morphology (e.g. a sphere) will be different if that sphere is changed into a different shape. This shape dependence allows a silver nanoparticle to experience optical enhancement at a range of different wavelengths, even by keeping the size relatively constant, just by changing its shape. The applications of this shape-exploited expansion of optical behavior range from developing more sensitive biosensors to increasing the longevity of textiles.

Triangular Nanoprisms

Triangular shaped nanoparticles are a canonical type of anisotropic morphology studied for both gold and silver.

Though many different techniques for silver nanoprism synthesis exist, several methods employ a seed-mediated approach, which involves first synthesizing small (3-5 nm diameter) silver nanoparticles that offer a template for shape-directed growth into triangular nanostructures.

The silver seeds are synthesized by mixing silver nitrate and sodium citrate in aqueous solution and then rapidly adding sodium borohydride. Additional silver nitrate is added to the seed solution at low temperature, and the prisms are grown by slowly reducing the excess silver nitrate using ascorbic acid.

With the seed-mediated approach to silver nanoprism synthesis, selectivity of one shape over another can in part be controlled by the capping ligand. Using essentially the same procedure above but changing citrate to poly (vinyl pyrrolidone) (PVP) yields cube and rod-shaped nanostructures instead of triangular nanoprisms.

In addition to the seed mediated technique, silver nanoprisms can also be synthesized using a photo-mediated approach, in which preexisting spherical silver nanoparticles are transformed into triangular nanoprisms simply by exposing the reaction mixture to high intensities of light.

Nanocubes

Silver nanocubes can be synthesized using ethylene glycol as a reducing agent and PVP as a capping agent, in a polyol synthesis reaction (vide supra). A typical synthesis using these reagents involves adding fresh silver nitrate and PVP to a solution of ethylene glycol heated at 140 °C.

This procedure can actually be modified to produce another anisotropic silver nanostructure, nanowires, by just allowing the silver nitrate solution to age before using it in the synthesis. By allowing the silver nitrate solution to age, the initial nanostructure formed during the synthesis is slightly different than that obtained with fresh silver nitrate, which influences the growth process, and therefore, the morphology of the final product.

Coating with Silica

In this method, polyvinylpyrrolidone (PVP) is dissolved in water by sonication and mixed with silver colloid particles. Active stirring ensures the PVP has adsorbed to the nanoparticle surface. Centrifuging separates the PVP coated nanoparticles which are then transferred to a solution of ethanol to be centrifuged further and placed in a solution of ammonia, ethanol and $Si(OEt_4)$ (TES). Stirring for twelve hours results in the silica shell being formed consisting of a surrounding layer of silicon oxide with an ether linkage available to add functionality. Varying the amount of TES allows for different thicknesses of shells formed. This technique is popular due to the ability to add a variety of functionality to the exposed silica surface.

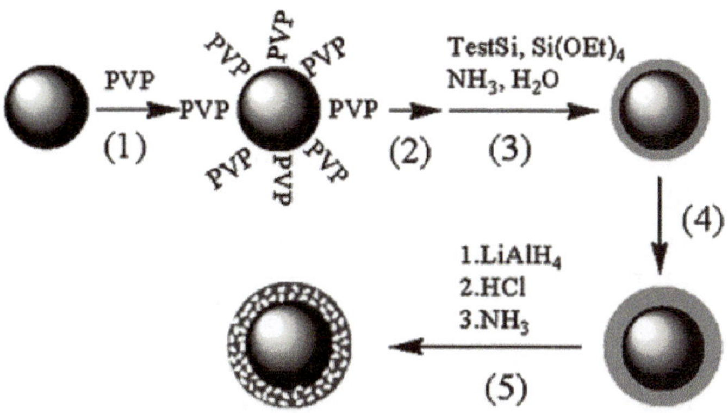

General procedure for coating colloid particles in silica. First PVP is absorbed onto the colloidal surface. These particles are put into a solution of ammonia in ethanol. the particle then begins to grow by addition of $Si(OET_4)$.

Use

Catalysis

Using silver nanoparticles for catalysis has been gaining attention in recent years. Although the most common applications are for medicinal or antibacterial purposes, silver nanoparticles have been demonstrated to show catalytic redox properties for dyes, benzene, carbon monoxide, and likely other compounds.

This paragraph is a general description of nanoparticle properties for catalysis; it is not exclusive to silver nanoparticles. The size of a nanoparticle greatly determines the properties that it exhibits due to various quantum effects. Additionally, the chemical environment of the nanoparticle plays a large role on the catalytic properties. With this in mind, it is important to note that heterogeneous catalysis takes place by adsorption of the reactant species to the catalytic substrate. When polymers, complex ligands, or surfactants are used to prevent coalescence of the nanoparticles, the catalytic ability is frequently hindered due to reduced adsorption ability. However, these compounds can also be used in such a way that the chemical environment enhances the catalytic ability.

Supported on Silica Spheres – Reduction of Dyes

Silver nanoparticles have been synthesized on a support of inert silica spheres. The support plays virtually no role in the catalytic ability and serves as a method of preventing coalescence of the silver nanoparticles in colloidal solution. Thus, the silver nanoparticles were stabilized and it was possible to demonstrate the ability of them to serve as an electron relay for the reduction of dyes by sodium borohydride. Without the silver nanoparticle catalyst, virtually no reaction occurs between sodium borohydride and the various dyes: methylene blue, eosin, and rose bengal.

Mesoporous Aerogel – Selective Oxidation of Benzene

Silver nanoparticles supported on aerogel are advantageous due to the higher number of active sites. The highest selectivity for oxidation of benzene to phenol was observed at low weight percent of silver in the aerogel matrix (1% Ag). This better selectivity is believed to be a result of the higher monodispersity within the aerogel matrix of the 1% Ag sample. Each weight percent solution formed different sized particles with a different width of size range.

Silver Alloy – Synergistic Oxidation of Carbon Monoxide

Au-Ag alloy nanoparticles have been shown to have a synergistic effect on the oxidation of carbon monoxide (CO). On its own, each pure-metal nanoparticle shows very poor catalytic activity for CO oxidation; together, the catalytic properties are greatly enhanced. It is proposed that the gold acts as a strong binding agent for the oxygen atom and the silver serves as a strong oxidizing catalyst, although the exact mechanism is still not completely understood. When synthesized in an Au/Ag ratio from 3:1 to 10:1, the alloyed nanoparticles showed complete conversion when 1% CO was fed in air at ambient temperature. Interestingly, the size of the alloyed particles did not play a big role in the catalytic ability. It is well known that gold nanoparticles only show catalytic properties for CO when they are ~3 nm in size, but alloyed particles up to 30 nm demonstrated excellent catalytic activity – catalytic activity better than that of gold nanoparticles on active support such as TiO_2, Fe_2O_3, etc.

Light-enhanced

Plasmonic effects have been studied quite extensively. Until recently, there have not been studies investigating the oxidative catalytic enhancement of a nanostructure via excitation of its surface plasmon resonance. The defining feature for enhancing the oxidative catalytic ability has been identified as the ability to convert a beam of light into the form of energetic electrons that can be transferred to adsorbed molecules. The implication of such a feature is that photochemical reactions can be driven by low-intensity continuous light can be coupled with thermal energy.

The coupling of low-intensity continuous light and thermal energy has been performed with silver nanocubes. The important feature of silver nanostructures that are enabling for photocatalysis is their nature to create resonant surface plasmons from light in the visible range.

The addition of light enhancement enabled the particles to perform to the same degree as particles that were heated up to 40 K greater. This is a profound finding when noting that a reduction in temperature of 25 K can increase the catalyst lifetime by nearly tenfold, when comparing the photothermal and thermal process.

Biological Research

Researchers have explored the use of silver nanoparticles as carriers for delivering various payloads such as small drug molecules or large biomolecules to specific targets. Once the AgNP has had sufficient time to reach its target, release of the payload could potentially be triggered by an internal or external stimulus. The targeting and accumulation of nanoparticles may provide high payload concentrations at specific target sites and could minimize side effects.

Chemotherapy

The introduction of nanotechnology into medicine is expected to advance diagnostic cancer imaging and the standards for therapeutic drug design. Nanotechnology may uncover insight about the structure, function and organizational level of the biosystem at the nanoscale.

Silver nanoparticles can undergo coating techniques that offer a uniform functionalized surface to which substrates can be added. When the nanoparticle is coated, for example, in silica the surface exists as silicic acid. Substrates can thus be added through stable ether and ester linkages that are not degraded immediately by natural metabolic enzymes. Recent chemotherapeutic applications have designed anti cancer drugs with a photo cleavable linker, such as an ortho-nitrobenzyl bridge, attaching it to the substrate on the nanoparticle surface. The low toxicity nanoparticle complex can remain viable under metabolic attack for the time necessary to be distributed throughout the bodies systems. If a cancerous tumor is being targeted for treatment, ultraviolet light can be introduced over the tumor region. The electromagnetic energy of the light causes the photo responsive linker to break between the drug and the nanoparticle substrate. The drug is now cleaved and released in an unaltered active form to act on the cancerous tumor cells. Advantages anticipated for this method is that the drug is transported without highly toxic compounds, the drug is released without harmful radiation or relying on a specific chemical reaction to occur and the drug can be selectively released at a target tissue.

A second approach is to attach a chemotherapeutic drug directly to the functionalized surface of the silver nanoparticle combined with a nucelophilic species to undergo a displacement reaction. For example, once the nanoparticle drug complex enters or is in the vicinity of the target tissue or cells, a glutathione monoester can be administered to the site. The nucleophilic ester oxygen will attach to the functionalized surface of the nanoparticle through a new ester linkage while the drug is released to its surroundings. The drug is now active and can exert its biological function on the cells immediate to its surroundings limiting non-desirable interactions with other tissues.

Multiple Drug Resistance

A major cause for the ineffectiveness of current chemotherapy treatments is multiple drug resistance which can arise from several mechanisms.

Nanoparticles can provide a means to overcome MDR. In general, when using a targeting agent to deliver nanocarriers to cancer cells, it is imperative that the agent binds with high selectivity to molecules that are uniquely expressed on the cell surface. Hence NPs can be designed with proteins that specifically detect drug resistant cells with overexpressed transporter proteins on their surface. A pitfall of the commonly used nano-drug delivery systems is that free drugs that are released from the nanocarriers into the cytosol get exposed to the MDR transporters once again, and are exported. To solve this, 8 nm nanocrystalline silver particles were modified by the addition of trans-activating transcriptional activator (TAT), derived from the HIV-1 virus, which acts as a cell penetrating peptide (CPP). Generally, AgNP effectiveness is limited due to the lack of efficient cellular uptake; however, CPP-modification has become one of the most efficient methods for improving intracellular delivery of nanoparticles. Once ingested, the export of the AgNP is prevented based on a size exclusion. The concept is simple: the nanoparticles are too large to be effluxed by the MDR transporters, because the efflux function is strictly subjected to the size of its substrates, which is generally limited to a range of 300-2000 Da. Thereby the nanoparticulates remain insusceptible to the efflux, providing a means to accumulate in high concentrations.

Antimicrobial

Introduction of silver into bacterial cells induces a high degree of structural and morphological changes, which can lead to cell death. As the silver nano particles come in contact with the bacteria, they adhere to the cell wall and cell membrane. Once bound, some of the silver passes through to the inside, and interacts with phosphate-containing compounds like DNA and RNA, while another portion adheres to the sulphur-containing proteins on the membrane. The silver-sulphur interactions at the membrane cause the cell wall to undergo structural changes, like the formation of pits and pores. Through these pores, cellular components are released into the extracellular fluid, simply due to the osmotic difference. Within the cell, the integration of silver creates a low molecular weight region where the DNA then condenses. Having DNA in a condensed state inhibits the cell's replication proteins contact with the DNA. Thus the introduction of silver nanoparticles inhibits replication and is sufficient to cause the death of the cell. Further increasing their effect, when silver comes in contact with fluids, it tends to ionize which increases the nanoparticles bactericidal activity. This has been correlated to the suppression of enzymes and inhibited expression of proteins that relate to the cell's ability to produce ATP.

Although it varies for every type of cell proposed, as their cell membrane composition varies greatly, It has been seen that in general, silver nano particles with an average size of 10 nm or less show electronic effects that greatly increase their bactericidal activity. This could also be partly due to the fact that as particle size decreases, reactivity increases due to the surface area to volume ratio increasing.

It has been noted that the introduction of silver nano particles has shown to have synergistic activity with common antibiotics already used today, such as; penicillin G, ampicillin, erythromycin, clindamycin, and vancomycin against E. coli and S. aureus.

Silver nanoparticles can prevent bacteria from growing on or adhering to the surface. This can be especially useful in surgical settings where all surfaces in contact with the patient must be sterile. Interestingly, silver nanoparticles can be incorporated on many types of surfaces including metals, plastic, and glass. In medical equipment, it has been shown that silver nano particles lower the bacterial count on devices used compared to old techniques. However, the problem arises when the procedure is over and a new one must be done. In the process of washing the instruments a large portion of the silver nano particles become less effective due to the loss of silver ions. They are more commonly used in skin grafts for burn victims as the silver nano particles embedded with the graft provide better antimicrobial activity and result in significantly less scarring of the victim. These new applications are direct decedents of older practices that used silver nitrate to treat conditions such as skin ulcers. Now, silver nanoparticles are used in bandages and patches to help heal certain burns and wounds.

They also show promising application as water treatment method to form clean potable water. This doesn't sound like much, but water contains numerous diseases and some parts of the world do not have the luxury of clean water, or any at all. It wasn't new to use silver for removing microbes, but this experiment used the carbonate in water to make microbes even more vulnerable to silver. First the scientists of the experiment use the nanopaticles to remove certain pesticides from the water, ones that prove fatal to people if ingested. Several other tests have shown that the silver nanoparticles were capable of removing certain ions in water as well, like iron, lead, and arsenic. But that is not the only reason why the silver nanoparticles are so appealing, they do not require any external force (no electricity of hydrolics) for the reaction to occur.

Consumer Goods

Household Applications

There are instances in which silver nanoparticles and colloidal silver are used in consumer goods. Samsung for example claimed that the use of silver nanoparticles in washing machines would help to sterilize clothes and water during the washing and rinsing functions, and allow clothes to be cleaned without the need for hot water. The nanoparticles in these appliances are synthesized using electrolysis. Through electrolysis, silver is extracted from metal plates and then turned into silver nanoparticles by a reduction agent. This method avoids the drying, cleaning and re-dispersion processes, which are generally required with alternative colloidal synthesis methods. Importantly, the electrolysis strategy also decreases the production cost of Ag nanoparticles, making these washing machines more affordable to manufacture. Samsung has described the system:

[A] grapefruit-sized device alongside the [washer] tub uses electrical currents to nanoshave two silver plates the size of large chewing gum sticks. Resulting in positively charged silver atoms-silver ions (Ag^{+})-are injected into the tub during the wash cycle.

Samsung's description of the Ag nanoparticle generating process seems to contradict its advertisement of silver nanoparticles. Instead, the statement indicates that laundry cycles. When clothes are run through the cycle, the intended mode of action is that bacteria contained in the water are sterilized as they interact with the silver present in the washing tub. As a result, these washing machines can provide antibacterial and sterilization benefits on top of conventional washing methods. Samsung has commented on the lifetime of these silver-containing washing machines. The electrolysis of silver generates over 400 billion silver ions during each wash cycle. Given the size of the silver source (two "gum-sized" plate of Ag), Samsung estimates that these plates can last up to 3000 wash cycles.

These plans by Samsung were note overlooked by regulatory agencies. Agencies investigating nanoparticle use include but are not limited to: the U.S. FDA, U.S. EPA, SIAA of Japan, and Korea's Testing and Research Institute for Chemical Industry and FITI Testing & Research Institute. These various agencies plan to regulate silver nanoparticles in appliances. These washing machines are some of the first cases in which the EPA has sought to regulate nanoparticles in consumer goods. Samsung stated that the silver gets washed away in the sewer and regulatory agencies worry over what that means for wastewater treatment streams. Currently, the EPA classifies silver nanoparticles as pesticides due to their use as antimicrobial agents in wastewater purification. The washing machines being developed by Samsung do contain a pesticide and have to be registered and tested for safety under the law, particularly the US Federal insecticide, fungicide and rodenticide act. The difficulty, however behind regulating nanotechnology in this manner is that there is no distinct way to measure toxicity.

Health and Safety

Although silver nanoparticles are widely used in a variety of commercial products, there has only recently been a major effort to study their effects on human health. There have been several studies that describe the *in vitro* toxicity of silver nanoparticles to a variety of different organs, including the lung, liver, skin, brain, and reproductive organs. The mechanism of the toxicity of silver nanoparticles to human cells appears to be derived from oxidative stress and inflammation that is caused by the generation of reactive oxygen species (ROS) stimulated by either the Ag NPs, Ag ions, or both. For example, Park *et al.* showed that exposure of a mouse peritoneal macrophage cell line (RAW267.7) to silver nanoparticles decreased the cell viability in a concentration- and time-dependent manner. They further showed that the intracellular reduced glutathionine (GSH), which is a ROS scavenger, decreased to 81.4% of the control group of silver nanoparticles at 1.6 ppm.

Modes of Toxicity

Since silver nanoparticles undergo dissolution releasing silver ions, which is well-documented to have toxic effects, there have been several studies that have been conducted to determine whether the toxicity of silver nanoparticles is derived from the release of silver ions or from the nanoparticle itself. Several studies suggest that the toxicity of silver nanoparticles is attributed to their release of silver ions in cells as both silver nanoparticles and silver ions have been reported to have similar cytotoxicity. For example, In some cases it is reported that silver nanoparticles facilitate the release of toxic free silver ions in cells via a "Trojan-horse type mechanism," where the particle enters cells and is then ionized within the cell. However, there have been reports that suggest that a combination of silver nanoparticles and ions is responsible for the toxic effect of silver nanoparticles. Navarro *et al.* using cysteine ligands as a tool to measure the concentration of free silver in solution, determined that although initially silver ions were 18 times more likely to inhibit the photosynthesis of an algae, Chlamydomanas reinhardtii, but after 2 hours of incubation it was revealed that the algae containing silver nanoparticles were more toxic than just silver ions alone. Furthermore, there are studies that suggest that silver nanoparticles induce toxicity independent of free silver ions. For example, Asharani *et al.* compared phenotypic defects observed in zebrafish treated with silver nanoparticles and silver ions and determined that the phenotypic defects observed with silver nanoparticle treatment was not observed with silver ion-treated embryos, suggesting that the toxicity of silver nanoparticles are independent of silver ions.

Protein channels and nuclear membrane pores can often be in the size range of 9 nm to 10 nm in diameter. Small silver nanoparticles constructed of this size have the ability to not only pass through the membrane to interact with internal structures but also to be become lodged within the membrane. Silver nanoparticle depositions in the membrane can impact regulation of solutes, exchange of proteins and cell recognition. Exposure to silver nanoparticles has been associated with "inflammatory, oxidative, genotoxic, and cytotoxic consequences"; the silver particulates primarily accumulate in the liver. but have also been shown to be toxic in other organs including the brain.

Nano-silver applied to tissue-cultured human cells leads to the formation of free radicals, raising concerns of potential health risks.

- Allergic reaction: There have been several studies conducted that show a precedence for allerginicity of silver nanoparticles.
- Argyria and staining: Ingested silver or silver compounds, including colloidal silver, can cause a condition called argyria, a discoloration of the skin and organs.In 2006, there was a case study of a 17-year-old man, who sustained burns to 30% of his body, and experienced a temporary bluish-grey hue after several days of treatment with Acticoat, a brand of wound dressing containing silver nanoparticles. Argyria is the deposition of silver in deep tissues, a condition that cannot happen on a temporary basis, raising the question of whether the cause of the man's discoloration was argyria or even a result of the silver treatment. Silver dressings are known to cause a "transient discoloration" that dissipates in 2–14 days, but not a permanent discoloration.
- Silzone heart valve: St. Jude Medical released a mechanical heart valve with a silver coated sewing cuff (coated using ion beam-assisted deposition) in 1997. The valve was designed to reduce the instances of endocarditis. The valve was approved for sale in Canada, Europe, the United States, and most other markets around the world. In a post-commercialization study, researchers showed that the valve prevented tissue ingrowth, created paravalvular leakage, valve loosening, and in the worst cases explantation. After 3 years on the market and 36,000 implants, St. Jude discontinued and voluntarily recalled the valve.

Platinum Nanoparticles

Platinum nanoparticles are usually in the form of a suspension or colloid of nanoparticles of platinum in a fluid, usually water. A colloid is technically defined as a stable dispersion of particles in a fluid medium (liquid or gas).

Spherical platinum nanoparticles can be made with sizes between about 2 and 100 nanometres (nm), depending on reaction conditions. Platinum nanoparticles are suspended in the colloidal solution of brownish-red or black color. Nanoparticles come in wide variety of shapes including spheres, rods, cubes, and tetrahedra.

Platinum nanoparticles are the subject of substantial research, with potential applications in a wide variety of areas. These include catalysis, medicine, and the synthesis of novel materials with unique properties.

Synthesis

Platinum nanoparticles are typically synthesized either by the reduction of platinum ion precursors in solution with a stabilizing or capping agent to form colloidal nanoparticles, or by the impregnation and reduction of platinum ion precursors in a micro-porous support such as alumina.

Some common examples of platinum precursors include potassium hexachloroplatinate (K_2PtCl_6) or platinous chloride ($PtCl_2$) Different combinations of precursors, such as ruthenium chloride ($RuCl_3$)

and chloroplatinic acid (H_2PtCl_6), have been used to synthesize mixed-metal nanoparticles Some common examples of reducing agents include hydrogen gas (H_2), sodium borohydride ($NaBH_4$) and ethylene glycol ($C_2H_6O_2$), although other alcohols and plant-derived compounds have also been used.

As the platinum metal precursor is reduced to neutral platinum metal (Pt^o), the reaction mixture becomes supersaturated with platinum metal and the Pt^o begins to precipitate in the form of nanoscale particles. A capping agent or stabilizing agent such as sodium polyacrylic acid or sodium citrate is often used to stabilize the nanoparticle surfaces, and prevents the aggregation and coalescence of the nanoparticles.

The size of nanoparticles synthesized colloidally may be controlled by changing the platinum precursor, the ratio of capping agent to precursor, and/or the reaction temperature. The size of the nanoparticles can also be controlled with small deviation by using a stepwise seed-mediated growth procedure as outlined by Bigall et al. (2008). The size of nanoparticles synthesized onto a substrate such as alumina depends on various parameters such as the pore size of the support.

Platinum nanoparticles can also be synthesized by decomposing $Pt_2(dba)_3$ (dba = dibenzylideneacetone) under a CO or H_2 atmosphere, in the presence of a capping agent. The size and shape distributions of the resulting nanoparticles depend on the solvent, the reaction atmosphere, the types of capping agents and their relative concentrations, the specific platinum ion precursor, as well at the temperature of the system and reaction time.

Shape and Size Control

Ramirez et al. reported the influence of ligand and solvent effects on the size and shape of platinum nanoparticles. Platinum nanoparticle seeds were prepared by the decomposition of $Pt_2(dba)_3$ in tetrahydrofuran (THF) under carbon monoxide (CO). These conditions produced Pt nanoparticles with weakly bound THF and CO ligands and an approximate diameter on 1.2 nm. Hexadecylamine (HDA) was added to the purified reaction mixture and allowed to displace the THF and CO ligands over the course of approximately seven days, producing monodispersed spherical crystalline Pt nanoparticles with an average diameter of 2.1 nm. After the seven-day period, an elongation of the Pt nanoparticles occurred. When the same procedure was followed using a stronger capping agent such as triphenyl phosphine or octanethiol, the nanoparticles remained spherical, suggesting that the HDA ligand affects particle shape.

When $Pt_2(dba)_3$ was decomposed in THF under hydrogen gas in the presence HDA, the reaction took much longer, and formed nanowires with diameters between 1.5 and 2 nm. Decomposition of $Pt_2(dba)_3$ under hydrogen gas in toluene yielded the formation of nanowires with 2-3 nm diameter independent of HDA concentration. The length of these nanowires was found to be inversely proportional to the concentration of HDA present in solution. When these nanowire syntheses were repeated using reduced concentrations of $Pt_2(dba)_3$, there was little effect on the size, length or distribution of the nanowires formed.

Platinum nanoparticles of controlled shape and size have also been accessed through varying the ratio of polymer capping agent concentration to precursor concentration. Reductive colloidal syntheses as such have yielded tetrahedral, cubic, irregular-prismatic, icosahedral, and cubo-octahedral nanoparticles, whose dispersity is also dependent on the concentration ratio of capping agent to precursor, and which may be applicable to catalysis. The precise mechanism of shape-controlled colloidal synthesis is not yet known; however, it is known that the relative growth rate of crystal facets within the growing nanostructure determines its final shape. Polyol syntheses of platinum nanoparticles, in which chloroplatinic acid is reduced to $PtCl_4^{2-}$ and Pt^o by ethylene glycol, have also been a means to shape-controlled fabrication. Addition of varying amounts of sodium nitrate to these reactions was shown to yield tetrahedra and octahedra at high concentration ratios of sodium nitrate to chloroplatinic acid. Spectroscopic studies suggest that nitrate is reduced to nitrite by $PtCl_4^{2-}$ early in this reaction, and that the nitrite may then coordinate both Pt(II) and Pt(IV), greatly slowing the polyol reduction and altering the growth rates of distinct crystal facets within the nanoparticles, ultimately yielding morphological differentiation.

Green Synthesis

An ecologically-friendly synthesis of platinum nanoparticles from chloroplatinic acid was achieved through the use of a leaf extract of *Diospyros kaki* as a reducing agent. Nanoparticles synthesized as such were spherical with an average diameter ranging from 2-12 nm depending on reaction temperature and concentration of leaf extract used. Spectroscopic analysis suggests that this reaction is not enzyme-mediated and proceeds instead through plant-derived reductive small molecules. Another eco-friendly synthesis from chloroplatinic acid was reported using leaf extract from *Ocimum sanctum* and tulsi as reducing agents. Spectroscopic analysis suggested that ascorbic acid, gallic acid, various terpenes, and certain amino acids were active in the reduction. Particles synthesized as such were shown through scanning electron microscopy to consist in aggregates with irregular shape. It has been shown that tea extracts with high polyphenol content may be used both as reducing agents and capping agents for platinum nanoparticle synthesis.

Properties

The chemical and physical properties of platinum nanoparticles (NP) make them applicable for a wide variety of research applications. Extensive experimentation has been done to create new species of platinum NPs, and study their properties. Platinum NP applications include electronics, optics, catalysts, and enzyme immobilization.

Catalytic Properties

Platinum NPs are used as catalysts for proton exchange membrane fuel cell (PEMFC), for industrial synthesis of nitric acid, reduction of exhaust gases from vehicles and as catalytic nucleating agents for synthesis of magnetic NPs. NPs can act as catalysts in homogeneous colloidal solution or as gas-phase catalysts while supported on solid state material. The catalytic reactivity of the NP is dependent on the shape, size and morphology of the particle.

One type of platinum NPs that have been researched on are colloidal platinum NPs. Monometallic and bimetallic colloids have been used as catalysts in a wide range of organic chemistry, including, oxidation of carbon monoxide in aqueous solutions, hydrogenation of alkenes in organic or biphasic solutions and hydrosilylation of olefins in organic solutions. Collodial platinum NPs protected by Poly(N-isopropylacrylamide) were synthesised and their catalytic properties measured. It was determined that they were more active in solution and inactive when phase separated due to its solubility being inversely proportional to temperature.

Optical Properties

Platinum NPs exhibit fascinating optical properties. Being a free electron metal NP like silver and gold, its linear optical response is mainly controlled by the surface plasmon resonance. Surface plasmon resonance occurs when the electrons in the metal surface are subject to an electromagnetic field that exerts a force on the electrons and cause them to displace from their original positions. The nuclei then exert a restoring force that results in oscillation of the electrons, which increase in strength when frequency of oscillations is in resonance with the incident electromagnetic wave.

The SPR of platinum nanoparticles is found in the ultraviolet range (215 nm), unlike the other noble metal nanoparticles which display SPR in the visible range. Experiments were done and the spectra obtained are similar for most platinum particles regardless of size. However, there is an exception. Platinum NPs synthesized via citrate reduction do not have a surface plasmon resonance peak around 215 nm. Through experimentation, the resonance peak only showed slight variations with the change of size and synthetic method (while maintaining the same shape), with the exception of those nanoparticles synthesized by citrate reduction, which did not exhibit and SPR peak in this region.

Through the control of percent composition of 2-5 nm platinum nanoparticles on SiO_2, Zhang et al. modeled distinct absorption peaks attributed to platinum in the visible range, distinct from the conventional SPR absorption. This research attributed these absorption features to the generation and transfer of hot electrons from the platinum nanoparticles to the semiconductive material. Interestingly, the addition of small platinum nanoparticles on semiconductors such as TiO_2 increases the photocatalytic oxidation activity under visible light irradiation. These concepts suggest the possible role of platinum nanoparticles in the development of solar energy conversion using metal nanoparticles. By changing the size, shape and environment of metal nanoparticles, their optical properties can be used for electrontic, catalytic, sensing, and photovoltaic applications.

Applications

Hydrogen Fuel Cells

Among the precious metals, platinum is the most active toward the hydrogen oxidation reaction that occurs at the anode in hydrogen fuel cells. In order to meet cost reductions of this magnitude, the Pt catalyst loading must be decreased. Two strategies have been investigated for reducing the Pt loading: the binary and ternary Pt-based alloyed nanomaterials and the dispersion of Pt-based nanomaterials onto high surface area substrates.

Methanol Fuel Cells

The methanol oxidation reaction occurs at the anode in direct methanol fuel cells (DMFCs). Platinum is the most promising candidate among pure metals for application in DMFCs. Platinum has the highest activity toward the dissociative adsorption of methanol. However, pure Pt surfaces are poisoned by carbon monoxide, a byproduct of methanol oxidation. Researchers have focused on dispersing nanostructured catalysts on high surface area supporting materials and the development of Pt-based nanomaterials with high electrocatalytic activity toward MOR to overcome the poisoning effect of CO.

Electrochemical Oxidation of Formic Acid

Formic acid is another attractive fuel for use in PEM-based fuel cells. The dehydration pathway produces adsorbed carbon monoxide. A number of binary Pt-based nanomaterial electrocatalysts have been investigated for enhanced electrocatalytic activity toward formic acid oxidation.

Modifying Conductivity of Zinc Oxide Materials

Platinum NPs can be used to dope zinc oxide (ZnO) materials to improve their conductivity. ZnO has several characteristics that allow it to be used in several novel devices such as development of light-emitting assemblies and solar cells. However, because ZnO is of slightly lower conductivity than metal and indium tin oxide (ITO), it can be doped and hybridized with metal NPs like platinum to improve its conductivity. A method to do so would be to synthesize ZnO NPs using methanol reduction and incorporate at 0.25 at.% platinum NPs. This boosts the electrical properties of ZnO films while preserving its transmittance for application in transparent conducting oxides.

Glucose Detection Applications

Enzymatic glucose sensors have drawbacks that originate from the nature of the enzyme. Nonenzymatic glucose sensors with Pt-based electrocatalysts offer several advantages, including high stability and ease of fabrication. Many novel Pt and binary Pt-based nanomaterials have been developed to overcome the challenges of glucose oxidation on Pt surfaces, such as low selectivity, poor sensitivity, and poisoning from interfering species.

Other Applications

Platinum catalysts are alternatives of automotive catalytic converters, carbon monoxide gas sensors, petroleum refining, hydrogen production, and anticancer drugs. These applications utilize platinum nanomaterials due to their catalytic ability to oxidize CO and NOx, dehydrogenate hydrocarbons, and electrolyze water and their ability to inhibit the division of living cells.

Biological Interactions

The increased reactivity of nanoparticles is one of their most useful properties and is leveraged in fields such as catalysis, consumer products, and energy storage. However, this high reactivity also means that a nanoparticle in a biological environment may have unintended impacts. For example, many nanoparticles such as silver, copper, and ceria interact with cells to produce reactive oxygen species or ROS which can cause premature cell death through apoptosis. Determining the toxicity of a specific nanoparticle requires knowledge of the particle's chemical composition, shape, size and is a field that is growing alongside advances in nanoparticle research.

Determining the impact of a nanoparticle on a living system is not straightforward. A multitude of in vivo and in vitro studies must be done to fully characterize reactivity. In vivo studies often use whole organisms such as mice or zebrafish to infer the interaction of the nanoparticle on a healthy human body. In vitro studies look at how nanoparticles interact with specific cell colonies, typically of human origin. Both types of experiments are needed for a complete understanding of nanoparticle toxicity, especially human toxicity, since no one model has complete human relevance.

Drug Delivery

A topic of research within the field of nanoparticles is how to use these small particles for drug delivery. Depending on particle properties, nanoparticle may move throughout the human body are promising as site-specific vehicles for the transport of medicine. Current research using platinum nanoparticles in drug delivery uses platinum-based carries to move antitumor medicine. In one study, platinum nanoparticles of diameter 58.3 nm were used to transport an anticancer drug to human colon carcinoma cells, HT-29. Uptake of the nanoparticles by the cell involves compartmentalization of the nanoparticles within lysosomes. The high acidity environment enables leaching of platinum ions from the nanoparticle, which the researchers identified as causing the increased effectiveness of the drug. In another study, a Pt nanoparticle of diameter 140 nm was encapsulated within a PEG nanoparticle to move an antitumor drug, Cisplatin, within a prostate cancer cell (LNCaP/PC3) population. Use of platinum in drug delivery hinges on its ability to not interact in a harmful manner in healthy portions of the body while also being able to release its contents when in the correct environment.

Toxicology

Toxicity stemming from platinum nanoparticles can take multiple forms. One possible interaction is cytotoxicity or the ability of the nanoparticle to cause cell death. A nanoparticle can also interact with the cell's DNA or genome to cause genotoxicity. These effects are seen in different levels of gene expression measured through protein levels. Last is the developmental toxicity that can occur as an organism grows. Developmental toxicity looks at the impact the nanoparticle has on the growth of an organism from an embryonic stage to a later set point. Most nanotoxicology research is done on cyto- and genotoxicity as both can easily be done in a cell culture lab.

Platinum nanoparticles have the potential to be toxic to living cells. In one case, 2 nm platinum nanoparticles were exposed to two different types of algae in order to understand how these nanoparticles interact with a living system. In both species of algae tested, the platinum nanoparticles inhibited growth, induced small amounts of membrane damage, and created a large amount of oxidative stress. In another study, researcher tested the effects of differently sized platinum nanoparticles on primary human keratinocytes. The authors tested 5.8 and 57.0 nm Pt nanoparticles. The 57 nm nanoparticles had some hazardous effects including decreased cell metabolism, but the effect of the smaller nanoparticles was much more damaging. The 5.8 nm nanoparticles exhibited a more deleterious effect on the DNA stability of the primary keratincoytes than did the larger nanoparticles. The damage to the DNA was measured for individual cells using single-gel electrophoresis via the comet assay.

Researchers have also compared the toxicity of Pt nanoparticles to other commonly used metallic nanoparticles. In one study, the authors compared the impact of different nanoparticle compositions on the red blood cells found in the human bloodstream. The study showed that 5-10 nm platinum nanoparticles and 20-35 nm gold nanoparticles have very little effect on the red blood cells. In the same study it was found that 5-30 nm silver nanoparticles caused membrane damage, detrimental morphological variation, and haemagglutination to the red blood cells.

In a recent paper published in Nanotoxicology, the authors found that between silver (Ag-NP, d=5-35 nm), gold (Au-NP, d=15-35 nm), and Pt (Pt-NP, d=3-10 nm) nanoparticles, the Pt nanoparticles were the second most toxic in developing zebrafish embryos, behind only the Ag-NPs. However, this work did not examine the size dependence of the nanoparticles on their toxicity or biocompatibility. Size-dependent toxicity was determined by researchers at the National Sun Yat - Sen University in Kaohsiung, Taiwan. This group's work showed that the toxicity of platinum nanoparticles in bacterial cells is strongly dependent on nanoparticle size and shape/morphology. Their conclusions were based on two major observations. First, the authors found that platinum nanoparticles with spherical morphologies and sizes less than 3 nm showed biologically toxic properties; measured in terms of mortality, hatching delay, phenotypic defects and metal accumulation. While those nanoparticles with alternative shapes—such as cuboidal, oval, or floral—and sizes of 5-18 nm showed biocompatibility and no biologically toxic properties. Secondly, out of the three varieties of platinum nanoparticles which exhibited biocompatibility, two showed an increase in bacterial cell growth.

The paper introduces many hypotheses for why these observations were made, but based on other works and basic knowledge of bacterial cell membranes, the reasoning behind the size dependent toxicity observation seems to be twofold. One: The smaller, spherically shaped nanoparticles are able to pass through cell membranes simply due to their reduced size, as well as their shape-compatibility with the typically spherical pores of most cell membranes. Although this hypothesis needs to be further supported by future work, the authors did cite another paper which tracked the respiratory intake of platinum nanoparticles. This group found that 10 μm platinum nanoparticles are absorbed by the mucus of the bronchi and trachea, and can travel no further through the respiratory tract. However, 2.5 μm particles showed an ability to pass through this mucus layer, and reach much deeper into the respiratory tract. Also the larger, uniquely shaped nanoparticles are too large to pass through the pores of the cell membrane, and/or have shapes which are incompatible with the more spherically shaped pores of the cellular membrane. In regards to the observation that the two largest platinum nanoparticles (6-8 nm oval, and 16-18 nm floral) actually increase bacterial cell growth, the explanation could originate in the findings of other works which have shown that platinum nanoparticles have demonstrated significant antioxidative capacity. However, it must be noted that in order for these antioxidative properties to be exploited, the platinum nanoparticles must first enter the cells, so perhaps there is another explanation for this observation of increased bacterial cell growth.

Most studies so far have been size based using an in vivo mouse model. In one study, researchers compared the effects of sun 1 nm and 15 nm platinum nanoparticles on mice. The 15 mg/kg dose of sub 1 nm platinum nanoparticles were found to cause liver damage while the larger particles had no effect. A similar study using a singular injection as an exposure source of platinum nanoparticles into the mouse model found necrosis of tubular epithelial cells for particles under 1 nm, but no effect with those particles of 8 nm. These in vivo studies show a trend that the toxicity of the platinum nanoparticles is size dependent, most likely due to the ability of the nanoparticle to get into a high impactful region within the body. A complete study analyzing the effect of varying sized platinum nanoparticles used both in vivo and in vitro models is used to gain a better understanding what impact these nanoparticles could have. Using mice as a model, they found retention of the platinum nanoparticles by the respiratory tract of the mouse. This was accompanied by a minor to mild inflammation of the surrounding lung tissue. However, their in vitro tests using human and lung epithelial cells found no cytotoxic or oxidative stress effects caused by the platinum nanoparticles despite clear evidence of cellular uptake.

Cobalt Oxide Nanoparticles

Though the amount of research on cobalt oxide nanoparticles is still relatively new, researchers have found that these particles have great potential for us in chemical and industrial areas.

Characteristics

Cobalt oxide nanoparticles appear white in color and are magnetic p-type semiconductors. It is interesting to note that the particles will convert into cobalt metal upon heating at 900 °C. Even though cobalt oxide nanoparticles have many beneficial uses, the substance is still toxic to humans and also aquatic life. It is harmful when swallowed and also acts as a skin irritant.

Chemical Properties	
Chemical Symbol	Co_3O_4
CAS Number	1308-06-1
Group	Cobalt 9 Oxygen 16
Electronic Configuration	Cobalt [Ar] $3d^7 4s^2$ Oxygen [He] $2s^2 2p^4$
Element	Content (%)
Cobalt	73.4
Oxygen	26.57

Physical Properties	
Properties	Metric
Density	6.11 g/cm^3
Molar Mass	74.93 g/mol

Thermal Properties	
Melting Point	895 °C
Boiling Point	900 °C

Applications

Although they have been observed to readily enter cells and may possibly be exploited for hyperthermic treatment, gene therapy and drug delivery, a protective and efficient molecular shield for the nanoparticles still needs to be developed. Cobalt nanoparticles are more commonly applied to devices for energy storage.

Lithium-ion Battery

Cobalt oxide nanoparticles are being investigated in optimizing high performance in lithium-ion batteries. This is due to their high surface-to-volume ratio and short path length for lithium cation transport, as opposed to nickel oxide and iron oxides. Some proposals include Co_3O_4 nanoparticles anchored on graphene as a source of possible lithium battery anodes. This is synthesized by first forming $Co(OH)_2$ on the sheet from Co^{2+} and NH_3H_2O, which is then heated to 450 °C for two hours to yield the cobalt oxide/graphene composite. The advantageous combination of the graphene sheet's flexibility and its strong interaction with Co_3O_4 nanoparticles prevent volume expansion/contraction and aggregation of Co_3O_4 during lithium charge and discharge processes.

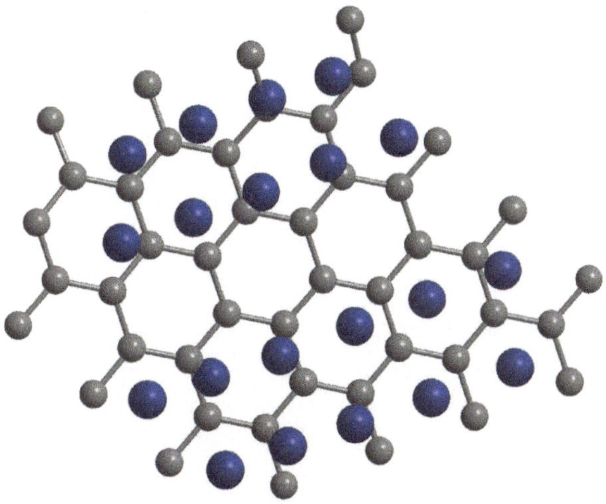

Cobalt oxide (Co_3O_4) nanoparticles anchored on a single sheet of graphene

Porous Co_3O_4 nanotubes are another good lithium battery anode candidate due to their higher surface-to-volume ratio than other one-dimensional nanostructures. These different types engineered nanostructures exhibited high reversible capacity and good cycle life. Diphenylalanine/cobalt oxide hybrid nanowires have a potential application as anodes for Li-ion batteries as well, also enhancing the specific capacity as an electrode.

Gas Sensor

The p-type Co_3O_4 hollow nanospheres have performed well as a gas sensor in sensing toluene, acetone, and other organic vapors with rapid response and high sensitivity to very low concentrations at low temperature. This serves to potentially monitor organic flammable and toxic gases. Co_3O_4 nanoparticles prepared on single-walled carbon nanotubes (SWCNTs) are being investigated in sensing NO_x and H_2 gases. This takes advantage of Co_3O_4's active reactivity with the gases, as well as the electrical continuum formed within the Co_3O_4 nanoparticle/SWCNT composite. With NO_x, the resistance of the sensor decreases because of an increase in the hole density in the composite. This is because both the Co_3O_4 nanoparticles and carbon nanotubes are p-type semiconductors and NO_x is electron-accepting. Thus, the adsorbed gas molecules continue to extract electrons from them. Oppositely, the resistance of the sensor increases for H_2 because it is electron donating. "The composite shows a highly enhanced response of ~200 % upon exposure to 4 % H_2 diluted in air compared with pure SWCNTs and Co_3O_4 film" at room temperature.

Synthesis

Hydrothermal Synthesis

Hydrothermal synthesis of metal oxide hollow spheres takes place in an autoclave and proceeds through a one-pot synthesis to increase yield while reducing time and resources. To initiate the procedure, carbohydrates and metal salts are dissolved in water at 100-200 °C. Carbon spheres are produced with metal ions integrated into the hydrophobic shell. Calcination allows for removal of the carbon from the product and results in hollow metal oxide spheres. Surface area and thickness of the shell can be manipulated by varying the carbohydrate to metal salt concentration. The completion time for the procedure varies from hours to days. Hydrothermal synthesis is generally simple and being able to control size and shape is beneficial for specificity. However, the drawback of this approach is a smaller yield compared to other methods. The pH of the reaction medium, temperature, pressure, and cations used in the solution all play a role in determining the nature of the solid. This approach allows for the production of a wide range of shapes and sizes for metal oxide hollow spheres.

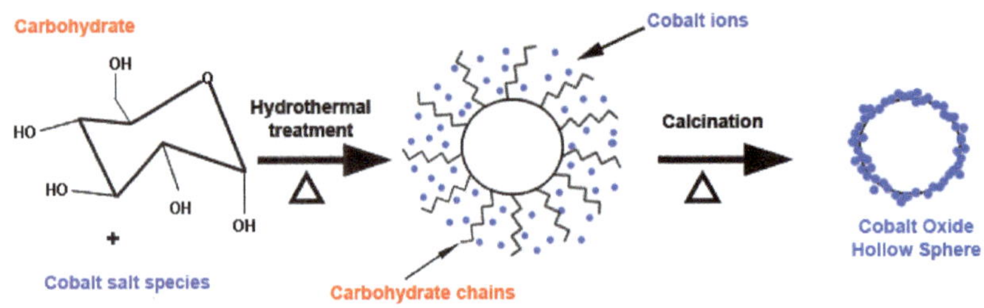

Hydrothermal Synthesis of Cobalt Oxide Hollow Sphere

Solvent Free Thermal Synthesis

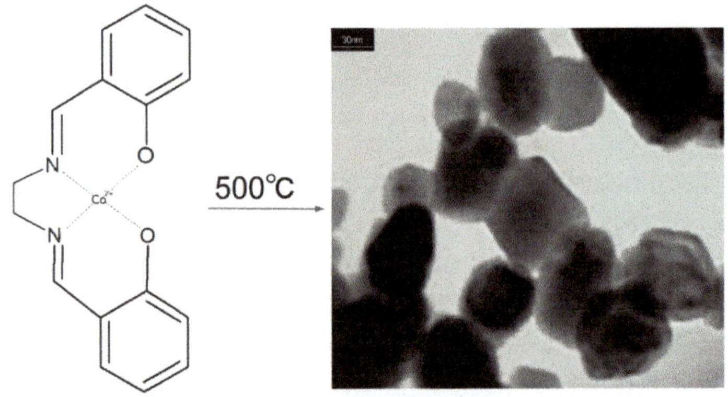

Organic precursor [Co-(salen)] TEM image of Cobalt Oxide Nanoparticles

TEM image of cobalt oxide nanoparticle produced by heating organometallic precursor [Co-(salen)] under high temperature.

Cobalt oxide (Co_3O_4) nanoparticles are generally synthesized at low temperature; However, the tendency of Co_3O_4 to grow into complex microstructure made the addition of organic surfactant necessary. Organic surfactant is both expensive and pollutant. Using organometallic molecular as a precursor to synthesize Co_3O_4 under high temperature can ensure purity, composition, phase purity, and microstructure. Using organometallic chemistry, Co_3O_4 is synthesized by heating 0.01 mol of [bis(salicylaldehyde)ethylenediiminecobalt(II], known as [Co-(salen)], to 500 °C in the air. The synthesis of the organometallic molecular precursor [Co-(salen)] is prepared by first making the bis(salicylaldehyde)ethylenediimine crystal. By dissolving salicylaldehyde in methanol with drops of 1,2- ethylenediamine, 92% of bis(salicylaldehyde)ethylenediimine crystal could be purified via reflux and filtration. [Bis(salicylaldehyde)ethylenediirninccobalt(II)] is then prepared by adding cobalt(II) acetate tetrahydrate in propanol and warmed to 50 °C under nitrogen atmosphere, followed by adding bis(salicylaldehyde)ethylenediimine and purification through reflux and recrystallization.

Nanofiber

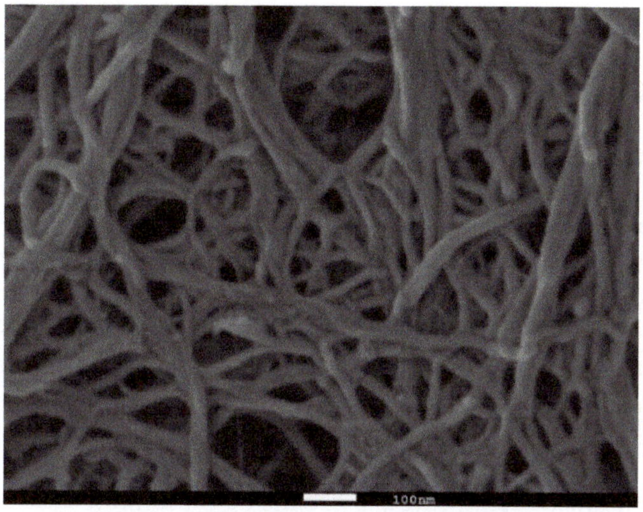

Example of a cellulose nanofiber network

Nanofibers are fibers with diameters in the nanometer range. Nanofibers can be generated from different polymers and hence have different physical properties and application potentials. Examples of natural polymers include collagen, cellulose, silk fibroin, keratin, gelatin and polysaccharides such as chitosan and alginate. Examples of synthetic polymers include poly(lactic acid) (PLA), polycaprolactone (PCL), polyurethane (PU), poly(lactic-co-glycolic acid) (PLGA), poly(3-hydroxybutyrate-co-3-hydroxyvalerate) (PHBV), and poly(ethylene-co-vinylacetate) (PEVA). Polymer chains are connected via covalent bonds. The diameters of nanofibers depend on the type of polymer used and the method of production. All polymer nanofibers are unique for their large surface area-to-volume ratio, high porosity, appreciable mechanical strength, and flexibility in functionalization compared to their microfiber counterparts.

There exist many different methods to make nanofibers, including drawing, electrospinning, self-assembly, template synthesis, and thermal-induced phase separation. Electrospinning is the most commonly used method to generate nanofibers because of the straightforward setup, the ability to mass-produce continuous nanofibers from various polymers, and the capability to generate ultrathin fibers with controllable diameters, compositions, and orientations. This flexibility allows for controlling the shape and arrangement of the fibers so that different structures (*i.e.* hollow, flat and ribbon shaped) can be fabricated depending on intended application purposes.

Nanofibers have many possible technological and commercial applications. They are used in tissue engineering, drug delivery, cancer diagnosis, lithium-air battery, optical sensors and air filtration.

History of Nanofiber Production

Nanofibers were first produced via electrospinning more than four centuries ago. Beginning with the development of the electrospinning method, English physicist William Gilbert (1544-1603) first documented the electrostatic attraction between liquids by preparing an experiment in which he observed a spherical water drop on a dry surface warp into a cone shape when it was held below an electrically charged amber. This deformation later came to be known as the Taylor cone. In 1882, English physicist Lord Rayleigh (1842-1919) analyzed the unstable states of liquid droplets that were electrically charged, and noted that the liquid was ejected in tiny jets when equilibrium was established between the surface tension and electrostatic force. In 1887, British physicist Charles Vernon Boys (1855-1944) published a manuscript about nanofiber development and production. In 1900, American inventor John Francis Cooley (1861-1903) filed the first modern electrospinning patent.

Anton Formhals was the first person to attempt nanofiber production between 1934 and 1944 and publish the first patent describing the experimental production of nanofibers. In 1966, Harold Simons published a patent for a device that could produce thin and light nanofiber fabrics with diverse motifs.

Only at the end of the 20th century have the words electrospinning and nanofiber become common language among scientists and researchers. Electrospinning continues to be developed today.

Synthesis Methods

Many chemical and mechanical techniques for preparing nanofibers exist.

Electrospinning

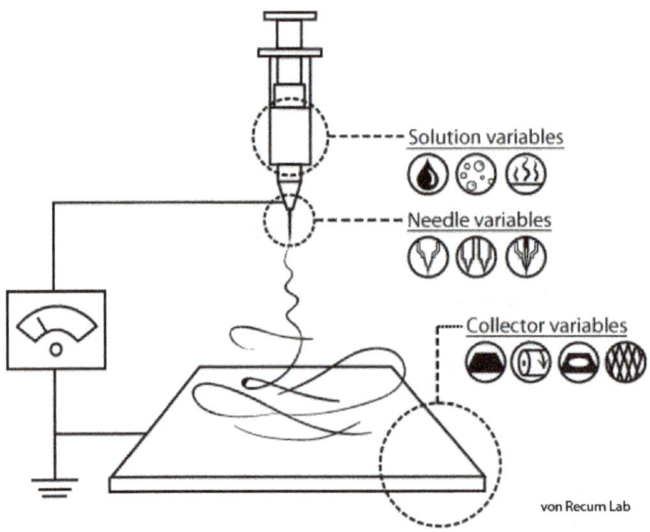

Diagram of a general set-up of electrospinning.

Electrospinning is the most commonly used method to fabricate nanofibers. The instruments necessary for electrospinning include a high voltage supplier, a capillary tube with a pipette or needle with a small diameter, and a metal collecting screen. One electrode is placed into the polymer solution and the other electrode is attached to the collector. An electric field is applied to the end of the capillary tube that contains the polymer solution held by its surface tension and forms a charge on the surface of the liquid. As the intensity of the electric field increases, the hemispherical surface of the fluid at the tip of the capillary tube elongates to form a conical shape known as the Taylor cone. A critical value is attained upon further increase in the electric field in which the repulsive electrostatic force overcomes the surface tension and the charged jet of fluid is ejected from the tip of the Taylor cone. The discharged polymer solution jet is unstable and elongates as a result, allowing the jet to become very long and thin. Charged polymer fibers solidifies with solvent evaporation. Randomly-oriented nanofibers are collected on the collector. Nanofibers can also be collected in a highly aligned fashion by using specialized collectors such as the rotating drum, metal frame, or a two-parallel plates system. Parameters such as jet stream movement and polymer concentration have to be controlled to produce nanofibers with uniform diameters and morphologies.

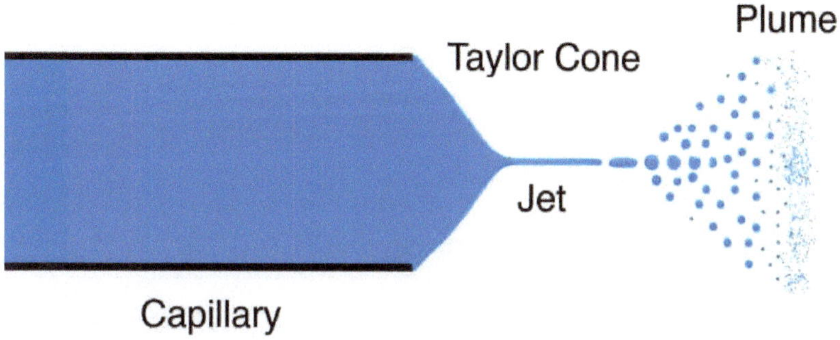

Taylor cone from which jet of polymer solution is ejected

The electrospinning technique transforms many types of polymers into nanofibers. An electrospun nanofiber network resembles the extracellular matrix (ECM) well. This resemblance is a major advantage of electrospinning because it opens up the possibility of mimicking the ECM with regards to fiber diameters, high porosity, and mechanical properties. Electrospinning is being further developed for mass production of one-by-one continuous nanofibers.

Thermal-induced Phase Separation

Thermal-induced phase separation separates a homogenous polymer solution into a multi-phase system via thermodynamic changes. The procedure involves five steps: polymer dissolution, liquid-liquid or liquid-solid phase separation, polymer gelation, extraction of solvent from the gel with water, and freezing and freeze-drying under vacuum. Thermal-induced phase separation method is widely used to generate scaffolds for tissue regeneration.

The homogenous polymer solution in the first step is thermodynamically unstable and tends to separate into polymer-rich and polymer-lean phases under appropriate temperature. Eventually after solvent removal, the polymer-rich phase solidifies to form the matrix and the polymer-lean phase develops into pores. Next, two types of phase separation can be carried out on the polymer solution depending on the desired pattern. Liquid-liquid separation is usually used to form bicontinuous phase structures while solid-liquid phase separation is used to form crystal structures. The gelation step plays a crucial role in controlling the porous morphology of the nanofibrous matrices. Gelation is influenced by temperature, polymer concentration, and solvent properties. Temperature regulates the structure of the fiber network: low gelation temperature results in formation of nanoscale fiber networks while high gelation temperature leads to the formation of a platelet-like structure. Polymer concentration affects fiber properties: an increase in polymer concentration decreases porosity and increases mechanical properties such as tensile strength. Solvent properties influence morphology of the scaffolds. After gelation, gel is placed in distilled water for solvent exchange. Afterwards, the gel is removed from the water and goes through freezing and freeze-drying. It is then stored in a desiccator until characterization.

Drawing

The drawing method makes long single strands of nanofibers one at a time. The pulling process is accompanied by solidification that converts the dissolved spinning material into a solid fiber. A cooling step is necessary in the case of melt spinning and evaporation of solvent in the case of dry spinning. A limitation, however, is that only a viscoelastic material that can undergo extensive deformations while possessing sufficient cohesion to survive the stresses developed during pulling can be made into nanofibers through this process.

Template Synthesis

The template synthesis method uses a nanoporous membrane template composed of cylindrical pores of uniform diameter to make fibrils (solid nanofiber) and tubules (hollow nanofiber). This method can be used to prepare fibrils and tubules of many types of materials, including metals, semiconductors and electronically conductive polymers. The uniform pores allow for control of the dimensions of the fibers so nanofibers with very small diameters can be produced through this method. However, a drawback of this method is that it cannot make continuous nanofibers one at a time.

Self-assembly

The self-assembly technique is used to generate peptide nanofibers and peptide amphiphiles. The method was inspired by the natural folding process of amino acid residues to form proteins with unique three-dimensional structures. The self-assembly process of peptide nanofibers involves various driving forces such as hydrophobic interactions, electrostatic forces, hydrogen bonding and van der Waals forces and is influenced by external conditions such as ionic strength and pH.

Polymer Materials

Due to their high porosity and large surface area-to-volume ratio, nanofibers are widely used to construct scaffolds for biological applications. Major examples of natural polymers used in scaffold production are collagen, cellulose, silk fibroin, keratin, gelatin and polysaccharides such as chitosan and alginate. Collagen is a natural extracellular component of many connective tissues. Its fibrillary structure, which varies in diameter from 50-500 nm, is important for cell recognition, attachment, proliferation and differentiation. Using type I collagen nanofibers produced via electrospinning, Shih et al. found that the engineered collagen scaffold showed an increase in cell adhesion and decrease in cell migration with increasing fiber diameter. Using silk scaffolds as a guide for growth for bone tissue regeneration, Kim et al. observed complete bone union after 8 weeks and complete healing of defects after 12 weeks whereas the control in which the bone did not have the scaffold displayed limited mending of defects in the same time period. Similarly, keratin, gelatin, chitosan and alginate demonstrate excellent biocompatibility and bioactivity in scaffolds.

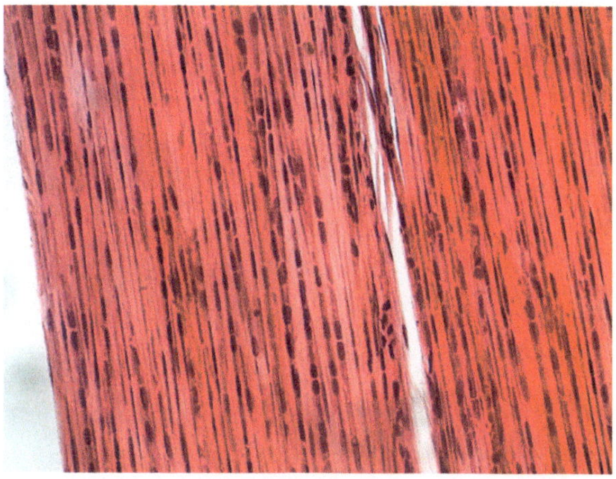

Collagen fibers in a cross-sectional area of dense connective tissue

However, cellular recognition of natural polymers can easily initiate an immune response. Consequently, synthetic polymers such as poly(lactic acid) (PLA), polycaprolactone (PCL), polyurethane (PU), poly(lactic-co-glycolic acid) (PLGA), poly(L-lactide) (PLLA), and poly(ethylene-co-vinylacetate) (PEVA) have been developed as alternatives for integration into scaffolds. Being biodegradable and biocompatible, these synthetic polymers can be used to form matrices with a fiber diameter within the nanometer range. Out of these synthetic polymers, PCL has generated considerable enthusiasm among researchers. PCL is a type of biodegradable polyester that can be prepared via ring-opening polymerization of ε-caprolactone using catalysts. It shows low toxicity, low cost and slow degradation. PCL can be combined with other materials such as gelatin, collagen, chitosan, and calcium phosphate to improve the differentiation and proliferation capacity (2, 17). PLLA is another popular synthetic polymer. PLLA is well known for its superior mechanical properties, biodegradability and biocompatibility. It shows efficient cell migration ability due to its high spatial interconnectivity, high porosity and controlled alignment. A blend of PLLA and PLGA scaffold matrix has shown proper biomimetic structure, good mechanical strength and favorable bioactivity.

Applications

Tissue Engineering

In tissue engineering, a highly porous artificial extracellular matrix is needed to support and guide cell growth and tissue regeneration. Natural and synthetic biodegradable polymers have been used to create such scaffolds.

Simon, in a 1988 NIH SBIR grant report, showed that electrospinning could be used to produced nano- and submicron-scale polystyrene and polycarbonate fibrous mats specifically intended for use as in vitro cell substrates. This early use of electrospun fibrous lattices for cell culture and tissue engineering showed that Human Foreskin Fibroblasts (HFF), transformed Human Carcinoma (HEp-2), and Mink Lung Epithelium (MLE) would adhere to and proliferate upon the fibers.

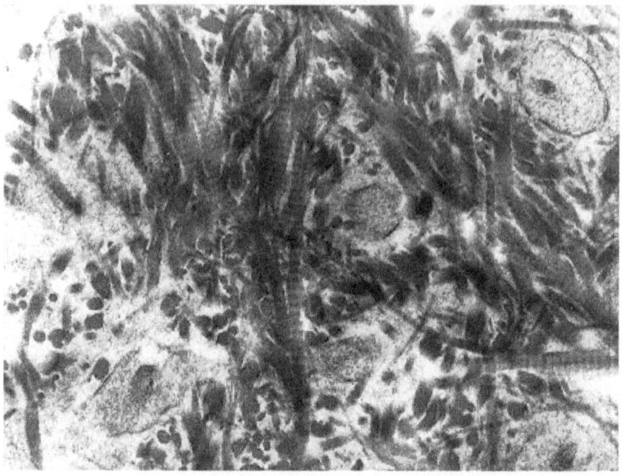

Bone matrix composed of collagen fibrils. Nanofiber scaffolds are able to mimic such structure

Nanofiber scaffolds are used in bone tissue engineering to mimic the natural extracellular matrix of the bones. The bone tissue is arranged either in a compact or trabecular pattern and composed of organized structures that vary in length from the centimeter range all the way to the nanometer scale. Nonmineralized organic component (i.e. type 1 collagen), mineralized inorganic component (i.e. hydroxyapatite), and many other noncollagenous matrix proteins (i.e. glycoproteins and proteoglycans) make up the nanocomposite structure of the bone ECM. The organic collagen fibers and the inorganic mineral salts provide flexibility and toughness, respectively, to ECM.

Although the bone is a dynamic tissue that can self-heal upon minor injuries, it cannot regenerate after experiencing large defects such as bone tumor resections and severe nonunion fractures because it lacks the appropriate template. Currently, the standard treatment is autografting which involves obtaining the donor bone from a non-significant and easily accessible site (i.e. iliac crest) in the patient own body and transplanting it into the defective site. Transplantation of autologous bone has the best clinical outcome because it integrates reliably with the host bone and can avoid complications with the immune system. But its use is limited by its short supply and donor site morbidity associated with the harvest procedure. Furthermore, autografted bones are avascular and hence are dependent on diffusion for nutrients, which affects their viability in the host. The grafts can also be resorbed before osteogenesis is complete due to high remodeling rates in the body. Another strategy for treating severe bone damage is allografting which transplants bones harvested from a human cadaver. However, allografts introduce the risk of disease and infection in the host.

Bone tissue engineering presents a versatile response to treat bone injuries and deformations. Nanofibers produced via electrospinning mimics the architecture and characteristics of natural extracellular matrix particularly well. These scaffolds can be used to deliver bioactive agents that promote tissue regeneration. These bioactive materials should ideally be osteoinductive, osteoconductive, and osseointegratable. Bone substitute materials intended to replace autologous or allogeneic bone consist of bioactive ceramics, bioactive glasses, and biological and synthetic polymers. The basis of bone tissue engineering is that the materials will be resorbed and replaced over time by the body's own newly regenerated biological tissue.

Tissue engineering is not only limited to the bone: a large amount of research is devoted to cartilage, ligament, skeletal muscle, skin, blood vessel, and neural tissue engineering as well.

Drug Delivery

Successful delivery of therapeutics to the intended target largely depends on the choice of the drug carrier. The criteria for an ideal drug carrier include maximum effect upon delivery of the drug to the target organ, evasion of the immune system of the body in the process of reaching the organ, retention of the therapeutic molecules from preparatory stages to the final delivery of the drug, and proper release of the drug for exertion of the intended therapeutic effect.

Nanofibers have grabbed the attention of researchers as a favorable carrier candidate. Natural polymers such as gelatin and alginate make for good fabrication biomaterials for carrier nanofibers because of their biocompatibility and biodegradability that result in no harm to the tissue of the host and no toxic accumulation in the human body, respectively. Additionally, the physical properties of the nanofibers are well suited for drug delivery system. Due to their cylindrical morphology, they possess a high surface area-to-volume ratio. As a result, the fibers possess high drug-loading capacity and can release therapeutic molecules over a large surface area in the medium. Whereas surface area to volume ratio can only be controlled by adjusting the radius for spherical vesicles, nanofibers have more degrees of freedom in controlling the ratio by varying both the length and the cross-sectional radius. This adjustability is important for their application in drug delivery system in which the functional parameters need to be precisely controlled.

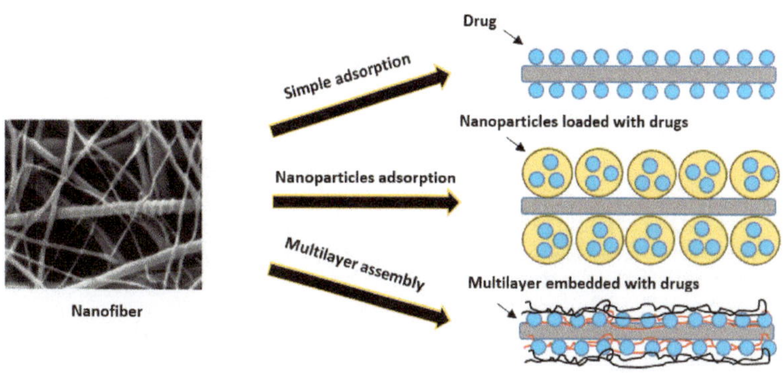

Drugs and biopolymers can be loaded onto nanofibers via simple adsorption, nanoparticles adsorption, and multilayer assembly.

Antibiotics and anticancer drugs have been successfully encapsulated in electrospun nanofibers by adding the drug into the polymer solution prior to electrospinning. Ignatova et al. showed that various antibiotics such as tetracycline hydrochloride, ciprofloxacin and levofloxacin can be incorporated into the synthetic polymer solution that can be electrospun to produce fibers that can be delivered to the site of interest. Drugs and other biopolymers such as DNA can be loaded onto the surfaces of nanofibers via simple adsorption, nanoparticles adsorption, and multilayer assembly. Surface-loaded nanofiber scaffolds are useful as adhesion barriers between internal organs and tissues post-surgery. Adhesion occurs during the healing process and can bring on complications such as chronic pain and reoperation failure. Nanofibers can separate the surgery-operated site from nearby tissues and organs while simultaneously deliver therapeutics such as antibiotics that are physically adsorbed onto their surfaces. Drugs can also be constructed in the form of nanoparticles and attached to the surface of nanofibers, allowing for loading of high concentration of the drugs in a given nanofiber. Rujitanaroj et al. showed that nanofibers functionalized with silver nanoparticles can be used as effective antibiotics against bacteria such as *E. coli*, *Pseudomonas aeruginosa*, *Staphylococcus aureus*, and *methicillin-resistant Staphylococcus aureus* found in burn wounds. Finally, charged biopolymers such as DNA can be incorporated into the multilayer assembly of nanofibers. Such modification can provide diverse drug releasing surface profiles of nanofibers for different medical contexts.

Cancer Diagnosis

Although pathologic examination is the current standard method for molecular characterization in testing for the presence of biomarkers in tumors, these single-sample analyses fail to account for the diverse genomic nature of tumors. Considering the invasive nature, psychological stress, and the financial burden resulting from repeated tumor biopsies in patients, biomarkers that could be judged through minimally invasive procedures, such as blood draws, constitute an opportunity for progression in precision medicine.

Liquid biopsy is an option that is becoming increasingly popular as an alternative to solid tumor biopsy. This is simply a blood draw that contains circulating tumor cells (CTCs) which are shed into the bloodstream from solid tumors. Patients with metastatic cancer are more likely to have detectable CTCs in the bloodstream but CTCs also exist in patients with localized diseases. It has been found that the number of CTCs present in the bloodstream of patients with metastatic prostate and colorectal cancer is prognostic of the overall survival of tumors. CTCs also have been demonstrated to inform prognosis in earlier stages of the disease.

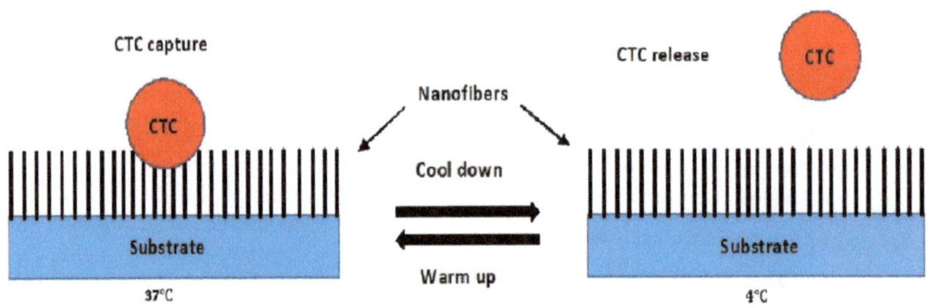

CTC capture and release mechanism of third generation Thermoresponsive Chip

Recently, Ke et al. developed a NanoVelcro chip that captures the CTCs from the blood samples. When blood is passed through the chip, the nanofibers coated with protein antibodies bind to the proteins expressed on the surface of cancer cells and act like Velcro to trap CTCs for analysis. The NanoVelcro CTC assays underwent three generations of development. The first generation NanoVelcro Chip was created for CTC enumeration for cancer prognosis, staging, and dynamic monitoring. The second generation NanoVelcro-LCM was developed for single-cell CTC isolation. The individually isolated CTCs can be subjected to single-CTC genotyping. The third generation Thermoresponsive Chip allowed for CTC purification. The nanofiber polymer brushes undergo temperature-dependent conformational changes to capture and release CTCs.

Lithium-air Battery

Among many advanced electrochemical energy storage devices, rechargeable lithium-air batteries are of particular interest due to their considerable energy storing capacities and high power densities. As the battery is being used, lithium ions combine with oxygen from the air to form particles of lithium oxides, which attach to carbon fibers on the electrode. During recharging, the lithium oxides separate again into lithium and oxygen which is released back into the atmosphere. This conversion sequence is highly inefficient because there is significant voltage difference of more than 1.2 volts between the output voltage and the charging voltage of the battery meaning that approximately 30% of the electrical energy is lost as heat when the battery is charging. Also the large volume changes resulting from continuous conversion of oxygen between its gaseous and solid state puts stress on the electrode and limits its lifetime.

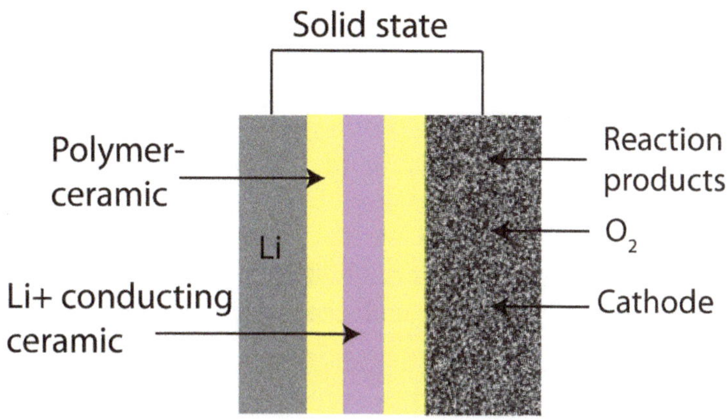

Schematic of a lithium-air battery. For the nanofiber-based lithium-air battery, the cathode would be made up of carbon nanofibers.

The performance of these batteries depends on the characteristics of the material that makes up the cathode. Carbon materials have been widely used as cathodes because of their excellent electrical conductivities, large surface areas, and chemical stability. Especially relevant for lithium-air batteries, carbon materials act as substrates for supporting metal oxides. Binder-free electrospun carbon nanofibers are particularly good potential candidates to be used in electrodes in lithium-oxygen batteries because they have no binders, have open macroporous structures, have carbons that support and catalyze the oxygen reduction reactions, and have versatility.

Zhu et al. developed a novel cathode that can store lithium and oxygen in the electrode they named nanolithia which is a matrix of carbon nanofibers periodically embedded with cobalt oxide. These cobalt oxides provide stability to the normally unstable superoxide-containing nanolithia. In this design, oxygen is stored as LiO_2 and does not convert between gaseous and solid forms during charging and discharging. When the battery is discharging, lithium ions in nanolithia and react with superoxide oxygen the matrix to form Li_2O_2, and Li_2O. The oxygen remains in its solid state as it transitions among these forms. The chemical reactions of these transitions provide electrical energy. During charging, the transitions occur in reverse.

Optical Sensors

Polymer optical fibers have generated increasing interest in recent years. Because of low cost, ease of handling, long wavelength transparency, great flexibility, and biocompatibility, polymer optical fibers show great potential for short-distance networking, optical sensing and power delivery.

Electrospun nanofibers are particularly well-suitable for optical sensors because sensor sensitivity increases with increasing surface area per unit mass. Optical sensing works by detecting ions and molecules of interest via fluorescence quenching mechanism. Wang et al. successfully developed nanofibrous thin film optical sensors for metal ion (Fe^{3+} and Hg^{2+}) and 2,4-dinitrotoluene (DNT) detection using the electrospinning technique.

Quantum dots show useful optical and electrical properties, including high optical gain and photochemical stability. A variety of quantum dots have been successfully incorporated into polymer nanofibers. Meng et al. showed that quantum dot-doped polymer nanofiber sensor for humidity detection shows fast response, high sensitivity, and long-term stability while requiring low power consumption.

Kelly et al. developed a sensor that warns first responders when the carbon filters in their respirators have become saturated with toxic fume particles. The respirators typically contain activated charcoal that traps airborne toxins. As the filters become saturated, chemicals begin to pass through and render the respirators useless. In order to easily determine when the filter is spent, Kelly and his team developed a mask equipped with a sensor composed of carbon nanofibers assembled into repeating structures called photonic crystals that reflect specific wavelengths of light. The sensors exhibit an iridescent color that changes when the fibers absorb toxins.

Air Filtration

Paints and protective coatings on furniture contain volatile organic compounds such as toluene and formaldehyde.

Electrospun nanofibers are useful for removing volatile organic compounds (VOC) from the atmopshere. Scholten et al. showed that adsorption and desorption of VOC by electrospun nanofibrous membrane were faster than the rates of conventional activated carbon.

Airborne contamination in the personnel cabins of mining equipment is of concern to the mining workers, mining companies, and government agencies such as the Mine Safety and Health Administration (MSHA). Recent work with mining equipment manufacturers and the MSHA has shown that nanofiber filter media can reduce cabin dust concentration to a greater extent compared to standard cellulose filter media.

Oil-water Separation

Nanofibre has the capabilities in oil–water separation, most particularly in sorption process when the material in use has the oleophilic and hydrophobic surfaces. These characteristic enable the nanofiber to be used as a tool to combat either oily waste- water from domestic household and industrial activities, or oily seawater due to the oil run down to the ocean from oil transportation activities and oil tank cleaning on a vessel.

Nanoscale Iron Particles

Nanoscale iron particles are sub-micrometer particles of iron metal. They are highly reactive because of their large surface area. In the presence of oxygen and water, they rapidly oxidize to form free iron ions. They are widely used in medical and laboratory applications and have also been studied for remediation of industrial sites contaminated with chlorinated organic compounds.

Synthesis

Iron nanoparticles can be synthesized by the reduction of Fe(II) or Fe(III) salt with sodium borohydride in an aqueous medium.

$$Fe(H_2O)_6^{3+} + 3BH_4^- + 3H_2O$$
$$\rightarrow Fe^0 + 3B(OH)_3 + 10.5H_2$$

Reactivity

When exposed to oxygen and water, iron oxidizes. This redox process can occur under either acidic or neutral/basic conditions:

$$2\ Fe^0(s) + 4\ H^+(aq) + O_2(aq) \rightarrow 2\ Fe^{2+}(aq) + 2\ H_2O(l)$$

$$Fe^0(s) + 2\ H_2O\ (aq) \rightarrow Fe^{2+}(aq) + H_2(g) + 2\ OH^-(aq)$$

Research

Research has shown that nanoscale iron particles can be effectively used to treat several forms of ground contamination, including grounds contaminated by polychlorinated biphenyls (PCBs), chlorinated organic solvents, and organochlorine pesticides. Nanoscale iron particle are easily transportable through ground water, allowing for in situ treatment. Additionally, the nanoparticle-water slurry can be injected into the contaminated area and stay there for long periods of time. These factors combine to make this method of cheaper than most currently used alternative.

Researchers have found that although metallic iron nanoparticles remediate contaminants well, they tend to agglomerate on the soil surfaces. In response, carbon nanoparticles and water-soluble polyelectrolytes have been used as supports to the metallic iron nanoparticles. The hydrophobic contaminants adsorb to these supports, improving permeability in sand and soil.

In field tests have generally confirmed lab findings. However, research is still ongoing and nanoscale iron particles are not yet commonly used for treating ground contamination.

Iron Oxide Nanoparticles

Iron oxide nanoparticles are iron oxide particles with diameters between about 1 and 100 nanometers. The two main forms are magnetite (Fe_3O_4) and its oxidized form maghemite (γ-Fe_2O_3). They have attracted extensive interest due to their superparamagnetic properties and their potential applications in many fields (although Co and Ni are also highly magnetic materials, they are toxic and easily oxidized).

Applications of iron oxide nanoparticles include terabit magnetic storage devices, catalysis, sensors, and high-sensitivity biomolecular magnetic resonance imaging (MRI) for medical diagnosis and therapeutics. These applications require coating of the nanoparticles by agents such as long-chain fatty acids, alkyl-substituted amines and diols.

Structure

Magnetite has an inverse spinel structure with oxygen forming a face-centered cubic crystal system. In magnetite, all tetrahedral sites are occupied by Fe^{3+} and octahedral sites are occupied by both Fe^{3+} nd Fe^{2+}. Maghemite differs from magnetite in that all or most of the iron is in the trivalent state (Fe^{3+}) and by the presence of cation vacancies in the octahedral sites. Maghemite has a cubic unit cell in which each cell contains 32 O ions, $21\frac{1}{3}$ Fe^{3+} ions and $2\frac{2}{3}$ vacancies. The cations are distributed randomly over the 8 tetrahedral and 16 octahedral sites.

Magnetic Properties

Due to its 4 unpaired electrons in 3d shell, an iron atom has a strong magnetic moment. Ions Fe^{2+}have also 4 unpaired electrons in 3d shell and Fe^{3+}have 5 unpaired electrons in 3d shell. Therefore, when crystals are formed from iron atoms or ions Fe^{2+} and Fe^{3+} they can be in ferromagnetic, antiferromagnetic or ferrimagnetic states.

In the paramagnetic state, the individual atomic magnetic moments are randomly oriented, and the substance has a zero net magnetic moment if there is no magnetic field. These materials have a relative magnetic permeability greater than one and are attracted to magnetic fields. The magnetic moment drops to zero when the applied field is removed. But in a ferromagnetic material, all the atomic moments are aligned even without an external field. A ferrimagnetic material is similar to a ferromagnet but has two different types of atoms with opposing magnetic moments. The material has a magnetic moment because the opposing moments have different strengths. If they have the same magnitude, the crystal is antiferromagnetic and possesses no net magnetic moment.

When an external magnetic field is applied to a ferromagnetic material, the magnetization (M) increases with the strength of the magnetic field (H) until it approaches saturation. Over some range of fields the magnetization has hysteresis because there is more than one stable magnetic state for each field. Therefore, a remanent magnetization will be present even after removing the external magnetic field.

A single domain magnetic material (ex: magnetic nanoparticles) that has no hysteresis loop is said to be superparamagnetic. The ordering of magnetic moments in ferromagnetic, antiferromagnetic, and ferrimagnetic materials decreases with increasing temperature. Ferromagnetic and ferrmagnetic materials become disordered and lose their magnetization beyond the Curie temperature "T_C" and antiferromagnetic materials lose their magnetization beyond the Néel temperature T_N. Magnetite is ferrimagnetic at room temperature and has a Curie temperature of 850 K. Maghemite is ferrimagnetic at room temperature, unstable at high temperatures, and loses its susceptibility with time. (Its Curie temperature is hard to determine). Both magnetite and maghemite nanoparticles are superparamagnetic at room temperature. This superparamagnetic behavior of iron oxide nanoparticles can be attributed to their size. When the size gets small enough (<10 nm), thermal fluctuations can change the direction of magnetization of the entire crystal. A material with many such crystals behaves like a paramagnet, except that the moments of entire crystals are fluctuating instead of individual atoms.

Synthesis

The preparation method has a large effect on shape, size distribution, and surface chemistry of the particles. It also determines to a great extent the distribution and type of structural defects or impurities in the particles. All these factors affect magnetic behavior. Recently, many attempts have been made to develop processes and techniques that would yield 'monodisperse colloids' consisting of nanoparticles uniform in size and shape.

Coprecipitation

By far the most employed method is coprecipitation. This method can be further divided into two types. In the first, ferrous hydroxide suspensions are partially oxidized with different oxidizing agents. For example, spherical magnetite particles of narrow size distribution with mean diameters between 30 and 100 nm can be obtained from a Fe(II) salt, a base and a mild oxidant (nitrate ions). The other method consists in ageing stoichiometric mixtures of ferrous and ferric hydroxides in aqueous media, yielding spherical magnetite particles homogeneous in size. In the second type, the following chemical reaction occurs:

$$2Fe^{3+} + Fe^{2+} + 8OH^{-} \dashrightarrow Fe_3O_4 + 4H_2O$$

Optimum conditions for this reaction are pH between 8 and 14, Fe^{3+}/Fe^{2+} ratio of 2:1 and a non-oxidizing environment. Being highly susceptibile to oxidation, magnetite (Fe_3O_4) is transformed to maghemite (γFe_2O_3) in the presence of oxygen:

$$2Fe_3O_4 + O_2 \rightarrow 2\gamma Fe_2O_3$$

The size and shape of the nanoparticles can be controlled by adjusting pH, ionic strength, temperature, nature of the salts (perchlorates, chlorides, sulfates, and nitrates), or the Fe(II)/Fe(III) concentration ratio.

Microemulsions

A microemulsion is a stable isotropic dispersion of 2 immiscible liquids consisting of nanosized domains of one or both liquids in the other stabilized by an interfacial film of surface-active molecules. Microemulsions may be categorized further as oil-in-water (o/w) or water-in-oil (w/o), depending on the dispersed and continuous phases. Water-in-oil is more popular for synthesizing many kinds of nanoparticles. The water and oil are mixed with an amphiphillic surfactant. The surfactant lowers the surface tension between water and oil, making the solution transparent. The water nanodroplets act as nanoreactors for synthesizing nanoparticles. The shape of the water pool is spherical. The size of the nanoparticles will depend on size of the water pool to a great extent. Thus, the size of the spherical nanoparticles can be tailored and tuned by changing the size of the water pool.

High-temperature Decomposition of Organic Precursors

The decomposition of iron precursors in the presence of hot organic surfactants results in samples with good size control, narrow size distribution (5-12 nm) and good crystallinity; and the nanoparticles are easily dispersed. For biomedical applications like magnetic resonance imaging, magnetic cell separation or magnetorelaxometry, where particle size plays a crucial role, magnetic nanoparticles produced by this method are very useful. Viable iron precursors include $Fe(Cup)_3$, $Fe(CO)_5$, or $Fe(acac)_3$ in organic solvents with surfactant molecules. A combination of Xylenes and Sodium Dodecylbenezensulfonate as a surfactant are used to create nanoreactors for which well dispersed iron(II) and iron (III) salts can react.

Biomedical Applications

Magnetite and maghemite are preferred in biomedicine because they are biocompatible and potentially non-toxic to humans. Iron oxide is easily degradable and therefore useful for in vivo applications. Results from exposure of a human mesothelium cell line and a murine fibroblast cell line to seven industrially important nanoparticles showed a nanoparticle specific cytotoxic mechanism for uncoated iron oxide. Solubility was found to strongly influence the cytotoxic response. Labelling cells (e.g. stem cells, dendritic cells) with iron oxide nanoparticles is an interesting new tool to monitor such labelled cells in real time by magnetic resonance tomography.

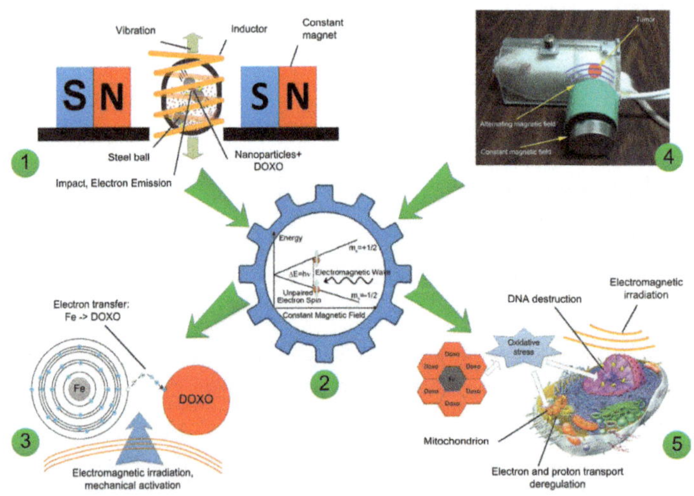

The magneto-mechano-chemical synthesis (1) is accompanied by splitting of electron energy levels (SEELs) and electron transfer in magnetic field (2) from nanoparticles Fe_3O_4 to doxorubicin. The concentration of paramagnetic centers (free radicals) is increased in the magneto-sensitive complex (MNC) (3). The local combined action of constant magnetic and electromagnetic fields and MNC in tumor (4) initiated SEELs, free radicals, leading to oxidative stress and electron and proton transport deregulation in the mitochondrion (5). Magnetic nanotherapy has more effectively inhibited the synthesis of ATP in mitochondria of tumor cell and induced the death of tumor cells compared to conventional doxorubicin.

Iron oxide nanoparticles are used in cancer magnetic nanotherapy that is based on the magneto-spin effects in free radical reactions and semiconductor material ability to generate oxygen radicals, furthermore, control oxidative stress in biological media under inhomogeneous electromagnetic radiation. The magnetic nanotherapy is remotely controlled by external electromagnetic field ROS (reactive oxygen species)-mediated local toxicity in the tumor during chemotherapy with antitumor magnetic complex and lesser side effects in normal tissues. Magnetic complexes that consist of iron oxide nanoparticles loaded with antitumor drug have additional advantages over conventional antitumor drugs due to their ability to be remotely controlled while targeting with a constant magnetic field and further strengthening of their antitumor activity by moderate inductive hyperthermia (below 40°C). The combined influence of inhomogeneous constant magnetic and electromagnetic fields during nanotherapy has initiated splitting of electron energy levels in magnetic complex and unpaired electron transfer from iron oxide nanoparticles to anticancer drug and tumor cells. In particular, anthracycline antitumor antibiotic doxorubicin, the native state of which is diamagnetic, acquires the magnetic properties of paramagnetic substances. Electromagnetic radiation at the hyperfine splitting frequency can increase time radical pairs are in the triplet state and hence the probability of dissociation and so the concentration of free radicals. The reactivity of magnetic particles depends on their spin state. The experimental data was received about correlation between the frequency of electromagnetic field radiation with magnetic properties and quantity paramagnetic centres of complex. It is possible to control the kinetics of free radical reactions by external magnetic fields and modulate the level of oxidative stress (local toxicity) in malignant tumor. Cancer cells are then particularly vulnerable to an oxidative assault and induction of high levels of oxidative stress locally in tumor tissue, that has the potential to destroy or arrest the growth of cancer cells and can be thought as therapeutic strategy against cancer. Multifunctional magnetic complexes can combine cancer magnetic nanotherapy, tumor targeting and medical imaging functionalities in theranostics approach for personalized cancer medicine.

Iron oxide nanoparticles also are using in magnetic hyperthermia as a cancer treatment method. In this method, ferrofluid which is contained Iron oxide is injected to the tumor and then heated up by an alternating high frequency magnetic field. Temperature distribution produced by this heat generation destroys cancerous cells inside tumor.

Iron-platinum Nanoparticles

Iron-platinum nanoparticles (FePt NPs) are 3D superlattices composed of an approximately equal atomic ratio of Fe and Pt. Under standard conditions, FePt NPs exist in the face centered cubic phase but can change to a chemically ordered face centered tetragonal phase as a result of thermal annealing. Currently there are many synthetic methods such as water-in-oil microemulsion, one-step thermal synthesis with metal precursors, and exchanged-coupled assembly for making FePt NPs. An important property of FePt NPs is their superparamagnetic character below 10 nanometers. The superparamagnetism of FePt NPs has made them attractive candidates to be used as MRI/CT scanning agents and a high-density recording material.

Properties

The various properties of iron-platinum nanoparticles allow them to function in multiple ways. In standard conditions, FePt NPs exist in the face centered cubic phase with a 3 to 10 nanometer diameter. However, once heat is added the structure becomes a face centered tetragonal and superparamagnetic. The nanoparticles become superparamagnetic because the addition of heat makes the particle smaller and iron rich, since it removes any impurities in the particles. As a result, the nanoparticles are used in CT or MRI scans.

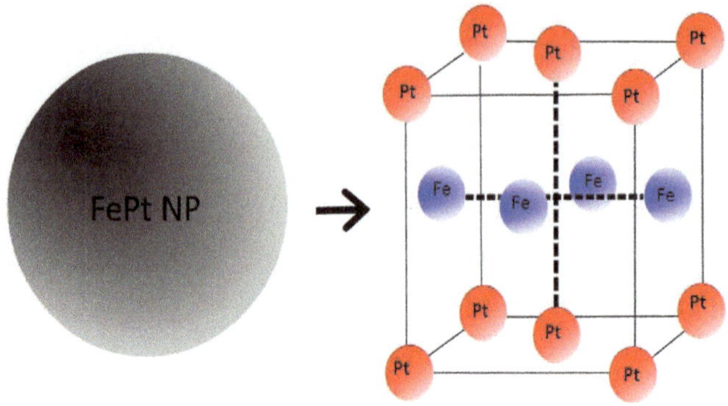

Iron-Platinum Nanoparticle Lattice

Plant viruses, known as Cowpea mosaic virus and Tobacco mosaic virus, enlarge the average radius of the FePt NPs through direct mineralization. The virus template acts as a natural and benign method to monodisperse the nanoparticles up to 30 nanometers in diameter. The size increase of the bimetallic nanoparticles entitles it to a wider range of biological applications.

Physical Properties	
Chemical Symbol	Fe_4Pt_{10}
Density (of a small particle around 7.4 nm diameter)	4.0 g/cm-^{3}
Density (of a large particles around 16nm diameter)	3.2 g/cm^3
Structure (in standard conditions)	Face Centered Cubic
Structure (after heating)	Face Centered Tetragonal
Chemical Properties	
Platinum (Pt)	71.43%
Iron (Fe)	28.57%

Physical properties of FePt NPs

Synthesis

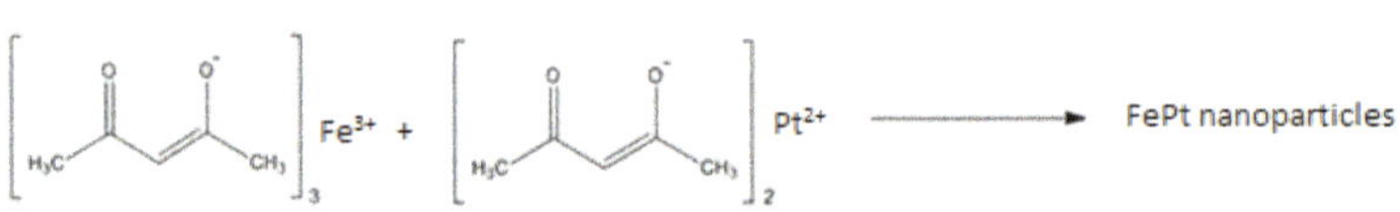

Synthesis of Iron-Platinum Nanoparticles

Platinum nanoparticles have a stronger chemical stability with the addition of iron, cobalt, or nickel. The platinum alloys also have a better detection range and catalytic activity than their counterpart. These magnetic metal additions to platinum reduce the overall sensitivity to oxidation while maintaining the desirable magnetic properties. Combined, FePt nanoparticles are synthesized for medicinal applications. One method of synthesis uses incident laser technology to irradiate solutions containing iron and platinum to combine the two alloys. A laser beam is emitted onto a 4:1

mixture of iron (III) acetylacetonate and platinum (II) acetylacetonate dissolved in methanol. The black precipitates are then washed and dried on silicon substrates to be characterized by transmission electron microscopy (TEM) and X-ray diffraction.

An alternative method of synthesis involves the coreduction of chloroplatinic acid (H_2PtCl_6) and iron (II) chloride in water-in-oil microemulsions. In this process, the normal face-centered cubic structure is transformed to a face-centered tetragonal configuration, offering a higher density product useful for many storage media applications.

Applications

FePt NPs are promising materials for ultra-high density magnetic recording media due to their high coercivity. Higher coercivity indicates the material cannot be demagnetized easily. After annealing at 700 °C, the film can have up to 14KOe coercivity compared to common hard drives that have 5KOe coercivity.

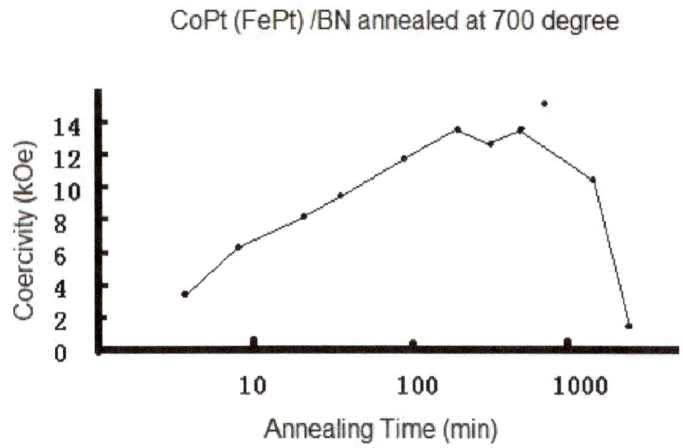

FePt Coercivity

Due to their superparamagnetism and controllable shape, size, and surface, iron-platinum nanoparticles have great potential for advancing medicine in many fields, including imagining, pathogen detection, and targeted cancer therapy. The NPs can be conjugated with antibodies for tissue-specific delivery, providing a systematic way to customize for either technology. FePt NPs are compatible for CT scans because of their strong ability to absorb x-rays. FePt NPs also provide a non-toxic, more persistent alternative to iodinated molecules that are harmful to the kidney and survive in the body for only a short time. The superparamagnetic properties of the nanoparticles and the systematic method for conjugating ligands to the FePt surface makes them viable vehicles for detection of pathogens such as gram-positive bacteria. Antibodies for the bacteria conjugated to the FePt NP bind to the bacteria and magnetic dipoles are used to detect the FePt NP-bacteria conjugate. By attaching peptides to the surface of the face centered cubic FePt NPs, cytotoxic iron can be delivered to specific locations and taken up with high selectivity. A phospholipid coating of the FCC-FePt prevents Fe release. Once in the cell, the low pH of lysosome's intracellular environments breaks down the phospholipid bilayer. Fe catalyzed decomposition of hydrogen peroxide into ROSs results in membrane lipid oxidation, damage to DNA and proteins, and tumor death.

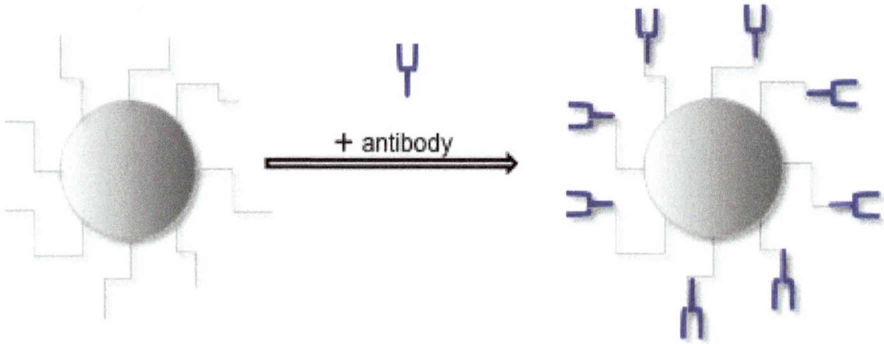

FePt-NP Antibody Application

Nanoceramic

Nanoceramic is a type of nanoparticle that is composed of ceramics, which are generally classified as inorganic, heat-resistant, nonmetallic solids made of both metallic and nonmetallic compounds. The material offers unique properties. Macroscale ceramics are brittle and rigid and break upon impact. However, nanoceramics take on a larger variety of functions, including dielectric, ferroelectric, piezoelectric, pyroelectric, ferromagnetic, magnetoresistive, superconductive and electro-optical.

Nanoceramics were discovered in the early 1980s. They were formed using a process called sol-gel which mixes nanoparticles within a solution and gel to form the nanoparticle. Later methods involved sintering (pressure and heat). The material is so small that it has basically no flaws. Larger scale materials have flaws that render them brittle.

In 2014 researchers announced a lasering process involving polymers and ceramic particles to form a nanotruss. This structure was able to recover its original form after repeated crushing.

Properties

Nanoceramics have unique properties because of their size and molecular structure. These properties are often shown in terms of various electrical and magnetic physics phenomenons which include:

- Dielectric - An electrical insulator that can be polarized (having electrons aligned so that there is a negative and positive side of the compound) by an electric field to shorten the distance of electron transfer in an electric current
- Ferroelectric - Dielectric materials that polarize in more than one direction (the negative and positive sides can be flipped via an electric field)
- Piezoelectric - Materials that accumulate an electrical charge under mechanical stress
- Pyroelectric - Material that can produce a temporary voltage given a temperature change
- Ferromagnetic - Materials that can to sustain a magnetic field after magnetization

- Magnetoresistive - Materials that change electrical resistance under an external magnetic field
- Superconductive - Materials that exhibit zero electric resistance when cooled to a critical temperature
- Electro-optical - Materials that change optical properties under an electric field

Nanotruss

Nanoceramic is more than 85% air and is very light, strong, flexible and durable. The fractal nanotruss is a nanostructure architecture made of alumina, or aluminum oxide. Its maximum compression is about 1 micron from a thickness of 50 nanometers. After its compression, it can revert to its original shape without any structural damage.

Synthesis

Sol-gel

One process for making nanoceramics varies is the sol-gel process, also known as chemical solution deposition. This involves a chemical solution, or the sol, made of nanoparticles in liquid phase and a precursor, usually a gel or polymer, made of molecules immersed in a solvent. The sol and gel are mixed to produce an oxide material which are generally a type of ceramic. The excess products (a liquid solvent) are evaporated. The particles desires are then heated in a process called densification to produce a solid product. This method could also be applied to produce a nanocomposite by heating the gel on a thin film to form a nanoceramic layer on top of the film.

Two-photon Lithography

This process uses a laser technique called two-photon lithography to etch out a polymer into a three-dimensional structure. The laser hardens the spots that it touches and leaves the rest unhardened. The unhardened material is then dissolved to produce a "shell". The shell is then coated with ceramic, metals, metallic glass, etc. In the finished state, the nanotruss of ceramic can be flattened and revert to its original state.

Sintering

In another approach sintering was used to consolidate nanoceramic powders using high temperatures. This resulted in a rough material that damages the properties of ceramics and requires more time to obtain an end product. This technique also limits the possible final geometries. Microwave sintering was developed to overcome such problems. Radiation is produced from a magnetron, which produces electromagnetic waves to vibrate and heat the powder. This method allows for heat to be instantly transferred across the entire volume of material instead of from the outside in.

The nanopowder is placed in an insulation box composed of low insulation boards to allow the microwaves to pass through it.The box increases temperature to aid absorption. Inside the boxes are suspectors that absorb microwaves at room temperature to initialize the sintering process. The microwave heats the suspectors to about 600 °C, sufficient to trigger the nanoceramics to absorb the microwaves.

History

In the early 1980s, the first nanoparticles, specifically nanoceramics were formed, using sol-gel. This process was replaced by sintering in the early 2000s and then by microwave sintering. None of these techniques proved suitable for large scale production.

In 2002, researchers tried to reverse engineer the microstructure of seashells to strengthen ceramics. They discovered that seashells' durability come from their "microarchitecture". Research began to focus on how ceramics could employ such an architecture.

In 2012 researchers replicated the sea sponge's structure using ceramics and the nanoarchitecture called nanotruss. As of 2015 the largest result is a 1mm cube. The lattice structure compresses up to 85% of its original thickness and can recover to its original form. These lattices are stabilized into triangles with cross-members for structural integrity and flexibility.

Applications

Medical technology used nanoceramics for bone repair. It has been suggested for areas including energy supply and storage, communication, transportation systems, construction and medical technology. Their electrical properties may allow energy to be transferred efficiencies approaching 100%. Nanotrusses may be eventually applicable for building materials, replacing concrete or steel.

Solid Lipid Nanoparticle

Solid lipid nanoparticles (SLNs) are a new pharmaceutical delivery system or pharmaceutical formulation.

The conventional approaches such as use of permeation enhancers, surface modification, prodrug synthesis, complex formation and colloidal lipid carrier based strategies have been developed for the delivery of drugs to intestinal lymphatics. In addition, polymeric nanoparticles, self-emulsifying delivery systems, liposomes, microemulsions, micellar solutions and recently solid lipid nanoparticles (SLN) have been exploited as probable possibilities as carriers for oral intestinal lymphatic delivery.

A solid lipid nanoparticle is typically spherical with -an average diameter between 10 and 1000 nanometers. Solid lipid nanoparticles possess a solid lipid core matrix that can solubilize lipophilic molecules. The lipid core is stabilized by surfactants (emulsifiers). The term lipid is used here in a broader sense and includes triglycerides (e.g. tristearin), diglycerides (e.g. glycerol bahenate), monoglycerides (e.g. glycerol monostearate), fatty acids (e.g. stearic acid), steroids (e.g. cholesterol), and waxes (e.g. cetyl palmitate). All classes of emulsifiers (with respect to charge and molecular weight) have been used to stabilize the lipid dispersion. It has been found that the combination of emulsifiers might prevent particle agglomeration more efficiently.

Development of solid lipid nanoparticles is one of the emerging fields of lipid nanotechnology with several potential applications in drug delivery, clinical medicine and research, as well as in other discipline. Due to their unique size-dependent properties, lipid nanoparticles offer the possibility to develop new therapeutics. The ability to incorporate drugs into nanocarriers offers a new prototype in drug delivery that could hold great promise for attaining the bioavailability enhancement along with controlled and site specific drug delivery. SLN's are also considered to well tolerated in general, due to their composition from physiologically similar lipids.

Use for Drug Delivery

Solid lipid nanoparticles have recently materialized as a novel approach to oral and parenteral drug delivery systems. SLNs combine the advantages of lipid emulsion and polymeric nanoparticle systems while overcoming the temporal and *in vivo* stability issues that troubles the conventional as well as polymeric nanoparticles drug delivery approaches (Mehnert *et al.*, 2001). It has been proposed that SLNs combine numerous advantages over the other colloidal carriers i.e. incorporation of lipophilic and hydrophilic drugs feasible, no biotoxicity of the carrier, avoidance of organic solvents, possibility of controlled drug release and drug targeting, increased drug stability and no problems with respect to large scale production. A recent study has demonstrated the use of solid lipid nanoparticles as a platform for oral delivery of the nutrient mineral iron, by incorporating the hydrophilic molecule ferrous sulphate ($FeSO_4$) in a lipid matrix composed of stearic acid. Carvedilol-loaded solid lipid nanoparticles were prepared using hot-homogenization technique for oral delivery using compritol and poloxamer 188 as a lipid and surfactant, respectively.

Solid Lipid Nanoparticles and its Potential Lymphatic Absorption Mechanism

Elucidation of intestinal lymphatic absorption mechanism from solid lipid nanoparticles using Caco-2 cell line as in vitro model was developed. Several researchers have shown the enhancement of oral bioavailibility of poorly water soluble drugs when encapsulated in solid lipid nanoparticle. This enhanced bioavailibility is achieved via lymphatic delivery. To elucidate the absorption mechanism, from solid lipid nanoparticle, human excised Caco-2 cell monolayer could be alternative tissue for development of an in-vitro model to be used as a screening tool before animal studies are undertaken. The results obtained in this model suggested that the main absorption mechanism of carvedilol loaded solid lipid nanoparticle could be endocytosis and, more specifically, clathrin-mediated endocytosis.

Characteristics and Production

An SLN is generally spherical in shape and consists of a solid lipid core stabilized by a surfactant. The core lipids can be fatty acids, acylglycerols, waxes, and mixtures of these surfactants. Biological membrane lipids such as phospholipids, sphingomyelins, bile salts (sodium taurocholate), and sterols (cholesterol) are utilized as stabilizers. Biological lipids having minimum carrier cytotoxicity and the solid state of the lipid permit better controlled drug release due to increased mass transfer resistance.

Advantages

Advantages of SLNs are the use of physiological lipids, the avoidance of organic solvents, a potential wide application spectrum (dermal, per os, intravenous) and the high pressure homogenization as an established production method. Additionally, improved bioavailability, protection of sensitive drug molecules from the outer environment water, light) and even controlled release characteristics were claimed by incorporation of poorly water-soluble drugs in the solid lipid matrix.

Nanotechnology and Ocular Drug Delivery

Nanotechnology is expected to revolutionize ocular drug delivery. Many nano-structured systems have been employed for ocular drug delivery and yielded some promising results. SLNs have been looked at as a potential drug carrier system since the 1990s. SLNs do not show biotoxicity as they are prepared from physiological lipids. SLNs are especially useful in ocular drug delivery as they can enhance the corneal absorption of drugs and improve the ocular bioavailability of both hydrophilic and lipophilic drugs. Solid lipid nanoparticles have another advantage of allowing autoclave sterilization, a necessary step towards formulation of ocular preparations.

References

- "Colloidal gold, a useful marker for transmission and scanning electron microscopy" by M Horisberger and J Rosset Journal of Histochemistry and Cytochemistry Volume 25, Issue 4, pp. 295–305, 4 January 1977
- Practical Electron Microscopy: A Beginner's Illustrated Guide, by Elaine Evelyn Hunter. Cambridge University Press; 2nd edition (September 24, 1993) ISBN 0-521-38539-3
- Berzelius, J. J. (1831). "Ueber den Cassius' schen Goldpurpur". Annalen der Physik. 98 (6): 306–308. Bibcode:1831AnP....98..306B. doi:10.1002/andp.18310980613
- Nune, Satish K., et al. "Green Nanotechnology from Tea: Phytochemicals in Tea as Building Blocks for Production of Biocompatible Gold Nanoparticles." Journal of materials chemistry 19.19 (2009): 2912–20. Print
- Faraday, M. (1857). "Experimental Relations of Gold (and Other Metals) to Light,". Philosophical Transactions of the Royal Society. 147: 145. doi:10.1098/rstl.1857.0011
- Jiang, Zhong-Jie; Liu, Chun-Yan; Sun, Lu-Wei (2005-02-01). "Catalytic Properties of Silver Nanoparticles Supported on Silica Spheres". The Journal of Physical Chemistry B. 109 (5): 1730–1735. ISSN 1520-6106. doi:10.1021/jp046032g
- Carlos, A; et al. (2011). Scaffolds for Tissue Engineering Via Thermally Induced Phase Separation, Advances in Regenerative Medicine. ISBN 978-953-307-732-1
- Mohan, R.R.; et., al. (June 2013). "BMP7 Gene Transfer via Gold Nanoparticles into Stroma Inhibits Corneal Fibrosis In Vivo". Plos One. 8 (6): 1–9. doi:10.1371/journal.pone.0066434
- Chuang; et al. "Allergenicity and toxicology of inhaled silver nanoparticles in allergen-provocation mice models". International Journal of Nanomedicine. 2013 (8): 4495–4506. doi:10.2147/IJN.S52239
- Teraoka, Iwao (2002). Polymer Solutions: An Introduction to Physical Properties. John Wiley & Sons, Inc. ISBN 0-471-22451-0
- Yin, Xi; Wu, Jianbo; Li, Panpan; Shi, Miao; Yang, Hong (January 2016). "Self-Heating Approach to the Fast Production of Uniform Metal Nanostructures". ChemNanoMat. 2 (1): 37–41. doi:10.1002/cnma.201500123
- Hirai; et al. (2014). "Silver nanoparticles induce silver nanoparticle-specific allergic responses (HYP6P.274)". The Journal of Immunology. 192 (118): 19

3

A Comprehensive Study of Nanostructure

A nanostructure is any structure whose size varies between microscopic and molecular structure. Nanomesh, gradient multilayer nanofilm, nanohole, nanosheet, nanostructured film, sculptured thin film, icosahedral twins, nano flake and nanofoam are important topics related to the study of nanostructures. Nanotechnology is best understood in confluence with the major topics listed in the following chapter.

Nanostructure

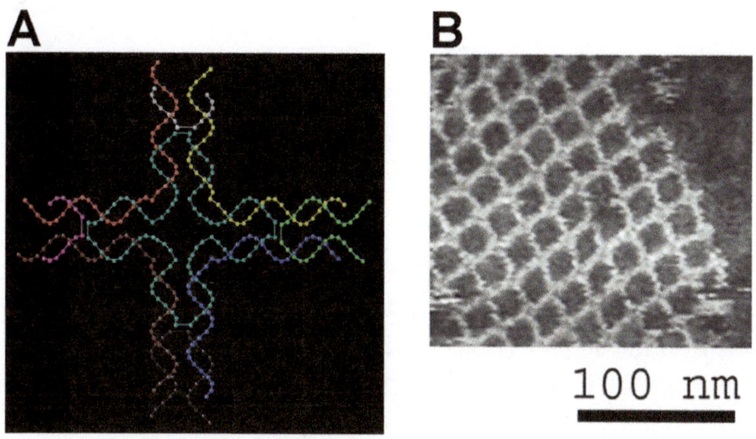

The DNA structure at left (schematic shown) will self-assemble into the structure visualized by atomic force microscopy at right. Image from Strong.

A nanostructure is a structure of intermediate size between microscopic and molecular structures. Nanostructural detail is microstructure at nanoscale.

In describing nanostructures, it is necessary to differentiate between the number of dimensions in the volume of an object which are on the nanoscale. Nanotextured surfaces have *one dimension* on the nanoscale, i.e., only the thickness of the surface of an object is between 0.1 and 100 nm. Nanotubes have *two dimensions* on the nanoscale, i.e., the diameter of the tube is between 0.1 and 100 nm; its length can be far more. Finally, spherical nanoparticles have *three dimensions* on the nanoscale, i.e., the particle is between 0.1 and 100 nm in each spatial dimension. The terms nanoparticles and ultrafine particles (UFP) are often used synonymously although UFP can reach into the micrometre range. The term *nanostructure* is often used when referring to magnetic technology.

Nanoscale structure in biology is often called ultrastructure.

Properties of nanoscale objects and ensembles of these objects are widely studied in physics.

Nanowire

A nanowire is a nanostructure, with the diameter of the order of a nanometer (10^{-9} meters). It can also be defined as the ratio of the length to width being greater than 1000. Alternatively, nanowires can be defined as structures that have a thickness or diameter constrained to tens of nanometers or less and an unconstrained length. At these scales, quantum mechanical effects are important — which coined the term "quantum wires". Many different types of nanowires exist, including superconducting (e.g. YBCO), metallic (e.g. Ni, Pt, Au), semiconducting (e.g. silicon nanowires (SiNWs), InP, GaN) and insulating (e.g. SiO_2, TiO_2). Molecular nanowires are composed of repeating molecular units either organic (e.g. DNA) or inorganic (e.g. $Mo_6S_{9-x}I_x$).

Overview

Crystalline 2×2-atom tin selenide nanowire grown inside a single-wall carbon nanotube (tube diameter ~1 nm)

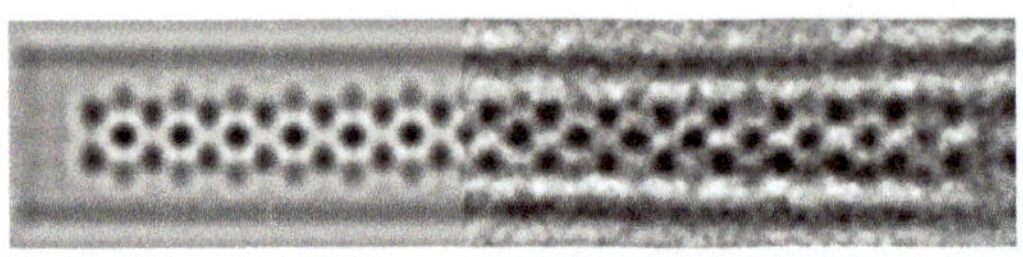

A noise-filtered HRTEM image of a HgTe extreme nanowire embedded down the central pore of a SWCNT. The image is also accompanied by a simulation of the crystal structure.

Typical nanowires exhibit aspect ratios (length-to-width ratio) of 1000 or more. As such they are often referred to as one-dimensional (1-D) materials. Nanowires have many interesting properties that are not seen in bulk or 3-D (three-dimensional) materials. This is because electrons in nanowires are quantum confined laterally and thus occupy energy levels that are different from the traditional continuum of energy levels or bands found in bulk materials.

Peculiar features of this quantum confinement exhibited by certain nanowires manifest themselves in discrete values of the electrical conductance. Such discrete values arise from a quantum mechanical restraint on the number of electrons that can travel through the wire at the nanometer scale. These discrete values are often referred to as the quantum of conductance and are integer multiples of

$$\frac{2e^2}{h} \simeq 77.41\,\mu S$$

They are inverse of the well-known resistance unit h/e^2, which is roughly equal to 25812.8 ohms, and referred to as the von Klitzing constant R_K (after Klaus von Klitzing, the discoverer of exact quantization). Since 1990, a fixed conventional value $R_{K\text{-}90}$ is accepted.

Examples of nanowires include inorganic molecular nanowires ($Mo_6S_{9-x}I_x$, $Li_2Mo_6Se_6$), which can have a diameter of 0.9 nm and be hundreds of micrometers long. Other important examples are based on semiconductors such as InP, Si, GaN, etc., dielectrics (e.g. SiO_2,TiO_2), or metals (e.g. Ni, Pt).

There are many applications where nanowires may become important in electronic, opto-electronic and nanoelectromechanical devices, as additives in advanced composites, for metallic interconnects in nanoscale quantum devices, as field-emitters and as leads for biomolecular nanosensors.

Synthesis of Nanowires

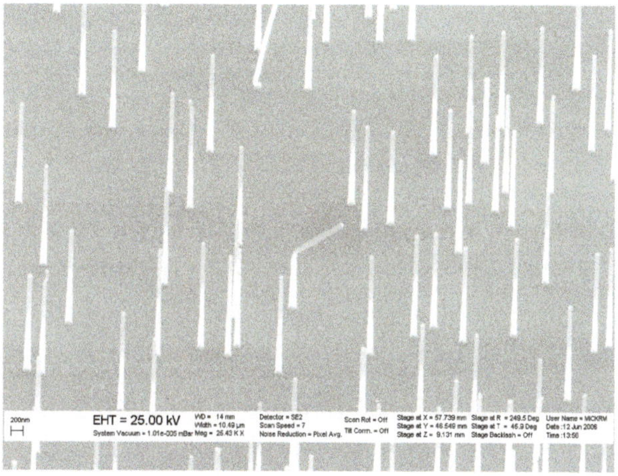

An SEM image of epitaxial nanowire heterostructures grown from catalytic gold nanoparticles

There are two basic approaches to synthesizing nanowires: top-down and bottom-up. A top-down approach reduces a large piece of material to small pieces, by various means such as lithography or electrophoresis. A bottom-up approach synthesizes the nanowire by combining constituent adatoms. Most synthesis techniques use a bottom-up approach. Initial synthesis via either method may often be followed by a nanowire thermal treatment step, often involving a form of self-limiting oxidation, to fine tune the size and aspect ratio of the structures.

Nanowire production uses several common laboratory techniques, including suspension, electrochemical deposition, vapor deposition, and VLS growth. Ion track technology enables growing homogeneous and segmented nanowires down to 8 nm diameter.

Suspension

A suspended nanowire is a wire produced in a high-vacuum chamber held at the longitudinal extremities. Suspended nanowires can be produced by:

- The chemical etching of a larger wire
- The bombardment of a larger wire, typically with highly energetic ions
- Indenting the tip of a STM in the surface of a metal near its melting point, and then retracting it

VLS Growth

A common technique for creating a nanowire is vapor-liquid-solid method (VLS). This process can produce high-quality crystalline nanowires of many semiconductor materials, for example, VLS–grown single crystalline silicon nanowires (SiNWs) with smooth surfaces could have excellent properties, such as ultra-large elasticity. This method uses a source material from either laser ablated particles or a feed gas such as silane.

VLS synthesis requires a catalyst. For nanowires, the best catalysts are liquid metal (such as gold) nanoclusters, which can either be self-assembled from a thin film by dewetting, or purchased in colloidal form and deposited on a substrate.

The source enters these nanoclusters and begins to saturate them. On reaching supersaturation, the source solidifies and grows outward from the nanocluster. Simply turning off the source can adjust the final length of the nanowire. Switching sources while still in the growth phase can create compound nanowires with super-lattices of alternating materials.

A single-step vapour phase reaction at elevated temperature synthesises inorganic nanowires such as $Mo_6S_{9-x}I_x$. From another point of view, such nanowires are cluster polymers.

Solution-phase Synthesis

Solution-phase synthesis refers to techniques that grow nanowires in solution. They can produce nanowires of many types of materials. Solution-phase synthesis has the advantage that it can produce very large quantities, compared to other methods. In one technique, the polyol synthesis, ethylene glycol is both solvent and reducing agent. This technique is particularly versatile at producing nanowires of gold, lead, platinum, and silver.

The supercritical fluid-liquid-solid growth method can be used to synthesize semiconductor nanowires, e.g., Si and Ge. By using metal nanocrystals as seeds, Si and Ge organometallic precursors are fed into a reactor filled with a supercritical organic solvent, such as toluene. Thermolysis results in degradation of the precursor, allowing release of Si or Ge, and dissolution into the metal nanocrystals. As more of the semiconductor solute is added from the supercritical phase (due to a concentration gradient), a solid crystallite precipitates, and a nanowire grows uniaxially from the nanocrystal seed.

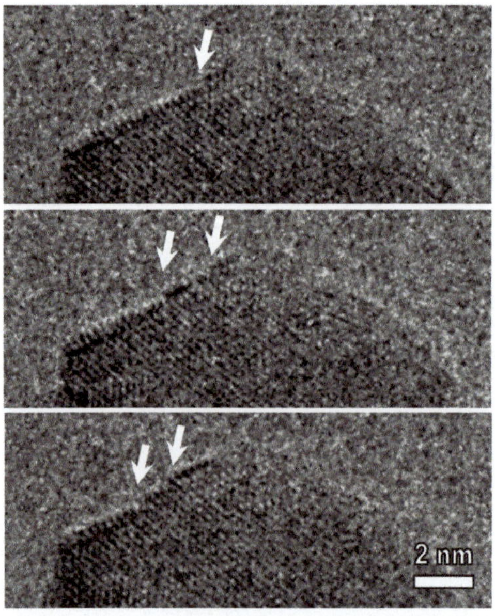

In situ observation of CuO nanowire growth

Non-catalytic Growth

Nanowires can be also grown without the help of catalysts, which gives an advantage of pure nanowires and minimizes the number of technological steps. The simplest methods to obtain metal oxide nanowires use ordinary heating of the metals, e.g. metal wire heated with battery, by Joule heating in air can be easily done at home. The vast majority of nanowire-formation mechanisms are explained through the use of catalytic nanoparticles, which drive the nanowire growth and are either added intentionally or generated during the growth. However the mechanisms for catalyst-free growth of nanowires (or whiskers) were known from 1950s. Spontaneous nanowire formation by non-catalytic methods were explained by the dislocation present in specific directions or the growth anisotropy of various crystal faces. More recently, after microscopy advancement, the nanowire growth driven by screw dislocations or twin boundaries were demonstrated. The picture on the right shows a single atomic layer growth on the tip of CuO nanowire, observed by in situ TEM microscopy during the non-catalytic synthesis of nanowire.

Physics of Nanowires

Conductivity of Nanowires

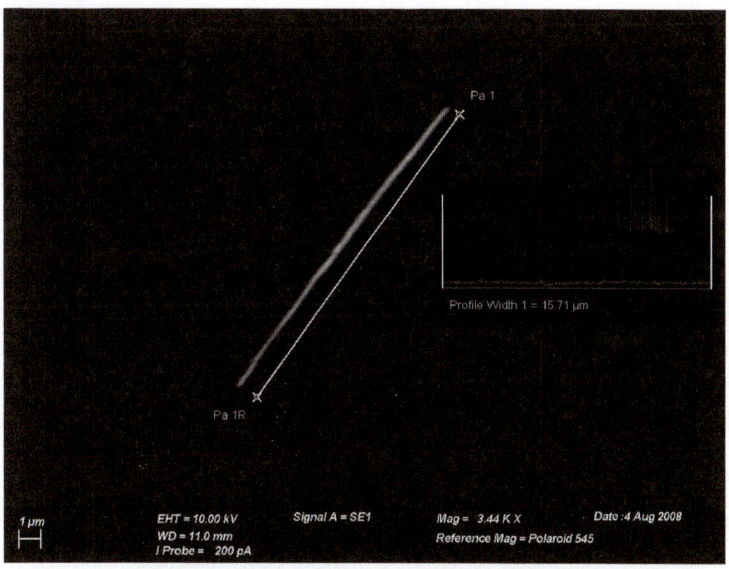

An SEM image of a 15 micrometer nickel wire.

Several physical reasons predict that the conductivity of a nanowire will be much less than that of the corresponding bulk material. First, there is scattering from the wire boundaries, whose effect will be very significant whenever the wire width is below the free electron mean free path of the bulk material. In copper, for example, the mean free path is 40 nm. Copper nanowires less than 40 nm wide will shorten the mean free path to the wire width.

Nanowires also show other peculiar electrical properties due to their size. Unlike single wall carbon nanotubes, whose motion of electrons can fall under the regime of ballistic transport (meaning the electrons can travel freely from one electrode to the other), nanowire conductivity is strongly influenced by edge effects. The edge effects come from atoms that lay at the nanowire surface and are not fully bonded to neighboring atoms like the atoms within the bulk of the nanowire. The unbonded atoms are often a source of defects within the nanowire, and may cause the nanowire to conduct electricity more poorly than the bulk material. As a nanowire shrinks in size, the surface atoms become more numerous compared to the atoms within the nanowire, and edge effects become more important.

Furthermore, the conductivity can undergo a quantization in energy: i.e. the energy of the electrons going through a nanowire can assume only discrete values, which are multiples of the conductance quantum $G = 2e^2/h$ (where e is the charge of the electron and h is the Planck constant).

The conductivity is hence described as the sum of the transport by separate *channels* of different quantized energy levels. The thinner the wire is, the smaller the number of channels available to the transport of electrons.

This quantization has been demonstrated by measuring the conductivity of a nanowire suspended between two electrodes while pulling it: as its diameter reduces, its conductivity decreases in a stepwise fashion and the plateaus correspond to multiples of G.

The quantization of conductivity is more pronounced in semiconductors like Si or GaAs than in metals, due to their lower electron density and lower effective mass. It can be observed in 25 nm wide silicon fins, and results in increased threshold voltage. In practical terms, this means that a MOSFET with such nanoscale silicon fins, when used in digital applications, will need a higher gate (control) voltage to switch the transistor on.

Welding Nanowires

To incorporate nanowire technology into industrial applications, researchers in 2008 developed a method of welding nanowires together: a sacrificial metal nanowire is placed adjacent to the ends of the pieces to be joined (using the manipulators of a scanning electron microscope); then an electric current is applied, which fuses the wire ends. The technique fuses wires as small as 10 nm.

For nanowires with diameters less than 10 nm, existing welding techniques, which require precise control of the heating mechanism and which may introduce the possibility of damage, will not be practical. Recently scientists discovered that single-crystalline ultrathin gold nanowires with diameters ~3–10 nm can be "cold-welded" together within seconds by mechanical contact alone, and under remarkably low applied pressures (unlike macro- and micro-scale cold welding process). High-resolution transmission electron microscopy and in situ measurements reveal that the welds are nearly perfect, with the same crystal orientation, strength and electrical conductivity as the rest of the nanowire. The high quality of the welds is attributed to the nanoscale sample dimensions, oriented-attachment mechanisms and mechanically assisted fast surface diffusion. Nanowire welds were also demonstrated between gold and silver, and silver nanowires (with diameters ~5–15 nm) at near room temperature, indicating that this technique may be generally applicable for ultrathin metallic nanowires. Combined with other nano- and microfabrication technologies, cold welding is anticipated to have potential applications in the future bottom-up assembly of metallic one-dimensional nanostructures.

Applications of Nanowires

Electronic Devices

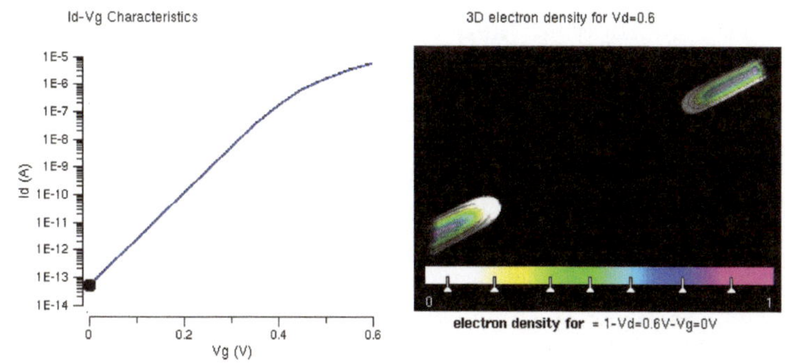

Atomistic simulation result for formation of inversion channel (electron density) and attainment of threshold voltage (IV) in a nanowire MOSFET. Note that the threshold voltage for this device lies around 0.45V.

Nanowires can be used for transistors. Transistors are used widely as fundamental building element in today's electronic circuits. As predicted by Moore's law, the dimension of transistors is shrinking smaller and smaller into nanoscale. One of the key challenges of building future nanoscale transistors is ensuring good gate control over the channel. Due to the high aspect ratio, if the gate dielectric is wrapped around the nanowire channel, we can get good control of channel electrostatic potential, thereby turning the transistor on and off efficiently.

To create active electronic elements, the first key step was to chemically dope a semiconductor nanowire. This has already been done to individual nanowires to create p-type and n-type semiconductors.

The next step was to find a way to create a p–n junction, one of the simplest electronic devices. This was achieved in two ways. The first way was to physically cross a p-type wire over an n-type wire. The second method involved chemically doping a single wire with different dopants along the length. This method created a p-n junction with only one wire.

After p-n junctions were built with nanowires, the next logical step was to build logic gates. By connecting several p-n junctions together, researchers have been able to create the basis of all logic circuits: the AND, OR, and NOT gates have all been built from semiconductor nanowire crossings.

In August 2012, researchers reported constructing the first NAND gate from undoped silicon nanowires. This avoids the problem of how to achieve precision doping of complementary nanocircuits, which is unsolved. They were able to control the Schottky barrier to achieve low-resistance contacts by placing a silicide layer in the metal-silicon interface.

It is possible that semiconductor nanowire crossings will be important to the future of digital computing. Though there are other uses for nanowires beyond these, the only ones that actually take advantage of physics in the nanometer regime are electronic.

In addition, nanowires are also being studied for use as photon ballistic waveguides as interconnects in quantum dot/quantum effect well photon logic arrays. Photons travel inside the tube, electrons travel on the outside shell.

When two nanowires acting as photon waveguides cross each other the juncture acts as a quantum dot.

Conducting nanowires offer the possibility of connecting molecular-scale entities in a molecular computer. Dispersions of conducting nanowires in different polymers are being investigated for use as transparent electrodes for flexible flat-screen displays.

Because of their high Young's moduli, their use in mechanically enhancing composites is being investigated. Because nanowires appear in bundles, they may be used as tribological additives to improve friction characteristics and reliability of electronic transducers and actuators.

Because of their high aspect ratio, nanowires are also uniquely suited to dielectrophoretic manipulation, which offers a low-cost, bottom-up approach to integrating suspended dielectric metal oxide nanowires in electronic devices such as UV, water vapor, and ethanol sensors.

Sensing of Proteins and Chemicals using Semiconductor Nanowires

In an analogous way to FET devices in which the modulation of conductance (flow of electrons/holes) in the semiconductor, between the input (source) and the output (drain) terminals, is controlled by electrostatic potential variation (gate-electrode) of the charge carriers in the device conduction channel, the methodology of a Bio/Chem-FET is based on the detection of the local change in charge density, or so-called "field effect", that characterizes the recognition event between a target molecule and the surface receptor.

This change in the surface potential influences the Chem-FET device exactly as a 'gate' voltage does, leading to a detectable and measurable change in the device conduction. When these devices are fabricated using semiconductor nanowires as the transistor element the binding of a chemical or biological species to the surface of the sensor can lead to the depletion or accumulation of charge carriers in the "bulk" of the nanometer diameter nanowire i.e. (small cross section available for conduction channels). Moreover, the wire, which serves as a tunable conducting channel, is in close contact with the sensing environment of the target, leading to a short response time, along with orders of magnitude increase in the sensitivity of the device as a result of the huge S/V ratio of the nanowires.

While several inorganic semiconducting materials such as Si, Ge, and metal oxides (e.g. In_2O_3, SnO_2, ZnO, etc.) have been used for the preparation of nanowires, Si is usually the material of choice when fabricating nanowire FET-based chemo/biosensors.

Several examples of the use of silicon nanowire(SiNW) sensing devices include the ultra sensitive, real-time sensing of biomarker proteins for cancer, detection of single virus particles, and the detection of nitro-aromatic explosive materials such as 2,4,6 Tri-nitrotoluene (TNT) in sensitives superior to these of canines. Silicon nanowires could also be used in their twisted form, as electromechanical devices, to measure intermolecular forces with great precision.

Limitations of Sensing with Silicon Nanowire FET Devices

Generally, the charges on dissolved molecules and macromolecules are screened by dissolved counterions, since in most cases molecules bound to the devices are separated from the sensor surface by approximately 2–12 nm (the size of the receptor proteins or DNA linkers bound to the sensor surface). As a result of the screening, the electrostatic potential that arises from charges on the analyte molecule decays exponentially toward zero with distance. Thus, for optimal sensing, the Debye length must be carefully selected for nanowire FET measurements. One approach of overcoming this limitation employs fragmentation of the antibody-capturing units and control over surface receptor density, allowing more intimate binding to the nanowire of the target protein. This approach proved useful for dramatically enhancing the sensitivity of cardiac biomarkers (e.g. Troponin) detection directly from serum for the diagnosis of acute myocardial infarction.

Silicon Nanowire

Silicon nanowires, also referred to as SiNWs, are a type of nanowire most often formed from a silicon precursor by etching of a solid or through catalyzed growth from a vapor or liquid phase. Initial synthesis is often accompanied by thermal oxidation steps to yield structures of accurately tailored size and morphology.

SiNWs have unique properties that are not seen in bulk (three-dimensional) Silicon materials. These properties arise from an unusual quasi one-dimensional electronic structure and are the subject of research across numerous disciplines and applications. The reason that SiNWs are considered as one of the most important one-dimensional materials is they could have a function as building blocks for nanoscale electronics assembled without the need for complex and costly fabrication facilities. SiNWs are frequently studied towards applications including photovoltaics, nanowire batteries, thermoelectrics and non-volatile memory.

Applications

Owing to their unique physical and chemical properties, silicon nanowires are a promising candidate for a wide range of applications that draw on their unique physico-chemical characteristics, which differ from those of bulk Silicon material.

SiNWs exhibit charge trapping behavior which renders such systems of value in applications necessitating electron hole separation such as photovoltaics, and photocatalysts.

Charge trapping behaviour and tuneable surface governed transport properties of SiNWs render this category of nanostructures of interest towards use as Metal Insulator Semiconductors and Field Effect Transistors, with further applications as nanoelectronic storage devices, in flash memory, logic devices as well as chemical and biological sensors.

The ability for lithium ions to intercalate into silicon structures renders various Si nanostructures of interest towards applications as anodes in Li-ion batteries (LiBs). SiNWs are of particular merit as such anodes as they exhibit the ability to undergo significant lithiation while maintaining structural integrity and electrical connectivity.

Synthesis

Several synthesis methods are known for SiNWs and these can be broadly divided into methods which start with bulk silicon and remove material to yield nanowires, also known as top-down synthesis, and methods which use a chemical or vapor precursor to build nanowires in a process generally considered to be bottom-up synthesis.

Top Down Synthesis Methods

These methods use material removal techniques to produce nanostructures from a bulk precursor

- Laser beam ablation
- Ion beam etching
- Thermal evaporation oxide-assisted growth (OAG)

Bottom-up Synthesis Methods

- Vapour liquid solid (VLS) growth - a type of catalysed CVD often using silane as Si precursor and gold nanoparticles as catalyst (or 'seed')
- Molecular beam epitaxy - a form of PVD applied in plasma environment
- Precipitation from a solution - A variation of the VLS method, aptly named Supercritical Fluid Liquid Solid (SFLS), that uses a supercritical fluid (e.g. organosilane at high temperature and pressure) as Si precursor instead of vapor. The catalyst would be a colloid in solution, such as colloidal gold nanoparticles, and the SiNWs are grown in this solution

Thermal Oxidation of Silicon Nanowires

Subsequent to physical or chemical processing, either top-down or bottom-up, to obtain initial silicon nanostructures, thermal oxidation steps are often applied in order to obtain materials with desired size and aspect ratio. Silicon nanowires exhibit a distinct and useful self-limiting oxidation behaviour whereby oxidation effectively ceases due to diffusion limitations, which can be modeled. This phenomenon allows accurate control of dimensions and aspect rations in SiNWs and has been used to obtain high aspect ratio SiNWs with diameters below 5 nm. The self-limiting oxidation of SiNWs is of value towards lithium ion battery materials.

Orientation of Nanowires

The orientation of SiNWs has profound influence on the overall properties of systems. For this reason several procedures have been proposed for the alignment of nanowires in chosen orientations. This includes the use of electric fields in polar alignment, electrophoresis, mircofluidic methods and contact printing.

Outlook

There is significant interest in SiNWs for their unique properties and the ability to control size and aspect ratio with great accuracy. As yet, limitations in large-scale fabrication impede the uptake of this material in the full range of investigated applications. Combined studies of synthesis methods, oxidation kinetics and properties of SiNW systems aim to overcome the present limitations and facilitate the implementation of SiNW systems, for example, high quality vapor-liquid-solid–grown SiNWs with smooth surfaces can be reversibly stretched with 10% or more elastic strain, approaching the theoretical elastic limit of silicon, which could open the doors for the emerging "elastic strain engineering" and flexible bio-/nano-electronics.

Nanomesh

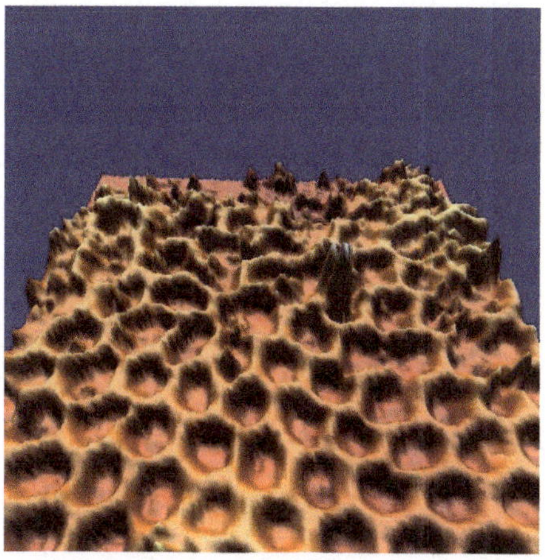

Perspective view of nanomesh, whose structure ends at the back of the figure.
The distance between two pore centers is 3.2nm, and the pores are 0.05nm deep.

The nanomesh is a new inorganic nanostructured two-dimensional material, similar to graphene. It was discovered in 2003 at the University of Zurich, Switzerland.

It consists of a single layer of boron (B) and nitrogen (N) atoms, which forms by self-assembly a highly regular mesh after high-temperature exposure of a clean rhodium or ruthenium surface to borazine under ultra-high vacuum.

The nanomesh looks like an assembly of hexagonal pores at the nanometer (nm) scale. The distance between 2 pore centers is only of 3.2 nm, whereas each pore has a diameter of about 2 nm and is 0.05 nm deep. The lowest regions bind strongly to the underlying metal, while the wires (highest regions) are only bound to the surface through strong cohesive forces within the layer itself.

The boron nitride nanomesh is not only stable under vacuum, air and some liquids, but also up to temperatures of 796°C (1070 K). In addition it shows the extraordinary ability to trap molecules and metallic clusters, which have similar sizes to the nanomesh pores, forming a well-ordered array. These characteristics promise interesting applications of the nanomesh in areas like nanocatalysis, surface functionalisation, spintronics, quantum computing and data storage media like hard drives.

Structure

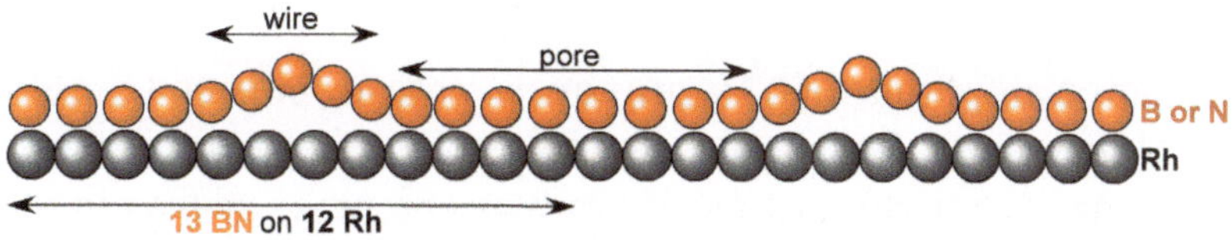

Cross-section of nanomesh on rhodium showing pore and wire regions.

h-BN nanomesh is a single sheet of hexagonal boron nitride, which forms on substrates like rhodium Rh(111) or ruthenium Ru(0001) crystals by a self-assembly process.

The unit cell of the h-BN nanomesh consists of 13x13 BN or 12x12 Rh atoms with a lattice constant of 3.2 nm. In a cross-section it means that 13 boron or nitrogen atoms are sitting on 12 rhodium atoms. This implies a modification of the relative positions of each BN towards the substrate atoms within a unit cell, where some bonds are more attractive or repulsive than other (site selective bonding), what induces the corrugation of the nanomesh.

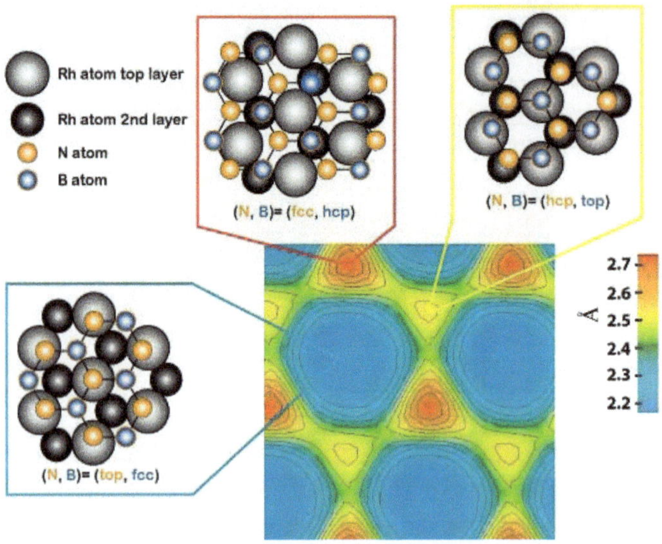

The image shows the boron nitride nanomesh measured by STM at 77K, where each "ball" represents one N atom. The center of each ring corresponds to the center of the pores.

The image is the theoretical calculation of the same area, where the N height relative to the underlying substrate is given. The exact arrangement of Rh, N and B atoms is given for three different areas (blue: pores, yellow-red: wires).

The nanomesh corrugation amplitude of 0.05 nm causes a strong effect on the electronic structure, where two distinct BN regions are observed. They are easily recognized in the lower right image, which is a scanning tunneling microscopy (STM) measurement, as well as in the lower left image representing a theoretical calculation of the same area. A strongly bounded region assigned to the pores is visible in blue in the left image below (center of bright rings in the above image) and a weakly bound region assigned to the wires appears yellow-red in the left image below (area in-between rings in the above image).

Properties

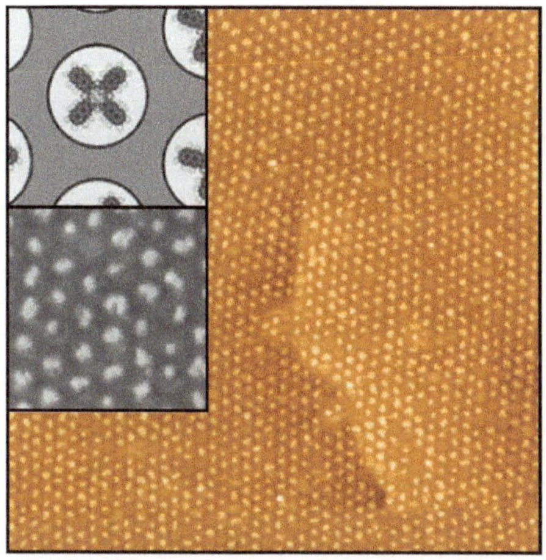

Naphthalocyanine molecules evaporated onto the nanomesh. They only adsorb in pores, forming a well-defined pattern.

The nanomesh is stable under a wide range of environments like air, water and electrolytes among others. It is also temperature resistant since it doesn't decompose up to 1275K under vacuum. In addition to these exceptional stabilities, the nanomesh shows the extraordinary ability to act as a scaffold for metallic nanoclusters and to trap molecules forming a well-ordered array.

In the case of gold (Au), its evaporation on the nanomesh leads to formation of well-defined round Au nanoparticles, which are centered at the nanomesh pores.

The STM figure shows Naphthalocyanine (Nc) molecules, which were vapor-deposited onto the nanomesh. These planar molecules have a diameter of about 2 nm, whose size is comparable to that of the nanomesh pores. It is spectacularly visible how the molecules form a well-ordered array with the periodicity of the nanomesh (3.22 nm). The lower inset shows a region of this substrate with higher resolution, where individual molecules are trapped inside the pores. In addition, the molecules seem to keep their native conformation, what means that their functionality is kept, which is nowadays a challenge in nanoscience.

Such systems with wide spacing between individual molecules/clusters and negligible intermolecular interactions might be interesting for applications such as molecular electronics and memory elements, in photochemistry or in optical devices.

Preparation and Analysis

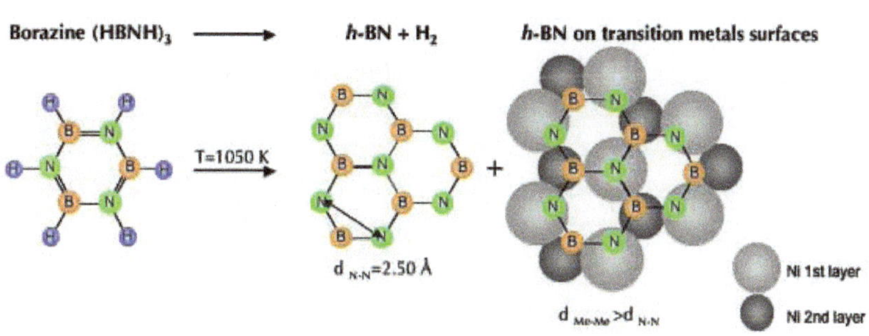

Decomposition of borazine on transition metal surfaces

Well-ordered nanomeshes are grown by thermal decomposition of borazine $(HBNH)_3$, a colorless substance that is liquid at room temperature. The nanomesh results after exposing the atomically clean Rh(111) or Ru(0001) surface to borazine by chemical vapor deposition (CVD).

The substrate is kept at a temperature of 796 °C (1070 K) when borazine is introduced in the vacuum chamber at a dose of about 40 L (1 Langmuir = 10^{-6} torr sec). A typical borazine vapor pressure inside the ultrahigh vacuum chamber during the exposure is $3x10^{-7}$ mbar.

After cooling down to room temperature, the regular mesh structure is observed using different experimental techniques. Scanning tunneling microscopy (STM) gives a direct look on the local real space structure of the nanomesh, while low energy electron diffraction (LEED) gives information about the surface structures ordered over the whole sample. Ultraviolet photoelectron spectroscopy (UPS) gives information about the electronic states in the outermost atomic layers of a sample, i.e. electronic information of the top substrate layers and the nanomesh.

Nanohole

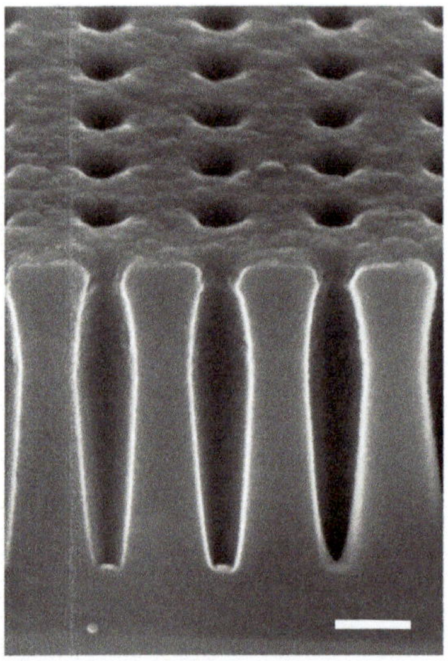

Angled cross-sectional scanning electron micrograph of a nanohole array etched in amorphous silicon, with a thin conductive polymer coating. Scale bar is 200 nm.

Nanoholes are a class of nanostructured material consisting of nanoscale voids in a surface of a material. It is different than nanofoam or nanoporous materials which support a network of voids permeating throughout the material (often in a disordered state), nanohole materials feature a regular hole pattern extending through a single surface. These can be thought of as the inverse of a nanopillar or nanowire structure.

Uses

Nanohole structures have been used for a variety of applications, ranging from superlenses produced from a metal nanohole array, to structured photovoltaic devices used to improve carrier extraction, and light absorption.

Nanohole structures are also extensively utilized for the creation of photonic crystals, particularly for creating photonic crystal waveguides.

Nanosheet

A nanosheet is a two-dimensional nanostructure with thickness in a scale ranging from 1 to 100 nm.

A typical example of a nanosheet is graphene, the thinnest two-dimensional material (0.34 nm) in the world. It consists of a single layer of carbon atoms with hexagonal lattices.

Examples and Applications

As of 2017 Silicon nanosheets are being used to prototype future generations of small (5 nm) transistors.

Carbon nanosheets (from hemp) may be an alternative to graphene as electrodes in supercapacitors.

Synthesis

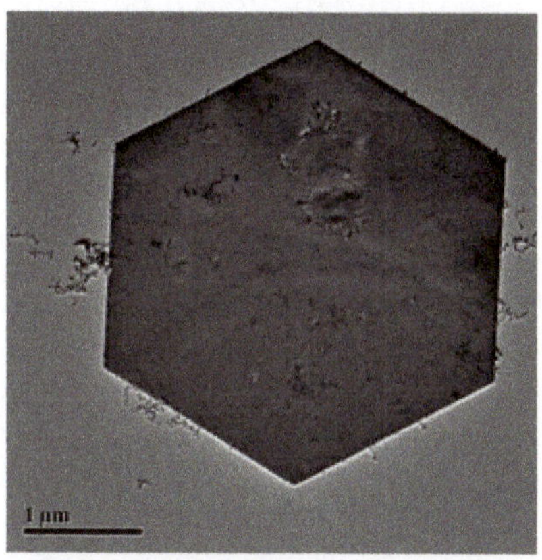

TEM image of PbO nanosheets with highly symmetric edge length. The edges of PbO nanosheets are surrounded with Au NPs seeds.

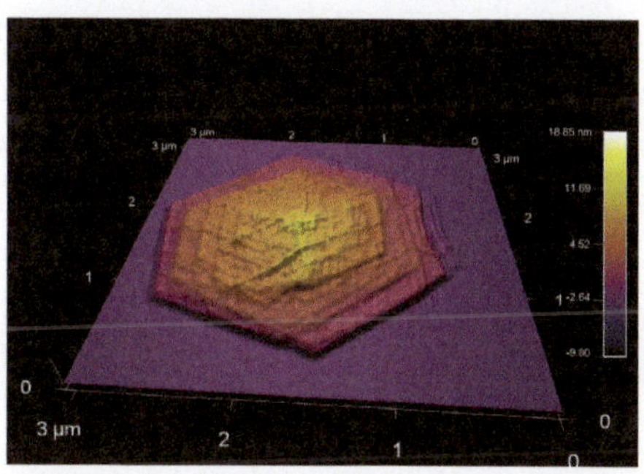

3D AFM topography image of multilayered palladium nanosheet on silicon wafer

The most commonly used nanosheet synthesis methods use a bottom-up approach, e.g., pre-organization and polymerization at interfaces like Langmuir–Blodgett films, solution phase synthesis and chemical vapor deposition (CVD). For example, CdTe (cadmium telluride) nanosheets could be synthesized by precipitating and aging CdTe nanoparticles in deionized water. The formation of free-floating CdTe nanosheets was due to directional hydrophobic attraction and anisotropic electrostatic interactions caused by dipole moment and small positive charges. Molecular simulations through a coarse-grained model with parameters from semi-empirical quantum mechanics calculations can be used to prove the experimental process.

Ultrathin single-crystal PbS (lead sulfur) sheets with micro scale in x-, y- dimensions can be obtained using a hot colloidal synthesis method. Compounds with linear chloroalkanes like 1,2-dichloroethane containing chlorine were used during the formation of PbS sheets. PbS ultrathin sheets probably resulted from the oriented attachment of the PbS nanoparticles in a two-dimensional fashion. The highly reactive facets were preferentially consumed in the growth process that led to the sheet-like PbS crystal growth.

Nanosheets can also be prepared at room temperature. For instance, hexagonal PbO (lead oxide)) nanosheets were synthesized using gold nanoparticles as seeds under room temperature. The size of the PbO nanosheet can be tuned by gold NPs and Pb^{2+} concentration in the growth solution. No organic surfactants were employed in the synthesis process. Oriented attachment, in which the sheets form by aggregation of small nanoparticles that each has a net dipole moment, and ostwald ripening are the two main reasons for the formation of the PbO nanosheets.

Carbon nanosheets have been produced using industrial hemp bast fibres with a technique that involves heating the fibres at over 350F (180C) for 24 hours. The result is then subjected to intense heat causing the fibers to exfoliate into a carbon nanosheet. This has been used to create an electrode for a supercapacitor with electrochemical qualities 'on a par with' devices made using graphene.

Metal nanosheets have also been synthesized from solution-based method by reducing metal precursors, including palladium, rhodium, and gold.

Nanopillar

Nanopillars is an emerging technology within the field of nanostructures. Nanopillars are pillar shaped nanostructures approximately 10 nanometers in diameter that can be grouped together in lattice like arrays. They are a type of metamaterial, which means that nanopillars get their attributes from being grouped into artificially designed structures and not their natural properties. Nanopillars set themselves apart from other nanostructures due to their unique shape. Each nanopillar has a pillar shape at the bottom and a tapered pointy end on top. This shape in combination with nanopillars' ability to be grouped together exhibits many useful properties. Nanopillars have many applications including efficient solar panels, high resolution analysis, and antibacterial surfaces.

Applications

Solar Panels

Due to their tapered ends, nanopillars are very efficient at capturing light. Solar collector surfaces coated with nanopillars are three times as efficient as nanowire solar cells. Less material is needed to build a solar cell out of nanopillars compared to regular semi conductive materials.

They also hold up well during the manufacturing process of solar panels. This durability allows manufacturers to use cheaper materials and less expensive methods to produce solar panels. Researchers are looking into putting dopants into the bottom of the nanopillars, to increase the amount of time photons will bounce around the pillars and thus the amount of light captured. As well as capturing light more efficiently, using nanopillars in solar panels will allow them to be flexible. The flexibility gives manufacturers more options on how they want their solar panels to be shaped as well as reduces costs in terms of how delicately the panels have to be handled. Although nanopillars are more efficient and cheaper than standard materials, scientists have not been able to mass-produce them yet. This is a significant drawback to using nanopillars as a part of the manufacturing process.

Antibacterial Surfaces

Nanopillars also have functions outside of electronics and can imitate natures defenses. Cicadas' wings are covered in tiny, nanopillar shaped rods. When bacteria hits a cicada's wing, it gets caught on the rods. The rods don't hold the bacteria up evenly so they get punctured and die. Since the rods on the cicadas are about the same size and shape as artificial nanopillars, it is possible for humans to copy this defense. A surface covered with nanopillars would immediately become germ free. If mass-produced and installed everywhere, nanopillars could eliminate much of the risk of transmitting diseases through touching infected surfaces.

High Resolution Molecular Analysis

Another use of nanopillars is observing cells. Nanopillars capture light so well that when lights hits them, the glow the nanopillars emit dies down at around 150 nanometers. Because this distance is less than the wavelength of light, it allows researchers to observe small objects without the interference of background light. This is especially useful in cellular analysis. The cells group around the nanopillars because of its small size and recognize it as an organelle. The nanopillars simply hold the cells in place while the cells are being observed.

History

In 2006, researchers at the University of Nebraska-Lincoln and the Lawrence Livermore National Laboratory developed a cheaper and more efficient way to create nanopillars. They used a combination of nanosphere lithography (a way of organizing the lattice) and reactive ion etching(molding the nanopillars to the right shape) to make large groups of silicon pillars with less than 500 nm diameters. Then, in 2010, researchers fabricated a way to manufacture nanopillars with tapered ends. The former design of a pillar with a flat blunt top reflected much of the light coming onto the pillars. The tapered tops allow light to enter the forest of nanopillars and the wider bottom absorbs almost all of the light that hits it. This design captures about 99% of the light whereas nanorods which have a uniform thickness only captured 85% of the light. After the introduction of tapered ends, researchers started to find many more applications for nanopillars.

Manufacturing Process

Constructing nanopillars is a simple but lengthy procedure that can take hours. The process to create nanopillars starts with anodizing a 2.5 mm thick aluminum foil mold. Anodizing the foil creates pores in the foil a micrometer deep and 60 nanometers wide. The next step is to treat the foil with phosphoric acid which expands the pores to 130 nanometers. The foil is anodized once more making its pores a micrometer deeper. Lastly, a small amount of gold is added to the pores to catalyze the reaction for the growth of the semiconductor material. When the aluminum is scraped away there is a forest of nanopillars left inside a casing of aluminum oxide.

Nanostructured Film

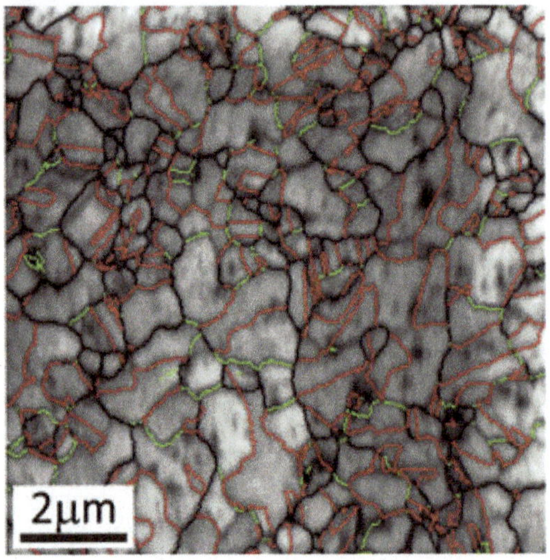

Surface of a nanotwinned copper film with highlighted Σ3 and low angle grain boundaries as imaged by EBSD. Image adapted from Zhao et al.

A nanostructured film is a film resulting from engineering of nanoscale features, such as dislocations, grain boundaries, defects, or twinning. In contrast to other nanostructures, such as nanoparticles, the film itself may be up to several microns thick, but possesses a large concentration of nanoscale features homogeneously distributed throughout the film. Like other nanomaterials, nanostructured films have sparked much interest as they possess unique properties not found in bulk, non-nanostructured material of the same composition. In particular, nanostructured films have been the subject of recent research due to their superior mechanical properties, including strength, hardness, and corrosion resistance compared to regular films of the same material. Examples of nanostructured films include those produced by grain boundary engineering, such as nano-twinned ultra-fine grain copper, or dual phase nanostructuring, such as crystalline metal and amorphous metallic glass nanocomposites.

Synthesis and Characterization

Nanostructured films are commonly created using magnetron sputtering from an appropriate target material. Films can be elemental in nature, formed by sputtering from a pure metal target such as copper, or composed of compound materials. Varying parameters such as the sputtering rate, substrate temperature, and sputtering interrupts allow the creation of films with a variety of different nanostructured elements. Control over nano-twinning, tailoring of specific types of grain boundaries, and restricting the movement and propagation of dislocations have been demonstrated using films produced via magnetron sputtering.

Methods used to characterize nanostructured films include transmission electron microscopy, scanning electron microscopy, electron backscatter diffraction, focused ion beam milling, and nanoindentation. These techniques are used as they allow imaging of nanoscale structures, including dislocations, twinning, grain boundaries, film morphology, and atomic structure.

Material Properties

Nanostructured films are of interest due to their superior mechanical and physical properties compared to their normal equivalent. Elemental nanostructured films composed of pure copper were found to possess good thermal stability due to the nano-twinned film possessing a higher fraction of grain boundaries. In addition to possessing higher thermal stability, copper films that were highly nano-twinned were found to have a better corrosion resistance than copper films with a low concentration of nano-twins. Control of the fraction of grains in a material with nano-twins present has great potential for less expensive alloys and coatings with a good degree of corrosion resistance.

Compound nanostructured films composed of crystalline $MgCu_2$ cores encapsulated by amorphous glassy shells of the same material were shown to possess a near-ideal mechanical strength. The crystalline $MgCu_2$ cores, typically less than 10 nm in size, were found to substantially strengthen the material by restricting the movement of dislocations and grains. The cores were also found to contribute to overall material strength by restricting the movement of shear bands in the material. This nanostructured film differs from both crystalline metals and amorphous metallic glasses, both of which exhibit behaviors such as the reverse Hall-Petch and shear-band softening effects that prevent them from reaching ideal strength values.

Applications

Nanostructured films with superior mechanical properties allow previously unusable materials to be utilized in new applications, enabling advances fields where coatings are heavily utilized, such as aerospace, energy, and other engineering fields. Production scalability of nanostructured films has already been demonstrated, and the ubiquity of sputtering techniques in industry is predicted to facilitate the incorporation of nanostructured films into existing applications.

Sculptured Thin Film

Sculptured thin films (STFs) are nanostructured materials with unidirectionally varying properties that can be designed and realized in a controllable manner using variants of physical vapor deposition. The ability to virtually instantaneously change the growth direction of their columnar morphology, through simple variations in the direction of the incident vapor flux, leads to a wide spectrum of columnar forms.

Forms

These forms can be:

1. two-dimensional, ranging from the simple slanted columns and chevrons to the more complex C- and S-shaped morphologies.
2. three-dimensional, including simple helixes and superhelixes.
3. combinations of two- and three-dimensional forms.

Properties

The column diameter and the column separation normal to the thickness direction of any STF are nominally constant. The column diameter can range from about 10 to 300 nm, while the density may lie between its theoretical maximum value to less than 20% thereof. The crystallinity must be at a scale smaller than the column diameter. The chemical composition is essentially unlimited, ranging from insulators to semiconductors to metals. Most recently, polymeric STFs have been deposited by combining physical and chemical vapor deposition processes; and deposition on micropatterned substrates has also been carried out.

Uses

To date, the chief applications of STFs are in optics as polarization filters, Bragg filters, and spectral hole filters. At visible and infrared wavelengths, a single-section STF is a unidirectionally nonhomogeneous continuum with direction-dependent properties. Several sections can be grown consecutively into a multisection STF, which can be conceived of as an optical circuit that can be integrated with electronic circuitry on a chip. Being porous, an STF can act as a sensor of fluids and can be impregnated with liquid crystals for switching applications too. Applications as low-permittivity barrier layers in electronic chips as well as solar cells have also been suggested. Biomedical applications such as tissue scaffolds, drug-delivery platforms, virus traps, and labs-on-a-chip are also in different stages of development.

Gradient Multilayer Nanofilm

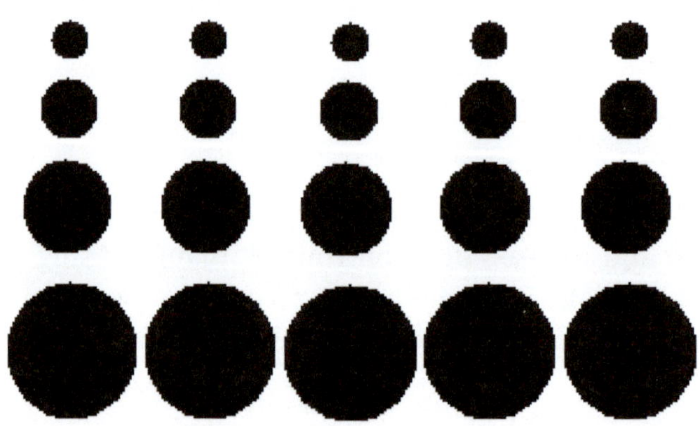

Schema of GML nanofilm

Gradient multilayer (GML) nanofilm is an assembly of quantum dot layers with a built-in gradient of nanoparticle size, composition or density.

Properties of such nanostructure are finding its applications in design of solar cells and energy storage devices.

The GML nanostructure can be embedded in the organic material (polymer), or can include quantum dots made of two or more types of material.

Photovoltaic Applications

The GML nanofilm only 100 nanometers thick can absorb the entire Sun spectrum (0.3–2.0+ eV). At the same time, gradient of the quantum dots size can create a gradient of the electrochemical potential, acting as an equivalent of built-in electric field inside a nanofilm. This enhances transport of electrons and holes, and improves internal quantum efficiency (IQE) and photocurrent.

Manufacturing

The industrial manufacturing of GML nanofilms represents a challenge. Traditional methods of building nanostructured materials (like spin coating) can't form GML nanostructures, while more effective methods like Atomic Layer Deposition (ALD) or Langmuir-Blodget "microchemical" method. are expensive.

Icosahedral Twins

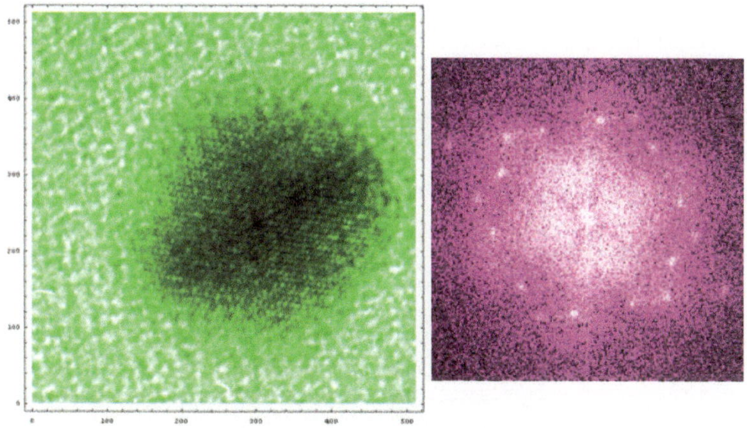

Lattice image of a Pd icotwin, with 10-fold symmetry in the power spectrum

An icosahedral twin is a twenty-face cluster made of ten interlinked dual-tetrahedron (bowtie) crystals, typically joined along triangular (e.g. cubic-111) faces having three-fold symmetry. One can think of their formation as a kind of nano-scale self-assembly.

A variety of nanostructures (e.g. condensing argon, metal atoms, and virus capsids) assume icosahedral form on size scales where surface forces eclipse those from the bulk. A twinned-form of these nanostructures is sometimes found to occur e.g. in face-centered-cubic (FCC) metal-atom clusters larger than 10 nm in diameter. This may occur when the building-blocks beneath each of the 20 facets of an initially icosahedral cluster make the case for conversion to a translationally symmetric crystalline form.

Causes

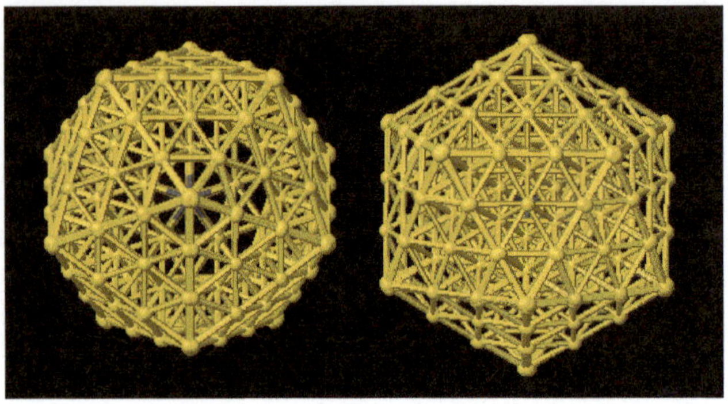

FCC icotwin model down 5-fold & 3-fold zones

When interatom bonding does not have strong directional preferences, it is not unusual for atoms to gravitate toward a kissing number of 12 nearest neighbors. The three most symmetric ways to do this are by icosahedral clustering, or by crystalline face-centered-cubic (cuboctahedral) and/or hexagonal (tri-orthobicupolar) close packing.

Icosahedral arrangements, perhaps because of their slightly smaller surface area, may be preferred for small clusters e.g. noble gas and metal atoms in condensed phases (both liquid and solid). However, the Achilles heel for icosahedral clustering about a single point is that it cannot fill space over large distances in a way that is translationally ordered.

Hence bulk atoms (i.e. sufficiently large clusters) generally revert to one of the crystalline close-packing configurations instead. In other words, when icosahedral clusters get sufficiently large, the bulk-atom vote wins out over the surface-atom vote, and the atoms beneath each of the 20 facets adopt a face-centered-cubic pyramidal arrangement with tetrahedral (111) facets. Thus icosahedral twins are born, with a certain amount of strain along the interfacial (111) planes.

Ubiquity

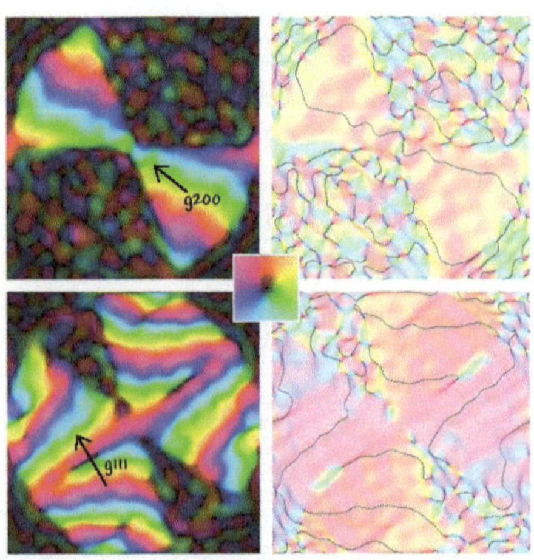

Darkfield analysis of dual-tetrahedron crystal pairs

Icosahedral-twinning has been seen in face-centered-cubic metal-nanoparticles that have nucleated: (i) by evaporation onto surfaces, (ii) out of solution, and (iii) by reduction in a polymer matrix. Icosahedral (or decahedral) twinning seems to be the result when clusters growing outward from an icoashedral-seed get larger than about 10[nm] in diameter.

Quasicrystals are *un-twinned* structures with long range rotational but not translational periodicity, that some initially tried to explain away as icosahedral twinning. Quasi-crystals let non-space-filling coordination persist to larger size scales. However, they generally form only when the compositional makeup (e.g. of two dissimilar metals like Ti and Mn) serves as an antagonist to formation of one of the more common close-packed space-filling but twinned crystalline forms.

Application

Face-centered-cubic noble-metal atom-clusters are important chemical-reaction nano-catalysts. One example of this is the platinum used in automobile catalytic converters. Icosahedral twinning makes it possible to cover the entire surface of a nanoparticle with {111} facets, in cases where those particular atomic-facets show favorable catalytic-activity.

Detection

Electron diffraction and high-resolution transmission electron microscope (TEM) imaging are two methods for identifying the icosahedral-twin structure of *individual* clusters. Digital darkfield analysis of lattice-fringe images shows promise for recognition of icosahedral twinning from most of the randomly oriented clusters in a microscope-image field of view.

Nano Flake

In a general meaning a nanoflake is a flake (that is, an uneven piece of material with one dimension substantially smaller than the other two) with at least one nanometric dimension (that is, between 1 and 100 nm). A flake is not necessarily perfectly flat but it is characterized by a plate-like form or structure. There are nanoflakes of all sorts of materials.

In a more restricted meaning, in the context of solar energy, Nano Flakes are a type of semiconductor that has potential for solar energy creation as the product itself is only in the prototype phase. With its crystalline structure the crystals are able to absorb light and harvest 30 percent of solar energy directed at its surface. These Nano Flakes can potentially also help out with economic and environmental problems associated with solar energy. Working on this Nano Flake Technology is Dr. Martin Aagesen at the Niels Bohr Institute at the University of Copenhagen with a PhD from the Nano Science Center. When Dr. Martin Aagesen discovered and published this new idea in 2007 there was much publicity about it and sparked many people to work on it. One major company that is working on these Nano structures is SunFlake which is a science company from the Nano Science Center with Dr. Martin Aagesen as Chief Executive Officer (CEO). The funding for the Nano flakes came from Danish Venture Capital fund SEED capital and University of Copenhagen.

Structure

The Nano flakes have a structure that contains tiny crystals in which millions of these crystals could fit into a single square centimeter. The tiny crystals absorb the sunlight and use the solar energy to convert it to electricity. This perfect crystalline structure is why this product can revolutionize solar energy. The large surface to volume ratio and the texture of the surface of this nano structure provides a larger absorption rate of the sun's light energy. Also researchers are working on trying to combine it with different semiconducting materials since the usual requirements of a need for a similar crystal structure for the carrier substrate is less stressed in the Nano flakes structure. The carrier substrate in the Nano flakes purpose is to permit growth of the nano structures and works as a contact for the Nano structures when they are actively absorbing the sun's energy.

Purpose

Solar energy obtained from the Nano flakes can help benefit in a couple of ways. Nano flakes can potentially help lower the cost of solar energy. Also since more solar energy can theoretically be obtained from Nano flakes, their use can potentially will keep the earth's environment cleaner by reducing the need for fossil fuels.

Cost

The high cost of solar energy stems from the difficulty of converting the solar energy into electricity for use, and less than 1 percent of the world's electricity comes from the sun because of this process. Nano flakes can potentially help with the economic issues of solar energy by lowering the cost due to an easier process and a better outcome of energy. Nano flake technology can potentially make it easier to convert solar energy into electricity estimated at twice the amount that today's solar cells can harvest. This new technology can also potentially lower the cost of solar energy because it allows for a reduction in expensive semiconducting silicon. Energy loss is also potentially reduced with a shorter distance of the solar energy transportation across smaller Nano flakes.

Environment

Nano Flake technology can also help keep the environment cleaner as the sun as the source it produces clean pure sustainable energy that can be converted into electricity. While fossil fuel is the primary energy source for electricity, using solar energy obtained from Nano flakes will lower dependence on fossil fuels. When fossil fuels are burned for use they release a toxic gas which has a huge impact on earth's pollution. Also the process of obtaining these fossil fuels is not good for the environment, whether it be mining for coal, drilling for oil, and hydraulic fracturing the earth's surface to reach the oil and gas.

Nanofoam

Nanofoams are a class of nanostructured, porous materials (foams) containing a significant population of pores with diameters less than 100 nm. Aerogels are one example of nanofoam.

Metal

In 2006, researchers produced metal nanofoams by igniting pellets of energetic metal bis(tetrazolato)amine complexes. Nanofoams of iron, cobalt, nickel, copper, silver, and palladium have been prepared through this technique. These materials exhibit densities as low as 11 mg/cm³, and surface areas as high as 258 m²/g. These foams are effective catalysts.

Carbon

Carbon nanofoam is an allotrope of carbon discovered in 1997. It consists of a cluster-assembly of carbon atoms strung together in a loose three-dimensional web. The material has a density of 2–10 mg/cm³ (0.0012 lb/ft³).

Glass

In 2014, researchers also fabricated glass nanofoam via femtosecond laser ablation. Their work consisted of raster scanning femtosecond laser pulses over the surface of glass to produce glass nanofoam with ~70 nm diameter wires.

Thermodynamics of Nanostructures

As the devices continue to shrink further into the sub-100 nm range following the trend predicted by Moore's law, the topic of thermal properties and transport in such nanoscale devices becomes increasingly important. Display of great potential by nanostructures for thermoelectric applications also motivates the studies of thermal transport in such devices. These fields, however, generate two contradictory demands: high thermal conductivity to deal with heating issues in sub-100 nm devices and low thermal conductivity for thermoelectric applications. These issues can be addressed with phonon engineering, once nanoscale thermal behaviors have been studied and understood.

The Effect of the Limited Length of Structure

In general two carrier types can contribute to thermal conductivity - electrons and phonons. In nanostructures phonons usually dominate and the phonon properties of the structure become of a particular importance for thermal conductivity. These phonon properties include: phonon group velocity, phonon scattering mechanisms, heat capacity, Grüneisen parameter. Unlike bulk materials, nanoscale devices have thermal properties which are complicated by boundary effects due to small size. It has been shown that in some cases phonon-boundary scattering effects dominate the thermal conduction processes, reducing thermal conductivity.

Depending on the nanostructure size, the phonon mean free path values (Λ) may be comparable or larger than the object size, L. When L is larger than the phonon mean free path, Umklapp scattering process limits thermal conductivity (regime of diffusive thermal conductivity). When L is comparable to or smaller than the mean free path (which is of the order 1 μm for carbon nanostructures), the continuous energy model used for bulk materials no longer applies and nonlocal and nonequilibrium aspects to heat transfer also need to be considered. In this case phonons in defectless structure could propagate without scattering and thermal conductivity becomes ballistic (similar to ballistic conductivity). More severe changes in thermal behavior are observed when the feature size *L* shrinks further down to the wavelength of phonons.

Nanowires

Thermal Conductivity Measurements

The first measurement of thermal conductivity in silicon nanowires was published in 2003. Two important features were pointed out: 1) The measured thermal conductivities are significantly lower than that of the bulk Si and, as the wire diameter is decreased, the corresponding thermal conductivity is reduced. 2) As the wire diameter is reduced, the phonon boundary scattering dominates over phonon–phonon Umklapp scattering, which decreases the thermal conductivity with an increase in temperature.

For 56 nm and 115 nm wires $k \sim T^3$ dependence was observed, while for 37 nm wire $k \sim T^2$ dependence and for 22 nm wire $k \sim T$ dependence were observed. Chen *et al.* has shown that the one-dimensional cross-over for 20 nm Si nanowire occurs around 8K, while the phenomenon was observed for temperature values greater than 20K. Therefore, the reason of such behaviour is not in the confinement experienced by phonons so that three-dimensional structures display two-dimensional or one-dimensional behavior.

Theoretical Models for Nanowires

Different Phonon Modes Contribution to Thermal Conductivity

Assuming that Boltzmann transport equation is valid, thermal conductivity can be written as:

$$k = \frac{1}{3} C v_g \Lambda = \frac{1}{3} C v_g^2 \tau$$

where C is the heat capacity, v_g is the group velocity and τ is the relaxation time. Note that this assumption breaks down when the dimensions of the system are comparable to or smaller than the wavelength of the phonons responsible for thermal transport. In our case, phonon wavelengths are generally in the 1 nm range and the nanowires under consideration are within tens of nanometers range, the assumption is valid.

Different phonon mode contributions to heat conduction can be extracted from analysis of the experimental data for silicon nanowires of different diameters to extract the $C \cdot v_g$ product for analysis. It was shown that all phonon modes contributing to thermal transport are excited well below the Si Debye temperature (645 K).

From the thermal conductivity equation, one can write the product $C \cdot v_g$ for each isotropic phonon branch *i*.

$$(C v_g)_i = \frac{k_B^4 T^3}{2\hbar^3 \pi^2} \int \frac{1}{v_{p,i}^2} \left[\frac{x^4 exp(x)}{(exp(x)-1)^2} \right] dx$$

where $x = h\omega / k_B T$ and $v_{p,i}$ is the phonon phase velocity, which is less sensitive to phonon dispersions than the group velocity v_g.

Many models of phonon thermal transport ignores the effects of transverse acoustic phonons (TA) at high frequency due to their small group velocity. (Optical phonon contributions are also ignored for the same reason.) However, upper branch of TA phonons have non-zero group velocity at the Brillouin zone boundary along the Γ-K direction and, in fact, behave similarly to the longitudinal acoustic phonons (LA) and can contribute to the heat transport.

Then, the possible phonon modes contributing to heat conduction are both LA and TA phonons at low and high frequencies. Using the corresponding dispersion curves, the $C \cdot v_g$ product can then be calculated and fitted to the experimental data. The best fit was found when contribution of high-frequency TA phonons is accounted as 70% of the product at room temperature. The remaining 30% is contributed by the LA and TA phonons at low-frequency.

Using Complete Phonon Dispersions

Thermal conductivity in nanowires can be computed based on complete phonon dispersions instead of the linearlized dispersion relations commonly used to calculate thermal conductivity in bulk materials.

Assuming the phonon transport is diffusive and Boltzmann transport equation (BTE) is valid, nanowire thermal conductance *G(T)* can be defined as:

$$G(T) \simeq \sum_{\alpha} \int \frac{\lambda_{\alpha}(k_z)}{L} \frac{\hbar\omega_{\alpha}(k_z)}{2\pi} \frac{df_b}{dT} v_z(\alpha, k_z) dk_z$$

where the variable α represents discrete quantum numbers associated with sub-bands found in one-dimensional phonon dispersion relations, f_B represents the Bose-Einstein distribution, v_z is the phonon velocity in the *z* direction and λ is the phonon relaxation length along the direction of the wire length. Thermal conductivity is then expressed as:

$$k(T) = \frac{1}{S} \sum \alpha \int_{0}^{\frac{\pi}{a_z}} \lambda_{\alpha}(k_z) \frac{\hbar\omega_{\alpha}(k_z)}{2\pi} \frac{df_B}{dT} v_z(\alpha, k_z) dk_z$$

where *S* is the cross sectional area of the wire, a_z is the lattice constant.

It was shown that, using this formula and atomistically computed phonon dispersions (with interatomic potentials developed in), it is possible to predictively calculate lattice thermal conductivity curves for nanowires, in good agreement with experiments. On the other hand, it was not possible to obtain correct results with the approximated Callaway formula. These results are expected to apply to "nanowhiskers" for which phonon confinement effects are unimportant. Si nanowires wider than ~35 nm are within this category.

Very Thin Nanowires

For large diameter nanowires, theoretical models assuming the nanowire diameters are comparable to the mean free path and that the mean free path is independent of phonon frequency have been able to closely match the experimental results. But for very thin nanowires whose dimensions are comparable to the dominant phonon wavelength, a new model is required. The study in has shown that in such cases, the phonon-boundary scattering is dependent on frequency. The new mean free path is then should be used:

$$l^{-1} = B\left(\frac{h}{d}\right)^2 \frac{1}{d}\left(\frac{\omega}{\omega_D}\right)^2 N(\omega)$$

Here, l is the mean free path (same as Λ). The parameter h is length scale associated with the disordered region, d is the diameter, $N(\omega)$ is number of modes at frequency ω, and B is a constant related to the disorder region.

Thermal conductance is then calculated using Landauer formula:

$$G(T) = \frac{1}{2\pi\hbar}\int_0^{\infty}\left(\frac{N_1(\omega)}{1+L/l(\omega)} + \frac{N_2(\omega)}{1+L/d}\right)\frac{\hbar^3\omega^2}{k_B T^2} \times \frac{e^{\frac{\hbar\omega}{k_B T}}}{(e^{\frac{\hbar\omega}{k_B T}}-1)^2}$$

Carbon Nanotubes

As nanoscale graphitic structures, carbon nanotubes are of great interest for their thermal properties. The low-temperature specific heat and thermal conductivity show direct evidence of 1-D quantization of the phonon band structure. Modeling of the low-temperature specific heat allows determination of the on-tube phonon velocity, the splitting of phonon subbands on a single tube, and the interaction between neighboring tubes in a bundle.

Thermal Conductivity Measurements

Measurements show a single-wall carbon nanotubes (SWNTs) room-temperature thermal conductivity about 3500 W/(m·K), and over 3000 W/(m·K) for individual multiwalled carbon nanotubes (MWNTs). It is difficult to replicate these properties on the macroscale due to imperfect contact between individual CNTs, and so tangible objects from CNTs such as films or fibres have reached only up to 1500 W/(m·K) so far. Addition of nanotubes to epoxy resin can double the thermal conductivity for a loading of only 1%, showing that nanotube composite materials may be useful for thermal management applications.

Theoretical Models for Nanotubes

Thermal conductivity in CNT is mainly due to phonons rather than electrons so the Wiedemann–Franz law is not applicable.

In general, the thermal conductivity is a tensor quality, but for this discussion, it is only important to consider the diagonal elements:

$$k_{zz} = \sum C v_z^2 \tau$$

where C is the specific heat, and v_z and τ are the group velocity and relaxation time of a given phonon state.

At low temperatures (T is far less than Debye temperature), the relaxation time is determined by scattering of fixed impurities, defects, sample boundaries, etc. and is roughly constant. Therefore, in ordinary materials, the low-temperature thermal conductivity has the same temperature dependence as the specific heat. However, in anisotropic materials, this relationship does not strictly hold. Because the contribution of each state is weighted by the scattering time and the square of the velocity, the thermal conductivity preferentially samples states with large velocity and scattering time. For instance, in graphite, the thermal conductivity parallel to the basal planes is only weakly dependent on the interlayer phonons. In SWNT bundles, it is likely that *k(T)* depends only on the on-tube phonons, rather than the intertube modes.

Thermal conductivity is of particular interest in low-dimensional systems. For CNT, represented as 1-D ballistic electronic channel, the electronic conductance is quantized, with a universal value of

$$G_0 = \frac{2e^2}{h}$$

Similarly, for a single ballistic 1-D channel, the thermal conductance is independent of materials parameters, and there exists a quantum of thermal conductance, which is linear in temperature:

$$G_{th} = \frac{\pi^2 k_B{}^2 T}{3h}$$

Possible conditions for observation of this quantum were examined by Rego and Kirczenow. In 1999, Keith Schwab, Erik Henriksen, John Worlock, and Michael Roukes carried out a series of experimental measurements that enabled first observation of the thermal conductance quantum. The measurements employed suspended nanostructures coupled to sensitive dc SQUID measurement devices. In 2008, a colorized electron micrograph of one of the Caltech devices was acquired for the permanent collection of the Museum of Modern Art in New York.

At high temperatures, three-phonon Umklapp scattering begins to limit the phonon relaxation time. Therefore, the phonon thermal conductivity displays a peak and decreases with increasing temperature. Umklapp scattering requires production of a phonon beyond the Brillouin zone boundary; because of the high Debye temperature of diamond and graphite, the peak in the thermal conductivity of these materials is near 100 K, significantly higher than for most other materials. In less crystalline forms of graphite, such as carbon fibers, the peak in *k(T)* occurs at higher temperatures, because defect scattering remains dominant over Umklapp scattering to higher temperature. In low-dimensional systems, it is difficult to conserve both energy and momentum for Umklapp processes, and so it may be possible that Umklapp scattering is suppressed in nanotubes relative to 2-D or 3-D forms of carbon.

Berber *et al.* have calculated the phonon thermal conductivity of isolated nanotubes. The value *k(T)* peaks near 100 K, and then decreases with increasing temperature. The value of *k(T)* at the peak (37,000 W/(m·K)) is comparable to the highest thermal conductivity ever measured (41,000

W/(m·K) for an isotopically pure diamond sample at 104 K). Even at room temperature, the thermal conductivity is quite high (6600 W/(m·K)), exceeding the reported room-temperature thermal conductivity of isotopically pure diamond by almost a factor of 2.

In graphite, the interlayer interactions quench the thermal conductivity by nearly 1 order of magnitude. It is likely that the same process occurs in nanotube bundles. Thus it is significant that the coupling between tubes in bundles is weaker than expected. It may be that this weak coupling, which is problematic for mechanical applications of nanotubes, is an advantage for thermal applications.

Phonon Density of States for Nanotubes

The phonon density of states is to calculated through band structure of isolated nanotubes, which is studied in Saito *et al.* and Sanchez-Portal *et al.* When a graphene sheet is "rolled" into a nanotube, the 2-D band structure folds into a large number of 1-D subbands. In a (10,10) tube, for instance, the six phonon bands (three acoustic and three optical) of the graphene sheet become 66 separate 1-D subbands. A direct result of this folding is that the nanotube density of states has a number of sharp peaks due to 1-D van Hove singularities, which are absent in graphene and graphite. Despite the presence of these singularities, the overall density of states is similar at high energies, so that the high temperature specific heat should be roughly equal as well. This is to be expected: the high-energy phonons are more reflective of carbon–carbon bonding than the geometry of the graphene sheet.

Thin Films

Thin films are prevalent in the micro and nanoelectronics industry for the fabrication of sensors, actuators and transistors; thus, thermal transport properties affect the performance and reliability of many structures such as transistors, solid-state lasers, sensors, and actuators. Although these devices are traditionally made from bulk crystalline material (silicon), they often contain thin films of oxides, polysilicon, metal, as well as superlattices such as thin-film stacks of GaAs/AlGaAs for lasers.

Single-crystal Thin Films

Silicon-on-insulator (SOI) films with silicon thicknesses of 0.05 μm to 10 μm above a buried silicon dioxide layer are increasingly popular for semiconductor devices due to the increased dielectric isolation associated with SOI/ SOI wafers contain a thin-layer of silicon on an oxide layer and a thin-film of single-crystal silicon, which reduces the effective thermal conductivity of the material by up to 50% as compared to bulk silicon, due to phonon-interface scattering and defects and dislocations in the crystalline structure. Previous studies by Asheghi *et al.*, show a similar trend. Other studies of thin-films show similar thermal effects.

Superlattices

Thermal properties associated with superlattices are critical in the development of semiconductor lasers. Heat conduction of superlattices is less understood than homogeneous thin films. It is theorized that superlattices have a lower thermal conductivity due to impurities from lattice mismatches and at the heterojunctions. Phonon-interface scattering at heterojunctions needs to be considered in this case; fully elastic scattering underestimates the heat conduction, while fully inelastic scattering overestimates the heat conduction. For example, a Si/Ge thin-film superlattice has a greater decrease in thermal conductivity than an AlAs/GaAs film stack due to increased lattice mismatch. A simple estimate of heat conduction of superlattices is:

$$k_n = \left(\frac{C_1 v_1 C_2 v_2}{C_1 v_1 + C_2 v_2} \right) \left(\frac{d_1 + d_2}{2} \right)$$

where C_1 and C_2 are the corresponding heat capacity of film1 and film2 respectively, v_1 and v_2 are the acoustic propagation velocities in film1 and film2, and *d1* and *d2* are the thicknesses of film1 and film2. This model neglects scattering within the layers and assumes fully diffuse, inelastic scattering.

Polycrystalline Films

Polycrystalline films are common in semiconductor devices, as the gate electrode of a field-effect transistor is often made of polycrystalline silicon. If the polysilicon grain sizes are small, internal scattering from grain boundaries can overwhelm the effects of film-boundary scattering. Also, grain boundaries contain more impurities, which result in impurity scattering. Likewise, disordered or amorphous films will experience a severe reduction of thermal conductivity, since the small grain size results in numerous grain-boundary scattering effects. Different deposition methods of amorphous films will result in differences in impurities and grain sizes.

The simplest approach to modeling phonon scattering at grain boundaries is to increase the scattering rate by introducing this equation:

$$\tau_G^{-1} = B \frac{v}{d_G}$$

where B is a dimensionless parameter that correlates with the phonon reflection coefficient at the grain boundaries, d_G is the characteristic grain size, and v is the phonon velocity through the material. A more formal approach to estimating the scattering rate is:

$$\tau_G^{-1} = \frac{2v}{\pi d_G} \left[1 - exp\left(-\frac{\pi^2}{4} v_G \right) \right]$$

where v_G is the dimensionless grain-boundary scattering strength, defined as

$$v_G = \sum_j \sigma_j v_j$$

Here σ_j is the cross-section of a grain-boundary area, and v_j is the density of the grain boundary area.

Measuring Thermal Conductivity of Thin Films

There are two approaches to experimentally determine the thermal conductivity of thin films. The goal of experimental metrology of thermal conductivity of thin films is to attain an accurate thermal measurement without disturbing the properties of the thin-film.

Electrical heating is used for thin films which have a lower thermal conductivity than the substrate; it is fairly accurate in measuring out-of-plane conductivity. Often, a resistive heater and thermistor is fabricated on the sample film using a highly conductive metal, such as aluminium. The most straightforward approach would be to apply a steady-state current and measure the change in temperature of adjacent thermistors. A more versatile approach uses an AC signal applied to the electrodes. The third harmonic of the AC signal reveals heating and temperature fluctuations of the material.

Laser heating is a non-contact metrology method, which uses picosecond and nanosecond laser pulses to deliver thermal energy to the substrate. Laser heating uses a pump-probe mechanism; the pump beam introduces energy to the thin-film, as the probe beam picks up the characteristics of how the energy propagates through the film. Laser heating is advantageous because the energy delivered to the film can be precisely controlled; furthermore, the short heating duration decouples the thermal conductivity of the thin film from the substrate.

References

- Liu, M.; Peng, J.; et al. (2016). "Two-dimensional modeling of the self-limiting oxidation in silicon and tungsten nanowires". Theoretical and Applied Mechanics Letters. 6 (5): 195–199. doi:10.1016/j.taml.2016.08.002
- "Could hemp nanosheets topple graphene for making the ideal supercapacitor?". acs.org. American Chemistry Society. Retrieved 14 August 2014
- Shao, M.; Duo Duo Ma, D.; Lee, ST (2010). "Silicon nanowires–synthesis, properties, and applications". European Journal of Inorganic Chemistry. 27: 4264–4278
- M. Corso; Greber, Thomas; Osterwalder, Jürg (2005). "h-BN on Pd(110): a tunable system for selfassembled nanostructures?". Surf. Sci. 577 (2–3): L78. Bibcode:2005SurSc.577L..78C. doi:10.1016/j.susc.2005.01.015
- Daniel, Shir; et al. (2006). "Oxidation of silicon nanowires.". Journal of Vacuum Science & Technology (3): 1333–1336
- Holm, Timothy P.; Cheng Chieh Chao; Prinz, Fritz B. (2009). "2009 34th IEEE Photovoltaic Specialists Conference (PVSC)": 000085. ISBN 978-1-4244-2949-3. doi:10.1109/PVSC.2009.5411731
- Geim, A. K. (2009). "Graphene: status and prospects". Science. 324 (5934): 1530–1534. PMID 19541989. doi:10.1126/science.1158877
- Harrison, Walter A. (1989). Electronic Structure and the Properties of Solids: The Physics of the Chemical Bond. Dover Publications. ISBN 0-486-66021-4
- "Simple fabrication method of hierarchical nano-pillars using aluminum anodizing processes". Current Applied Physics. 9: e81–e85. doi:10.1016/j.cap.2008.12.034
- "Nano-flake structure boosts solar cell performance - Eco-innovation Action Plan - European Commission". Eco-innovation Action Plan. Retrieved 2016-02-11
- R Saito; G Dresselhaus; M S Dresselhaus (22 July 1998). Physical Properties of Carbon Nanotubes. World Scientific. pp. 229–. ISBN 978-1-78326-241-0
- "Synthesis of Single-Crystal Gold Nanosheets of Large Size in Ionic Liquids". The Journal of Physical Chemistry B. 109: 14445–14448. doi:10.1021/jp0520998
- Callaway, J. (1959). "Model for Lattice Thermal Conductivity at Low Temperatures". Phys. Rev. Lett. 113 (4): 1046–51. Bibcode:1959PhRv..113.1046C. doi:10.1103/PhysRev.113.1046

4

An Integrated Study of Carbon Nanotube

Carbon nanotubes are carbon allotropes that have a cylindrical nanostructure. They posses exceptional stiffness and strength. The different types of carbon nanotubes that exist are single-walled, junctions and crosslinking, multi-walled, etc. The chapter strategically encompasses and incorporates the major components and key concepts of carbon nanotubes, providing a complete understanding.

Carbon Nanotube

Carbon nanotubes (CNTs) are allotropes of carbon with a cylindrical nanostructure. These cylindrical carbon molecules have unusual properties, which are valuable for nanotechnology, electronics, optics and other fields of materials science and technology. Owing to the material's exceptional strength and stiffness, nanotubes have been constructed with length-to-diameter ratio of up to 132,000,000:1, significantly larger than for any other material.

In addition, owing to their extraordinary thermal conductivity, mechanical, and electrical properties, carbon nanotubes find applications as additives to various structural materials. For instance, nanotubes form a tiny portion of the material(s) in some (primarily carbon fiber) baseball bats, golf clubs, car parts or damascus steel.

Nanotubes are members of the fullerene structural family. Their name is derived from their long, hollow structure with the walls formed by one-atom-thick sheets of carbon, called graphene. These sheets are rolled at specific and discrete ("chiral") angles, and the combination of the rolling angle and radius decides the nanotube properties; for example, whether the individual nanotube shell is a metal or semiconductor. Nanotubes are categorized as single-walled nanotubes (SWNTs) and multi-walled nanotubes (MWNTs). Individual nanotubes naturally align themselves into "ropes" held together by van der Waals forces, more specifically, pi-stacking.

Applied quantum chemistry, specifically, orbital hybridization best describes chemical bonding in nanotubes. The chemical bonding of nanotubes involves entirely sp^2-hybrid carbon atoms. These bonds, which are similar to those of graphite and stronger than those found in alkanes and diamond (which employ sp^3 -hybrid carbon atoms), provide nanotubes with their unique strength.

Types of Carbon Nanotubes and Related Structures

There is no consensus on some terms describing carbon nanotubes in scientific literature: both "-wall" and "-walled" are being used in combination with "single", "double", "triple" or "multi", and the letter C is often omitted in the abbreviation; for example, multi-walled carbon nanotube (MWNT).

Single-walled

Armchair (n,n) i.e.: m=n

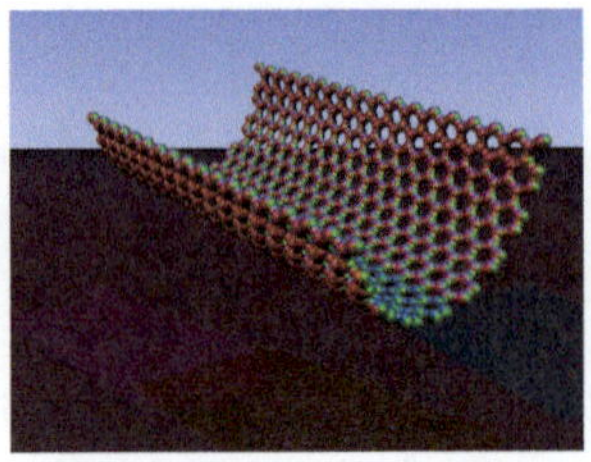
The translation vector is bent, while the chiral vector stays straight

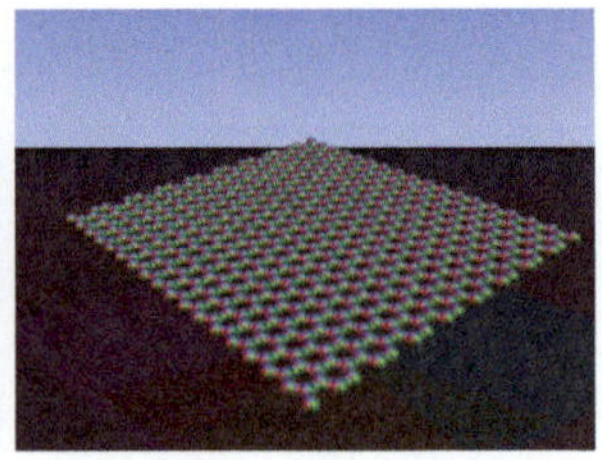
Graphene nanoribbon

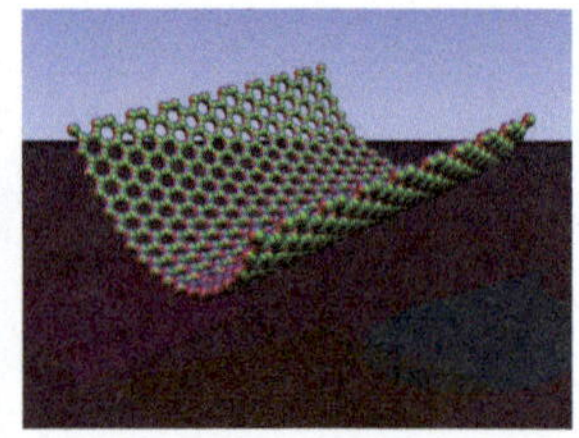
The chiral vector is bent, while the translation vector stays straight

Zigzag $(n,0)$

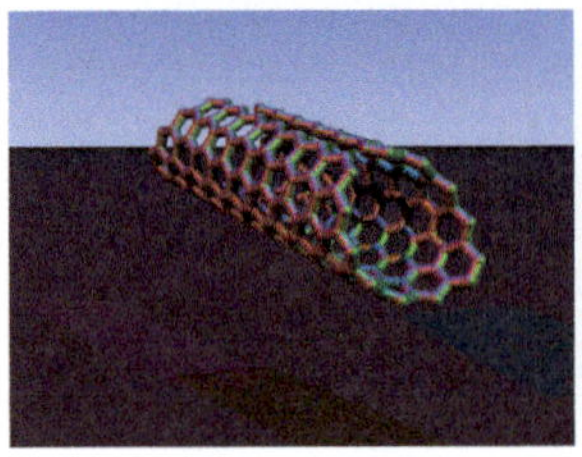
Chiral (n,m)

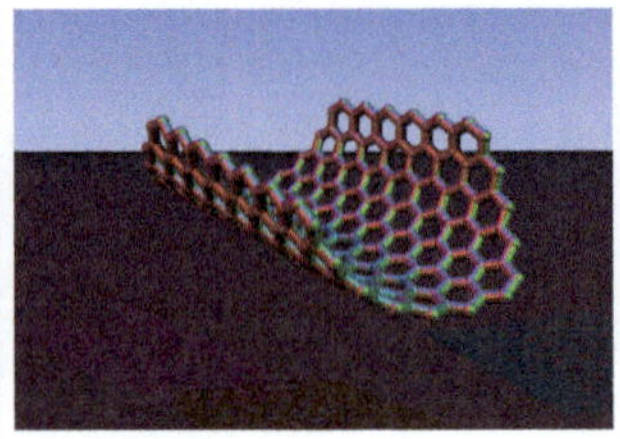
n and m can be counted at the end of the tube

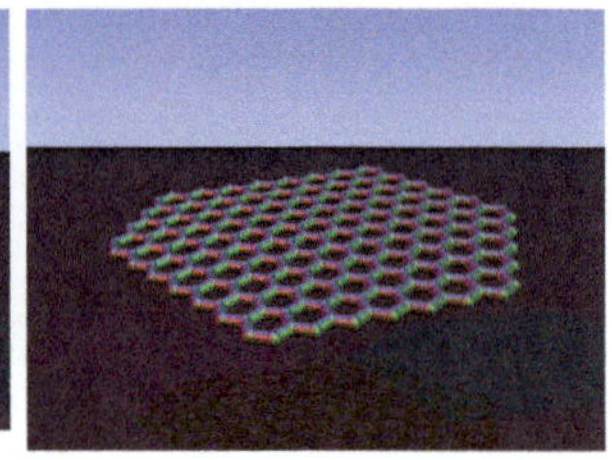
Graphene nanoribbon

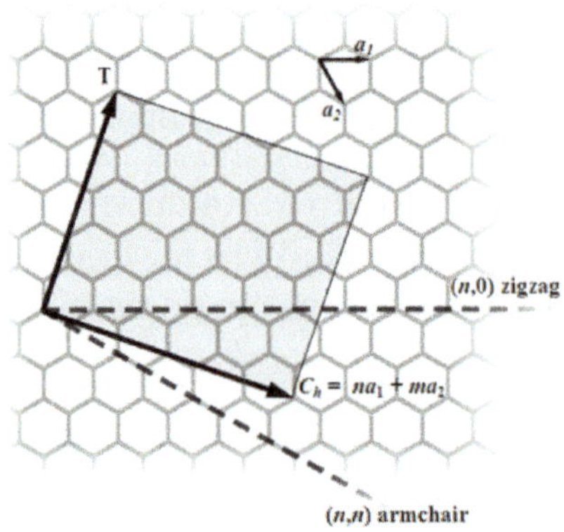

The (n,m) nanotube naming scheme can be thought of as a vector (C_h) in an infinite graphene sheet that describes how to “roll up” the graphene sheet to make the nanotube. T denotes the tube axis, and a_1 and a_2 are the unit vectors of graphene in real space.

A scanning tunneling microscopy image of single-walled carbon nanotube

Most single-walled nanotubes (SWNTs) have a diameter of close to 1 nanometer, and can be many millions of times longer. The structure of a SWNT can be conceptualized by wrapping a one-atom-thick layer of graphite called graphene into a seamless cylinder. The way the graphene sheet is wrapped is represented by a pair of indices (n,m). The integers n and m denote the number of unit vectors along two directions in the honeycomb crystal lattice of graphene. If $m = 0$, the nanotubes are called zigzag nanotubes, and if n = m, the nanotubes are called armchair nanotubes. Otherwise, they are called chiral. The diameter of an ideal nanotube can be calculated from its (n,m) indices as follows

$$d = \frac{a}{\pi}\sqrt{(n^2 + nm + m^2)} = 78.3\sqrt{((n+m)^2 - nm)}\,pm,$$

where a = 0.246 nm. Most single-walled nanotubes (SWNTs) have a diameter of close to 1 nanometer, and can be many millions of times longer. The structure of a SWNT can be conceptualized by wrapping a one-atom-thick layer of graphite called graphene into a seamless cylinder. The way the graphene sheet is wrapped is represented by a pair of indices (n,m). The integers n and m denote the number of unit vectors along two directions in the honeycomb crystal lattice of graphene. If m = 0, the nanotubes are called zigzag nanotubes, and if n = m, the nanotubes are called armchair nanotubes. Otherwise, they are called chiral. The diameter of an ideal nanotube can be calculated from its (n,m) indices as follows

$$d = \frac{a}{\pi}\sqrt{(n^2 + nm + m^2)} = 78.3\sqrt{((n+m)^2 - nm)}\,pm,$$

where a = 0.246 nm.

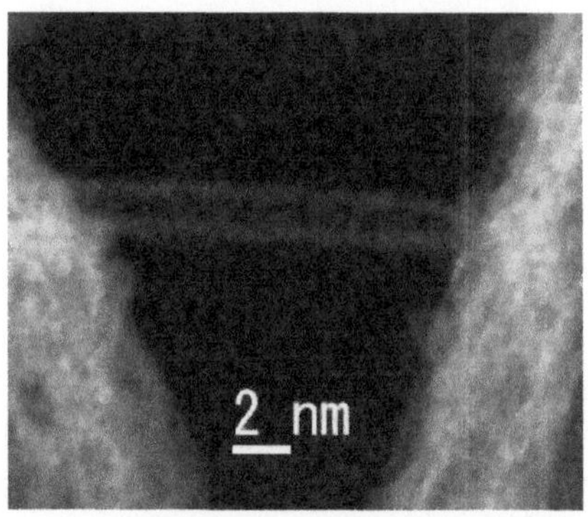

A transmission electron microscopy image of a single-walled carbon nanotube

SWNTs are an important variety of carbon nanotube because most of their properties change significantly with the (n,m) values, and this dependence is non-monotonic. In particular, their band gap can vary from zero to about 2 eV and their electrical conductivity can show metallic or semiconducting behavior. Single-walled nanotubes are likely candidates for miniaturizing electronics. The most basic building block of these systems is the electric wire, and SWNTs with diameters of an order of a nanometer can be excellent conductors. One useful application of SWNTs is in the development of the first intermolecular field-effect transistors (FET). The first intermolecular logic gate using SWCNT FETs was made in 2001. A logic gate requires both a p-FET and an n-FET. Because SWNTs are p-FETs when exposed to oxygen and n-FETs otherwise, it is possible to expose half of an SWNT to oxygen and protect the other half from it. The resulting SWNT acts as a *not* logic gate with both p and n-type FETs in the same molecule.

Prices for single-walled nanotubes declined from around $1500 per gram as of 2000 to retail prices of around $50 per gram of as-produced 40–60% by weight SWNTs as of March 2010. As of

2016 the retail price of as-produced 75% by weight SWNTs were $2 per gram, cheap enough for widespread use. SWNTs are forecast to make a large impact in electronics applications by 2020 according to *The Global Market for Carbon Nanotubes* report.

Multi-walled

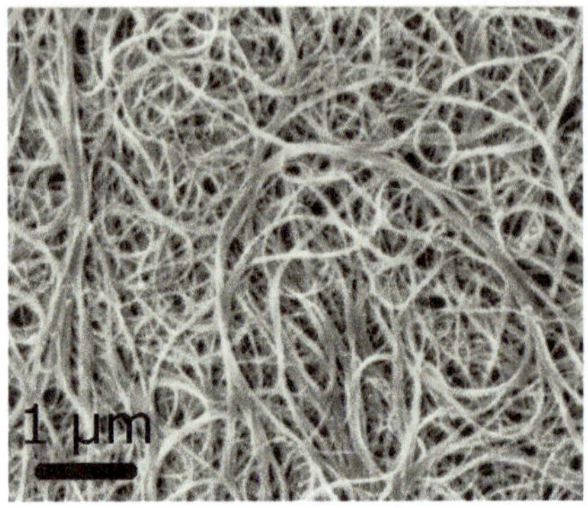

A scanning electron microscopy image of carbon nanotubes bundles

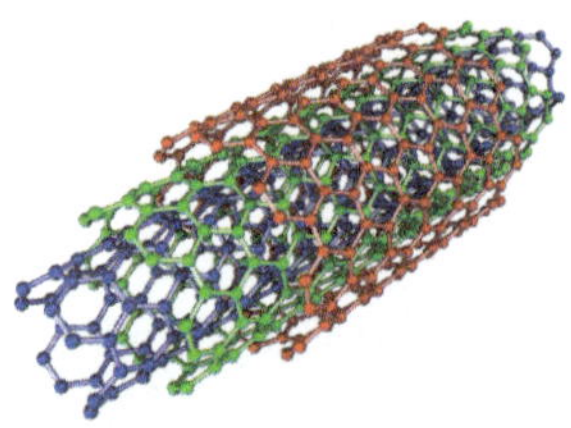

Triple-walled armchair carbon nanotube

Multi-walled nanotubes (MWNTs) consist of multiple rolled layers (concentric tubes) of graphene. There are two models that can be used to describe the structures of multi-walled nanotubes. In the *Russian Doll* model, sheets of graphite are arranged in concentric cylinders, e.g., a (0,8) single-walled nanotube (SWNT) within a larger (0,17) single-walled nanotube. In the *Parchment* model, a single sheet of graphite is rolled in around itself, resembling a scroll of parchment or a rolled newspaper. The interlayer distance in multi-walled nanotubes is close to the distance between graphene layers in graphite, approximately 3.4 Å. The Russian Doll structure is observed more commonly. Its individual shells can be described as SWNTs, which can be metallic or semiconducting. Because of statistical probability and restrictions on the relative diameters of the individual tubes, one of the shells, and thus the whole MWNT, is usually a zero-gap metal.

Double-walled carbon nanotubes (DWNTs) form a special class of nanotubes because their morphology and properties are similar to those of SWNTs but they are more resistant to chemicals. This is especially important when it is necessary to graft chemical functions to the surface of the nanotubes (functionalization) to add properties to the CNT. Covalent functionalization of SWNTs will break some C=C double bonds, leaving "holes" in the structure on the nanotube, and thus modifying both its mechanical and electrical properties. In the case of DWNTs, only the outer wall is modified. DWNT synthesis on the gram-scale was first proposed in 2003 by the CCVD technique, from the selective reduction of oxide solutions in methane and hydrogen.

The telescopic motion ability of inner shells and their unique mechanical properties will permit the use of multi-walled nanotubes as main movable arms in coming nanomechanical devices. Retraction force that occurs to telescopic motion caused by the Lennard-Jones interaction between shells and its value is about 1.5 nN.

Junctions and Crosslinking

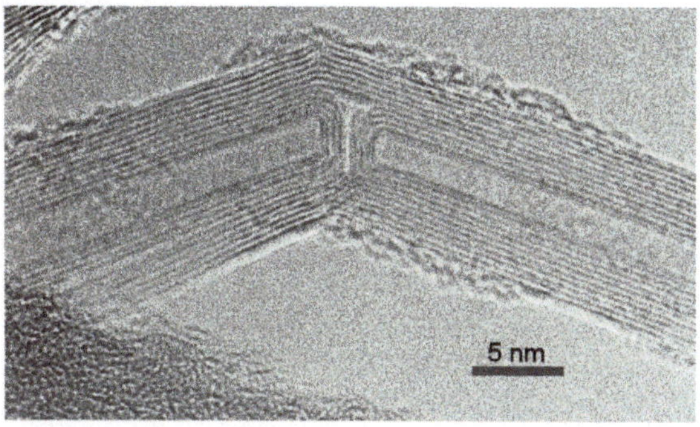

Transmission electron microscope image of carbon nanotube junction

Junctions between 2 or more nanotubes have been widely discussed theoretically. Such junctions are quite frequently observed in samples prepared by arc discharge as well as by chemical vapor deposition. The electronic properties of such junctions were first considered theoretically by Lambin et al., who pointed out that a connection between metallic tube and a semiconducting one would represent a nanoscale heterojunction. Such a junction could therefore form a component of a nanotube-based electronic circuit. The image on the right shows a junction between two multi-walled nanotubes. Junctions between nanotubes and graphene have been considered theoretically, but not widely studied experimentally. Such junctions form the basis of pillared graphene, in which parallel graphene sheets are separated by short nanotubes. Pillared graphene represents a class of three-dimensional carbon nanotube architectures.

3D carbon scaffolds

Recently, several studies have highlighted the prospect of using carbon nanotubes as building blocks to fabricate three-dimensional macroscopic (>100 nm in all three dimensions) all-carbon devices. Lalwani et al. have reported a novel radical initiated thermal crosslinking method to fabricate macroscopic, free-standing, porous, all-carbon scaffolds using single- and multi-walled carbon nanotubes as building blocks. These scaffolds possess macro-, micro-, and nano- structured pores and the porosity can be tailored for specific applications. These 3D all-carbon scaffolds/architectures may be used for the fabrication of the next generation of energy storage, supercapacitors, field emission transistors, high-performance catalysis, photovoltaics, and biomedical devices and implants.

Other Morphologies

A stable nanobud structure

Carbon nanobuds are a newly created material combining two previously discovered allotropes of carbon: carbon nanotubes and fullerenes. In this new material, fullerene-like "buds" are covalently bonded to the outer sidewalls of the underlying carbon nanotube. This hybrid material has useful properties of both fullerenes and carbon nanotubes. In particular, they have been found to be exceptionally good field emitters. In composite materials, the attached fullerene molecules may function as molecular anchors preventing slipping of the nanotubes, thus improving the composite's mechanical properties.

A carbon peapod is a novel hybrid carbon material which traps fullerene inside a carbon nanotube. It can possess interesting magnetic properties with heating and irradiation. It can also be applied as an oscillator during theoretical investigations and predictions.

In theory, a nanotorus is a carbon nanotube bent into a torus (doughnut shape). Nanotori are predicted to have many unique properties, such as magnetic moments 1000 times larger than previously expected for certain specific radii. Properties such as magnetic moment, thermal stability, etc. vary widely depending on radius of the torus and radius of the tube.

Graphenated carbon nanotubes are a relatively new hybrid that combines graphitic foliates grown along the sidewalls of multiwalled or bamboo style CNTs. The foliate density can vary as a function of deposition conditions (e.g. temperature and time) with their structure ranging from few layers of graphene (< 10) to thicker, more graphite-like. The fundamental advantage of an integrated graphene-CNT structure is the high surface area three-dimensional framework of the CNTs coupled with the high edge density of graphene. Depositing a high density of graphene foliates along the length of aligned CNTs can significantly increase the total charge capacity per unit of nominal area as compared to other carbon nanostructures.

Cup-stacked carbon nanotubes (CSCNTs) differ from other quasi-1D carbon structures, which normally behave as quasi-metallic conductors of electrons. CSCNTs exhibit semiconducting behaviors due to the stacking microstructure of graphene layers.

Extreme Carbon Nanotubes

The observation of the *longest* carbon nanotubes grown so far are over 1/2 m (550 mm long) was reported in 2013. These nanotubes were grown on Si substrates using an improved chemical vapor deposition (CVD) method and represent electrically uniform arrays of single-walled carbon nanotubes.

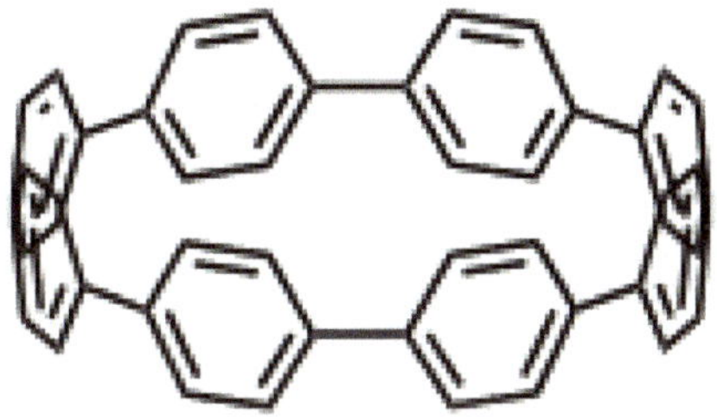

Cycloparaphenylene

The *shortest* carbon nanotube is the organic compound cycloparaphenylene, which was synthesized in 2008.

The *thinnest* carbon nanotube is the armchair (2,2) CNT with a diameter of 0.3 nm. This nanotube was grown inside a multi-walled carbon nanotube. Assigning of carbon nanotube type was done by a combination of high-resolution transmission electron microscopy (HRTEM), Raman spectroscopy and density functional theory (DFT) calculations.

The *thinnest freestanding* single-walled carbon nanotube is about 0.43 nm in diameter. Researchers suggested that it can be either (5,1) or (4,2) SWCNT, but the exact type of carbon nanotube remains questionable. (3,3), (4,3) and (5,1) carbon nanotubes (all about 0.4 nm in diameter) were unambiguously identified using aberration-corrected high-resolution transmission electron microscopy inside double-walled CNTs.

The *highest density* of CNTs was achieved in 2013, grown on a conductive titanium-coated copper surface that was coated with co-catalysts cobalt and molybdenum at lower than typical temperatures of 450 °C. The tubes averaged a height of 380 nm and a mass density of 1.6 g cm^{-3}. The material showed ohmic conductivity (lowest resistance 22 kΩ).

Properties

Mechanical

Carbon nanotubes are the strongest and stiffest materials yet discovered in terms of tensile strength and elastic modulus respectively. This strength results from the covalent sp^2 bonds formed between the individual carbon atoms. In 2000, a multi-walled carbon nanotube was tested to have a tensile strength of 63 gigapascals (9,100,000 psi). (For illustration, this translates into the ability to endure tension of a weight equivalent to 6,422 kilograms-force (62,980 N; 14,160 lbf) on a cable with cross-section of 1 square millimetre (0.0016 sq in).) Further studies, such as one conducted in 2008, revealed that individual CNT shells have strengths of up to ≈100 gigapascals (15,000,000 psi), which is in agreement with quantum/atomistic models. Since carbon nanotubes have a low density for a solid of 1.3 to 1.4 g/cm^3, its specific strength of up to 48,000 kN·m·kg^{-1} is the best of known materials, compared to high-carbon steel's 154 kN·m·kg^{-1}.

Although the strength of individual CNT shells is extremely high, weak shear interactions between adjacent shells and tubes lead to significant reduction in the effective strength of multi-walled carbon nanotubes and carbon nanotube bundles down to only a few GPa. This limitation has been recently addressed by applying high-energy electron irradiation, which crosslinks inner shells and tubes, and effectively increases the strength of these materials to ≈60 GPa for multi-walled carbon nanotubes and ≈17 GPa for double-walled carbon nanotube bundles. CNTs are not nearly as strong under compression. Because of their hollow structure and high aspect ratio, they tend to undergo buckling when placed under compressive, torsional, or bending stress.

On the other hand, there was evidence that in the radial direction they are rather soft. The first transmission electron microscope observation of radial elasticity suggested that even the van der Waals forces can deform two adjacent nanotubes. Later, nanoindentations with atomic force microscope were performed by several groups to quantitatively measure radial elasticity of multi-walled carbon nanotubes and tapping/contact mode atomic force microscopy was also performed on single-walled carbon nanotubes. Young's modulus of on the order of several GPa showed that CNTs are in fact very soft in the radial direction.

Electrical

Unlike graphene, which is a two-dimensional semimetal, carbon nanotubes are either metallic or semiconducting along the tubular axis. For a given (n,m) nanotube, if $n = m$, the nanotube is metallic; if $n - m$ is a multiple of 3 and $n \neq m$ and $nm \neq 0$, then the nanotube is quasi-metallic with a very small band gap, otherwise the nanotube is a moderate semiconductor. Thus all armchair ($n = m$) nanotubes are metallic, and nanotubes (6,4), (9,1), etc. are semiconducting. Carbon nanotubes are not semimetallic because the degenerate point (that point where the π [bonding] band meets the π [anti-bonding] band, at which the energy goes to zero) is slightly shifted away from the K point in the Brillouin zone due to the curvature of the tube surface, causing hybridization between the σ* and π* anti-bonding bands, modifying the band dispersion.

The rule regarding metallic versus semiconductor behavior has exceptions, because curvature effects in small diameter tubes can strongly influence electrical properties. Thus, a (5,0) SWCNT that should be semiconducting in fact is metallic according to the calculations. Likewise, zigzag and chiral SWCNTs with small diameters that should be metallic have a finite gap (armchair nanotubes remain metallic). In theory, metallic nanotubes can carry an electric current density of 4×10^9 A/cm², which is more than 1,000 times greater than those of metals such as copper, where for copper interconnects current densities are limited by electromigration. Carbon nanotubes are thus being explored as interconnects, conductivity enhancing components in composite materials and many groups are attempting to commercialize highly conducting electrical wire assembled from individual carbon nanotubes. There are significant challenges to be overcome, however, such as undesired current saturation under voltage, the much more resistive nanotube-to-nanotube junctions and impurities, all of which lower the electrical conductivity of the macroscopic nanotube wires by orders of magnitude, as compared to the conductivity of the individual nanotubes.

Because of its nanoscale cross-section, electrons propagate only along the tube's axis. As a result, carbon nanotubes are frequently referred to as one-dimensional conductors. The maximum electrical conductance of a single-walled carbon nanotube is $2G_0$, where $G_0 = 2e^2/h$ is the conductance of a single ballistic quantum channel.

Due to the role of the π-electron system in determining the electronic properties of graphene, doping in carbon nanotubes differs from that of bulk crystalline semiconductors from the same group of the periodic table (e.g. silicon). Graphitic substitution of carbon atoms in the nanotube wall by boron or nitrogen dopants leads to p-type and n-type behavior, respectively, as would be expected in silicon. However, some non-substitutional (intercalated or adsorbed) dopants introduced into a carbon nanotube, such as alkali metals as well as electron-rich metallocenes, result in n-type conduction because they donate electrons to the π-electron system of the nanotube. By contrast, π-electron acceptors such as $FeCl_3$ or electron-deficient metallocenes function as p-type dopants since they draw π-electrons away from the top of the valence band.

Intrinsic superconductivity has been reported, although other experiments found no evidence of this, leaving the claim a subject of debate.

Optical

Carbon nanotubes have useful absorption, photoluminescence (fluorescence), and Raman spectroscopy properties. Spectroscopic methods offer the possibility of quick and non-destructive characterization of relatively large amounts of carbon nanotubes. There is a strong demand for such characterization from the industrial point of view: numerous parameters of the nanotube synthesis can be changed, intentionally or unintentionally, to alter the nanotube quality. As shown below, optical absorption, photoluminescence and Raman spectroscopies allow quick and reliable characterization of this "nanotube quality" in terms of non-tubular carbon content, structure (chirality) of the produced nanotubes, and structural defects. Those features determine nearly any other properties such as optical, mechanical, and electrical properties.

Carbon nanotubes are unique "one-dimensional systems" which can be envisioned as rolled single sheets of graphite (or more precisely graphene). This rolling can be done at different angles and curvatures resulting in different nanotube properties. The diameter typically varies in the range 0.4–40 nm (i.e. "only" ~100 times), but the length can vary ~10,000 times, reaching 55.5 cm. The nanotube aspect ratio, or the length-to-diameter ratio, can be as high as 132,000,000:1, which is unequalled by any other material. Consequently, all the properties of the carbon nanotubes relative to those of typical semiconductors are extremely anisotropic (directionally dependent) and tunable.

Whereas mechanical, electrical and electrochemical (supercapacitor) properties of the carbon nanotubes are well established and have immediate applications, the practical use of optical properties is yet unclear. The aforementioned tunability of properties is potentially useful in optics and photonics. In particular, light-emitting diodes (LEDs) and photo-detectors based on a single nanotube have been produced in the lab. Their unique feature is not the efficiency, which is yet relatively low, but the narrow selectivity in the wavelength of emission and detection of light and the possibility of its fine tuning through the nanotube structure. In addition, bolometer and optoelectronic memory devices have been realised on ensembles of single-walled carbon nanotubes.

Crystallographic defects also affect the tube's electrical properties. A common result is lowered conductivity through the defective region of the tube. A defect in armchair-type tubes (which can conduct electricity) can cause the surrounding region to become semiconducting, and single monatomic vacancies induce magnetic properties.

Thermo

All nanotubes are expected to be very good thermal conductors along the tube, exhibiting a property known as "ballistic conduction", but good insulators lateral to the tube axis. Measurements show that an individual SWNT has a room-temperature thermal conductivity along its axis of about 3500 $W \cdot m^{-1} \cdot K^{-1}$; compare this to copper, a metal well known for its good thermal conductivity, which transmits 385 $W \cdot m^{-1} \cdot K^{-1}$. An individual SWNT has a room-temperature thermal conductivity across its axis (in the radial direction) of about 1.52 $W \cdot m^{-1} \cdot K^{-1}$, which is about as thermally conductive as soil. Macroscopic assemblies of nanotubes such as films or fibres have reached up to 1500 $W \cdot m^{-1} \cdot K^{-1}$ so far. The temperature stability of carbon nanotubes is estimated to be up to 2800 °C in vacuum and about 750 °C in air.

Crystallographic defects strongly affect the tube's thermal properties. Such defects lead to phonon scattering, which in turn increases the relaxation rate of the phonons. This reduces the mean free path and reduces the thermal conductivity of nanotube structures. Phonon transport simulations indicate that substitutional defects such as nitrogen or boron will primarily lead to scattering of high-frequency optical phonons. However, larger-scale defects such as Stone Wales defects cause phonon scattering over a wide range of frequencies, leading to a greater reduction in thermal conductivity.

Synthesis

Techniques have been developed to produce nanotubes in sizable quantities, including arc discharge, laser ablation, high-pressure carbon monoxide disproportionation, and chemical vapor deposition (CVD). Most of these processes take place in a vacuum or with process gases. The CVD growth method is popular, as it yields high purity and has a high degree of control over diameter, length and morphology. Using particulate catalysts, large quantities of nanotubes can be synthesized by these methods; advances in catalysis and continuous growth are making CNTs more commercially viable.

Vertically aligned carbon nanotube arrays are also grown by thermal chemical vapor deposition. A substrate (quartz, silicon, stainless steel, etc.) is coated with a catalytic metal (Fe, Co, Ni) layer. Typically that layer is iron, and is deposited via sputtering to a thickness of 1–5 nm. A 10–50 nm underlayer of alumina is often also put down on the substrate first. This imparts controllable wetting and good interfacial properties. When the substrate is heated to the growth temperature (~700 °C), the continuous iron film breaks up into small islands... each island then nucleates a carbon nanotube. The sputtered thickness controls the island size, and this in turn determines the nanotube diameter. Thinner iron layers drive down the diameter of the islands, and they drive down the diameter of the nanotubes grown. The amount of time that the metal island can sit at the growth temperature is limited, as they are mobile, and can merge into larger (but fewer) islands. Annealing at the growth temperature reduces the site density (number of CNT/mm^2) while increasing the catalyst diameter.

Chemical Modification

Carbon nanotubes can be functionalized to attain desired properties that can be used in a wide variety of applications. The two main methods of carbon nanotube functionalization are covalent and non-covalent modifications. Because of their hydrophobic nature, carbon nanotubes tend to agglomerate hindering their dispersion in solvents or viscous polymer melts. The resulting nanotube bundles or aggregates reduce the mechanical performance of the final composite. The surface of the carbon nanotubes can be modified to reduce the hydrophobicity and improve interfacial adhesion to a bulk polymer through chemical attachment.

Applications

Current

Current use and application of nanotubes has mostly been limited to the use of bulk nanotubes, which is a mass of rather unorganized fragments of nanotubes. Bulk nanotube materials may never achieve a tensile strength similar to that of individual tubes, but such composites may, nevertheless, yield strengths sufficient for many applications. Bulk carbon nanotubes have already been used as composite fibers in polymers to improve the mechanical, thermal and electrical properties of the bulk product.

- Easton-Bell Sports, Inc. have been in partnership with Zyvex Performance Materials, using CNT technology in a number of their bicycle components—including flat and riser handlebars, cranks, forks, seatposts, stems and aero bars.
- Zyvex Technologies has also built a 54' maritime vessel, the Piranha Unmanned Surface Vessel, as a technology demonstrator for what is possible using CNT technology. CNTs help improve the structural performance of the vessel, resulting in a lightweight 8,000 lb boat that can carry a payload of 15,000 lb over a range of 2,500 miles.
- Amroy Europe Oy manufactures Hybtonite carbon nanoepoxy resins where carbon nanotubes have been chemically activated to bond to epoxy, resulting in a composite material that is 20% to 30% stronger than other composite materials. It has been used for wind turbines, marine paints and a variety of sports gear such as skis, ice hockey sticks, baseball bats, hunting arrows, and surfboards.

Other current applications include:

- tips for atomic force microscope probes
- in tissue engineering, carbon nanotubes can act as scaffolding for bone growth

There is also ongoing research in using carbon nanotubes as a scaffold for diverse microfabrication techniques.

Potential

The strength and flexibility of carbon nanotubes makes them of potential use in controlling other nanoscale structures, which suggests they will have an important role in nanotechnology engineering. The highest tensile strength of an individual multi-walled carbon nanotube has been tested to be 63 GPa. Carbon nanotubes were found in Damascus steel from the 17th century, possibly helping to account for the legendary strength of the swords made of it. Recently, several studies have highlighted the prospect of using carbon nanotubes as building blocks to fabricate three-dimensional macroscopic (>1mm in all three dimensions) all-carbon devices. Lalwani et al. have reported a novel radical initiated thermal crosslinking method to fabricated macroscopic, free-standing, porous, all-carbon scaffolds using single- and multi-walled carbon nanotubes as building blocks. These scaffolds possess macro-, micro-, and nano- structured pores and the porosity can be tailored for specific applications. These 3D all-carbon scaffolds/architectures may be used for the fabrication of the next generation of energy storage, supercapacitors, field emission transistors, high-performance catalysis, photovoltaics, and biomedical devices and implants.

Large quantities of pure CNTs can be made into a fiber (a.k.a. filament) by wet spinning. The fiber is either directly spun from the synthesis pot or spun from pre-made dissolved CNTs. Individual fibers can be turned into a yarn. Apart from its strength and flexibility, the main advantage is making an electrically conducting yarn. The electronic properties of individual CNT fibers (i.e. bundle of individual CNT) is governed by the two-dimensional structure of CNTs.. At 300 K, CNT fibers have a conductivity one order of magnitude higher than the best electrical conductors.

CNT-based yarns are suitable for applications in energy and electrochemical water treatment when coated with an ion-exchange membrane. Also, CNT-based yarns could replace copper as a winding material. Pyrhönen et al. (2015) have built a motor using CNT winding.

Safety and Health

The National Institute for Occupational Safety and Health (NIOSH) is the leading United States federal agency conducting research and providing guidance on the occupational safety and health implications and applications of nanotechnology. Early scientific studies have indicated that some of these nanoscale particles may pose a greater health risk than the larger bulk form of these materials. In 2013, NIOSH published a Current Intelligence Bulletin detailing the potential hazards and recommended exposure limit for carbon nanotubes and fibers.

As of October 2016, single wall carbon nanotubes have been registered through the European Union's Registration, Evaluation, Authorization and Restriction of Chemicals (REACH) regulations, based on evaluation of the potentially hazardous properties of SWCNT. Based on this registration, SWCNT commercialization is allowed in the EU up to 10 metric tons. Currently, the type of SWCNT registered through REACH is limited to the specific type of single wall carbon nanotubes manufactured by OCSiAl, which submitted the application.

History

The true identity of the discoverers of carbon nanotubes is a subject of some controversy. A 2006 editorial written by Marc Monthioux and Vladimir Kuznetsov in the journal *Carbon* described the interesting and often-misstated origin of the carbon nanotube. A large percentage of academic and popular literature attributes the discovery of hollow, nanometer-size tubes composed of graphitic carbon to Sumio Iijima of NEC in 1991. He published a paper describing his discovery which initiated a flurry of excitement and could be credited by inspiring the many scientists now studying applications of carbon nanotubes. Though Iijima has been given much of the credit for discovering carbon nanotubes, it turns out that the timeline of carbon nanotubes goes back much further than 1991.

In 1952, L. V. Radushkevich and V. M. Lukyanovich published clear images of 50 nanometer diameter tubes made of carbon in the Soviet *Journal of Physical Chemistry*. This discovery was largely unnoticed, as the article was published in Russian, and Western scientists' access to Soviet press was limited during the Cold War. Monthioux and Kuznetsov mentioned in their *Carbon* editorial:

The fact is, Radushkevich and Lukyanovich [..] should be credited for the discovery that carbon filaments could be hollow and have a nanometer- size diameter, that is to say for the discovery of carbon nanotubes.

In 1976, Morinobu Endo of CNRS observed hollow tubes of rolled up graphite sheets synthesised by a chemical vapour-growth technique. The first specimens observed would later come to be known as single-walled carbon nanotubes (SWNTs). Endo, in his early review of vapor-phase-grown carbon fibers (VPCF), also reminded us that he had observed a hollow tube, linearly extended with parallel carbon layer faces near the fiber core. This appears to be the observation of multi-walled carbon nanotubes at the center of the fiber. The mass-produced MWCNTs today are strongly related to the VPGCF developed by Endo. In fact, they call it the "Endo-process", out of respect for his early work and patents.

In 1979, John Abrahamson presented evidence of carbon nanotubes at the 14th Biennial Conference of Carbon at Pennsylvania State University. The conference paper described carbon nanotubes as carbon fibers that were produced on carbon anodes during arc discharge. A characterization of these fibers was given as well as hypotheses for their growth in a nitrogen atmosphere at low pressures.

In 1981, a group of Soviet scientists published the results of chemical and structural characterization of carbon nanoparticles produced by a thermocatalytical disproportionation of carbon monoxide. Using TEM images and XRD patterns, the authors suggested that their "carbon multi-layer tubular crystals" were formed by rolling graphene layers into cylinders. They speculated that by rolling graphene layers into a cylinder, many different arrangements of graphene hexagonal nets are possible. They suggested two possibilities of such arrangements: circular arrangement (armchair nanotube) and a spiral, helical arrangement (chiral tube).

In 1987, Howard G. Tennent of Hyperion Catalysis was issued a U.S. patent for the production of "cylindrical discrete carbon fibrils" with a "constant diameter between about 3.5 and about 70 nanometers..., length 10^2 times the diameter, and an outer region of multiple essentially continuous layers of ordered carbon atoms and a distinct inner core...."

Iijima's discovery of multi-walled carbon nanotubes in the insoluble material of arc-burned graphite rods in 1991 and Mintmire, Dunlap, and White's independent prediction that if single-walled carbon nanotubes could be made, then they would exhibit remarkable conducting properties helped create the initial buzz that is now associated with carbon nanotubes. Nanotube research accelerated greatly following the independent discoveries by Bethune at IBM and Iijima at NEC of *single-walled* carbon nanotubes and methods to specifically produce them by adding transition-metal catalysts to the carbon in an arc discharge. The arc discharge technique was well-known to produce the famed Buckminster fullerene on a preparative scale, and these results appeared to extend the run of accidental discoveries relating to fullerenes. The discovery of nanotubes remains a contentious issue. Many believe that Iijima's report in 1991 is of particular importance because it brought carbon nanotubes into the awareness of the scientific community as a whole.

Optical Properties of Carbon Nanotubes

Within materials science, the optical properties of carbon nanotubes refer specifically to the absorption, photoluminescence (fluorescence), and Raman spectroscopy of carbon nanotubes. Spectroscopic methods offer the possibility of quick and non-destructive characterization of relatively large amounts of carbon nanotubes. There is a strong demand for such characterization from the industrial point of view: numerous parameters of the nanotube synthesis can be changed, intentionally or unintentionally, to alter the nanotube quality. As shown below, optical absorption, photoluminescence and Raman spectroscopies allow quick and reliable characterization of this "nanotube quality" in terms of non-tubular carbon content, structure (chirality) of the produced nanotubes, and structural defects. Those features determine nearly any other properties such as optical, mechanical, and electrical properties.

Carbon nanotubes are unique "one-dimensional systems" which can be envisioned as rolled single sheets of graphite (or more precisely graphene). This rolling can be done at different angles and curvatures resulting in different nanotube properties. The diameter typically varies in the range 0.4–40 nm (i.e. "only" ~100 times), but the length can vary ~10,000 times, reaching 55.5 cm. The nanotube aspect ratio, or the length-to-diameter ratio, can be as high as 132,000,000:1, which is unequalled by any other material. Consequently, all the properties of the carbon nanotubes relative to those of typical semiconductors are extremely anisotropic (directionally dependent) and tunable.

Whereas mechanical, electrical and electrochemical (supercapacitor) properties of the carbon nanotubes are well established and have immediate applications, the practical use of optical properties is yet unclear. The aforementioned tunability of properties is potentially useful in optics and photonics. In particular, light-emitting diodes (LEDs) and photo-detectors based on a single nanotube have been produced in the lab. Their unique feature is not the efficiency, which is yet relatively low, but the narrow selectivity in the wavelength of emission and detection of light and the possibility of its fine tuning through the nanotube structure. In addition, bolometer and optoelectronic memory devices have been realised on ensembles of single-walled carbon nanotubes.

Terminology

This section uses the following abbreviations:

- Carbon nanotube (CNT)
- Single wall carbon nanotube (SWCNT)
- Multiwall carbon nanotube (MWCNT)

However, C is often omitted in scientific literature, so NT, SWNT and MWNT are more commonly used. Also, "wall" is often exchanged with "walled".

Electronic Structure of Carbon Nanotube

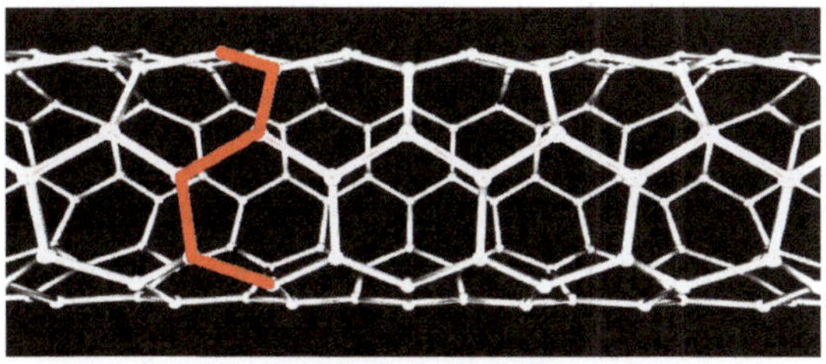

Armchair nanotube

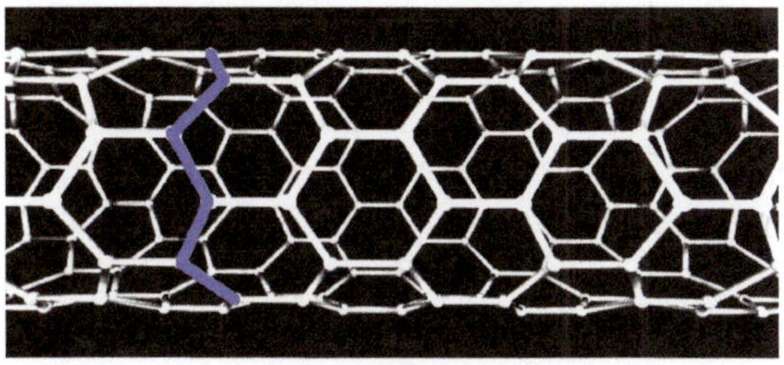

Zigzag nanotube

A single-wall carbon nanotube can be imagined as graphene sheet rolled at a certain "chiral" angle with respect to a plane perpendicular to the tube's long axis. Consequently, SWCNT can be defined by its diameter and chiral angle. The chiral angle can range from 0 to 30 degrees.

However, more conveniently, a pair of indices (n, m) is used instead. The indices refer to equally long unit vectors at 60° angles to each other across a single 6-member carbon ring. Taking the origin as carbon number 1, the a_1 unit vector may be considered the line drawn from carbon 1 to carbon 3, and the a_2 unit vector is then the line drawn from carbon 1 to carbon 5. To visualize a CNT with indices (n, m), draw n a_1 unit vectors across the graphene sheet, then draw m a_2 unit vectors at a 60° angle to the a_1 vectors, then add the vectors together. The line representing the sum of the vectors will define the circumference of the CNT along the plane perpendicular to its long axis, connecting one end to the other. In the diagram C_h is a (4, 2) vector: the sum of 4 unit vectors from the origin directly to the right, then 2 unit vectors at a 60° angle down and to the right.

Tubes having $n = m$ (chiral angle = 30°) are called "armchair" and those with $m = 0$ (chiral angle = 0°) "zigzag". Those indices uniquely determine whether CNT is a metal, semimetal or semiconductor, as well as its band gap: when $|m - n| = 3k$ (k is integer), the tube is metallic; but if $|m - n| = 3k \pm 1$, the tube is semiconducting. The nanotube diameter d is related to m and n as

$$\sum_j d = \frac{a}{\pi}\sqrt{(n^2 + nm + m^2)}.$$

In this equation, $a = 0.246$ nm is the magnitude of either unit vector a_1 or a_2.

The situation in multi-wall CNTs is complicated as their properties are determined by contribution of all individual shells; those shells have different structures, and, because of the synthesis, are usually more defective than SWCNTs. Therefore, optical properties of MWCNTs will not be considered here.

Van Hove Singularities

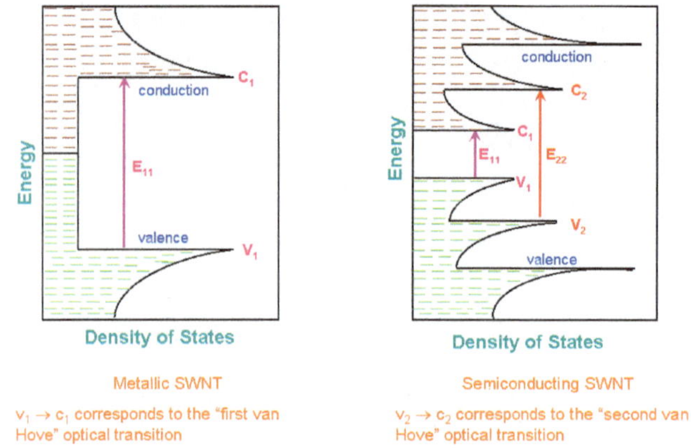

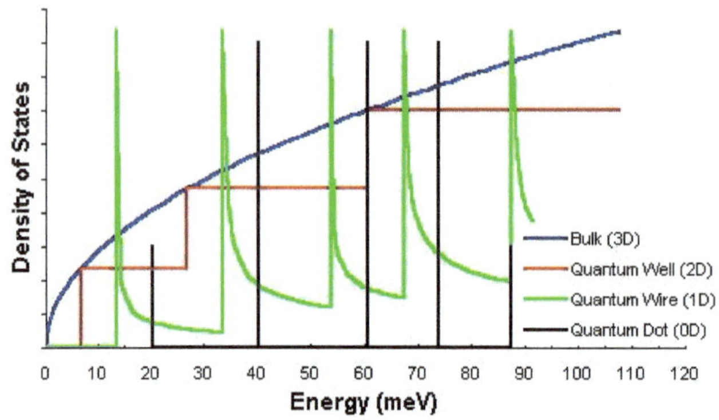

A bulk 3D material (blue) has continuous DOS, but a 1D wire (green) has Van Hove singularities

Optical properties of carbon nanotubes derive from electronic transitions within one-dimensional density of states (DOS). A typical feature of one-dimensional crystals is that their DOS is not a continuous function of energy, but it descends gradually and then increases in a discontinuous spike. In contrast, three-dimensional materials have continuous DOS. The sharp peaks found in one-dimensional materials are called Van Hove singularities.

Van Hove singularities result in the following remarkable optical properties of carbon nanotubes:

- Optical transitions occur between the $v_1 - c_1$, $v_2 - c_2$, etc., states of semiconducting or metallic nanotubes and are traditionally labeled as S_{11}, S_{22}, M_{11}, etc., or, if the "conductivity" of the tube is unknown or unimportant, as E_{11}, E_{22}, etc. Crossover transitions $c_1 - v_2$, $c_2 - v_1$, etc., are dipole-forbidden and thus are extremely weak, but they were possibly observed using cross-polarized optical geometry.

- The energies between the Van Hove singularities depend on the nanotube structure. Thus by varying this structure, one can tune the optoelectronic properties of carbon nanotube. Such fine tuning has been experimentally demonstrated using UV illumination of polymer-dispersed CNTs.

- Optical transitions are rather sharp (~10 meV) and strong. Consequently, it is relatively easy to selectively excite nanotubes having certain (*n*, *m*) indices, as well as to detect optical signals from individual nanotubes.

Kataura Plot

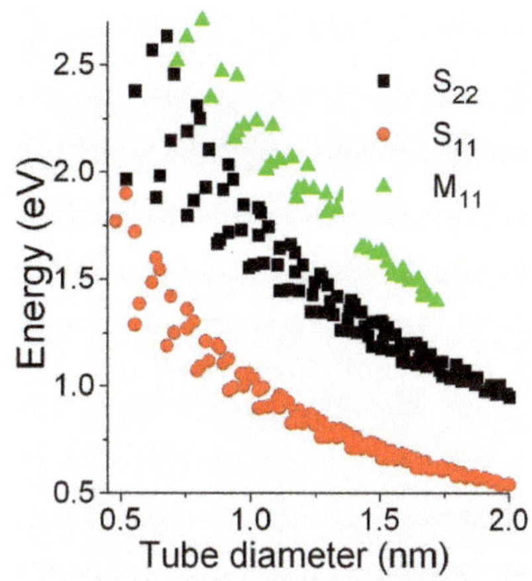

In this Kataura plot, the energy of an electronic transition decreases as the diameter of the nanotube increases

The band structure of carbon nanotubes having certain (n, m) indexes can be easily calculated. A theoretical graph based on this calculations was designed in 1999 by Hiromichi Kataura to rationalize experimental findings. A Kataura plot relates the nanotube diameter and its bandgap energies for all nanotubes in a diameter range. The oscillating shape of every branch of the Kataura plot reflects the intrinsic strong dependence of the SWCNT properties on the (n, m) index rather than on its diameter. For example, (10, 1) and (8, 3) tubes have almost the same diameter, but very different properties: the former is a metal, but the latter is a semiconductor.

Optical Absorption

Optical absorption in carbon nanotubes differs from absorption in conventional 3D materials by presence of sharp peaks (1D nanotubes) instead of an absorption threshold followed by an absorption increase (most 3D solids). Absorption in nanotubes originates from electronic transitions from the v_2 to c_2 (energy E_{22}) or v_1 to c_1 (E_{11}) levels, etc. The transitions are relatively sharp and can be used to identify nanotube types. Note that the sharpness deteriorates with increasing energy, and that many nanotubes have very similar E_{22} or E_{11} energies, and thus significant overlap occurs in absorption spectra. This overlap is avoided in photoluminescence mapping measurements, which instead of a combination of overlapped transitions identifies individual (E_{22}, E_{11}) pairs.

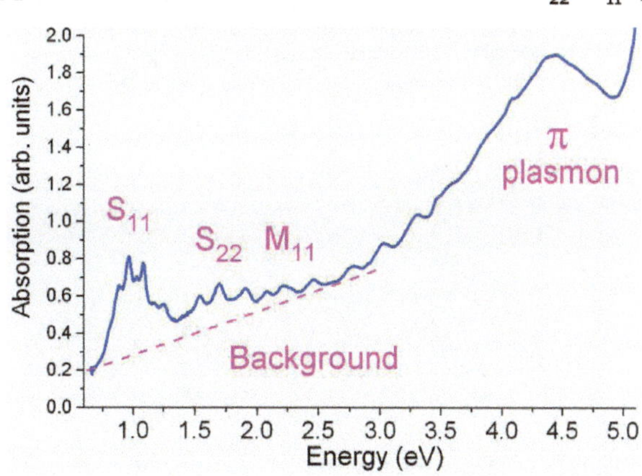

Optical absorption spectrum from dispersed single-wall carbon nanotubes

Interactions between nanotubes, such as bundling, broaden optical lines. While bundling strongly affects photoluminescence, it has much weaker effect on optical absorption and Raman scattering. Consequently, sample preparation for the latter two techniques is relatively simple.

Optical absorption is routinely used to quantify quality of the carbon nanotube powders.

The spectrum is analyzed in terms of intensities of nanotube-related peaks, background and pi-carbon peak; the latter two mostly originate from non-nanotube carbon in contaminated samples. However, it has been recently shown that by aggregating nearly single chirality semiconducting nanotubes into closely packed Van der Waals bundles the absorption background can be attributed to free carrier transition originating from intertube charge transfer.

Carbon Nanotubes as a Black Body

An ideal black body should have emissivity or absorbance of 1.0, which is difficult to attain in practice, especially in a wide spectral range. Vertically aligned "forests" of single-wall carbon nanotubes can have absorbances of 0.98–0.99 from the far-ultraviolet (200 nm) to far-infrared (200 μm) wavelengths.

These SWNT forests (buckypaper) were grown by the super-growth CVD method to about 10 μm height. Two factors could contribute to strong light absorption by these structures: (i) a distribution of CNT chiralities resulted in various bandgaps for individual CNTs. Thus a compound material was formed with broadband absorption. (ii) Light might be trapped in those forests due to multiple reflections.

Reflectance Measurements			
	UV-to-near IR	Near-to-mid IR	Mid-to-far IR
Wavelength, μm	0.2-2	2–20	25–200
Incident angle, °	8	5	10
Reflection	Hemispherical-directional	Hemispherical-directional	Specular
Reference	White reflectance standard	Gold mirror	Aluminum mirror
Average reflectance	0.0160	0.0097	0.0017
Standard deviation	0.0048	0.0041	0.0027

Luminescence

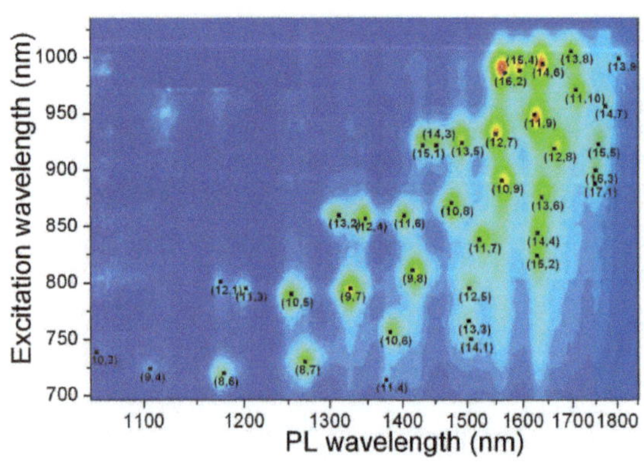

Photoluminescence map from single-wall carbon nanotubes. (n, m) indexes identify certain semiconducting nanotubes. Note that PL measurements do not detect nanotubes with $n = m$ or $m = 0$.

Photoluminescence (Fluorescence)

Semiconducting single-walled carbon nanotubes emit near-infrared light upon photoexcitation, described interchangeably as fluorescence or photoluminescence (PL). The excitation of PL usually occurs as follows: an electron in a nanotube absorbs excitation light via S_{22} transition, creating an electron-hole pair (exciton). Both electron and hole rapidly relax (via phonon-assisted processes) from c_2 to c_1 and from v_2 to v_1 states, respectively. Then they recombine through a $c_1 - v_1$ transition resulting in light emission.

No excitonic luminescence can be produced in metallic tubes. Their electrons can be excited, thus resulting in optical absorption, but the holes are immediately filled by other electrons out of the many available in the metal. Therefore, no excitons are produced.

Salient Properties

- Photoluminescence from SWCNT, as well as optical absorption and Raman scattering, is linearly polarized along the tube axis. This allows monitoring of the SWCNTs orientation without direct microscopic observation.
- PL is quick: relaxation typically occurs within 100 picoseconds.
- PL efficiency was first found to be low (~0.01%), but later studies measured much higher quantum yields. By improving the structural quality and isolation of nanotubes, emission efficiency increased. A quantum yield of 1% was reported in nanotubes sorted by diameter and length through gradient centrifugation, and it was further increased to 20% by optimizing the procedure of isolating individual nanotubes in solution.
- The spectral range of PL is rather wide. Emission wavelength can vary between 0.8 and 2.1 micrometers depending on the nanotube structure.
- Excitons are apparently delocalized over several nanotubes in single chirality bundles as the photoluminescence spectrum displays a splitting consistent with intertube exciton tunneling.
- Interaction between nanotubes or between a nanotube and another material may quench or increase PL. No PL is observed in multi-walled carbon nanotubes. PL from double-wall carbon nanotubes strongly depends on the preparation method: CVD grown DWCNTs show emission both from inner and outer shells. However, DWCNTs produced by encapsulating fullerenes into SWCNTs and annealing show PL only from the outer shells. Isolated SWCNTs lying on the substrate show extremely weak PL which has been detected in few studies only. Detachment of the tubes from the substrate drastically increases PL.
- Position of the (S_{22}, S_{11}) PL peaks depends slightly (within 2%) on the nanotube environment (air, dispersant, etc.). However, the shift depends on the (n, m) index, and thus the whole PL map not only shifts, but also warps upon changing the CNT medium.

Applications

- Photoluminescence is used for characterization purposes to measure the quantities of semiconducting nanotube species in a sample. Nanotubes are isolated (dispersed) using an appropriate chemical agent ("dispersant") to reduce the intertube quenching. Then PL is measured, scanning both the excitation and emission energies and thereby producing a PL map. The ovals in the map define (S_{22}, S_{11}) pairs, which unique identify (*n*, *m*) index of a tube. The data of Weisman and Bachilo are conventionally used for the identification.

- Nanotube fluorescence has been investigated for the purposes of imaging and sensing in biomedical applications.

Sensitization

Optical properties, including the PL efficiency, can be modified by encapsulating organic dyes (carotene, lycopene, etc.) inside the tubes. Efficient energy transfer occurs between the encapsulated dye and nanotube — light is efficiently absorbed by the dye and without significant loss is transferred to the SWCNT. Thus potentially, optical properties of a carbon nanotube can be controlled by encapsulating certain molecule inside it. Besides, encapsulation allows isolation and characterization of organic molecules which are unstable under ambient conditions. For example, Raman spectra are extremely difficult to measure from dyes because of their strong PL (efficiency close to 100%). However, encapsulation of dye molecules inside SWCNTs completely quenches dye PL, thus allowing measurement and analysis of their Raman spectra.

Cathodoluminescence

Cathodoluminescence (CL) — light emission excited by electron beam — is a process commonly observed in TV screens. An electron beam can be finely focused and scanned across the studied material. This technique is widely used to study defects in semiconductors and nanostructures with nanometer-scale spatial resolution. It would be beneficial to apply this technique to carbon nanotubes. However, no reliable CL, i.e. sharp peaks assignable to certain (*n*, *m*) indices, has been detected from carbon nanotubes yet.

Electroluminescence

If appropriate electrical contacts are attached to a nanotube, electron-hole pairs (excitons) can be generated by injecting electrons and holes from the contacts. Subsequent exciton recombination results in electroluminescence (EL). Electroluminescent devices have been produced from single nanotubes and their macroscopic assemblies. Recombination appears to proceed via triplet-triplet annihilation giving distinct peaks corresponding to E_{11} and E_{22} transitions.

Raman Scattering

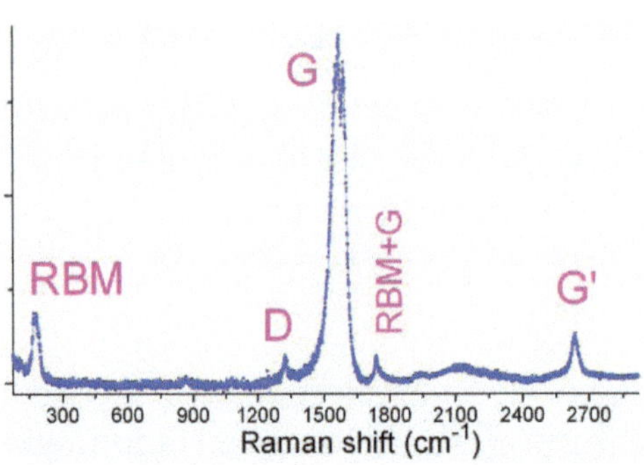

Raman spectrum of single-wall carbon nanotubes

Raman spectroscopy has good spatial resolution (~0.5 micrometers) and sensitivity (single nanotubes); it requires only minimal sample preparation and is rather informative. Consequently, Raman spectroscopy is probably the most popular technique of carbon nanotube characterization. Raman scattering in SWCNTs is resonant, i.e., only those tubes are probed which have one of the bandgaps equal to the exciting laser energy. Several scattering modes dominate the SWCNT spectrum, as discussed below.

Similar to photoluminescence mapping, the energy of the excitation light can be scanned in Raman measurements, thus producing Raman maps. Those maps also contain oval-shaped features uniquely identifying (*n*, *m*) indices. Contrary to PL, Raman mapping detects not only semiconducting but also metallic tubes, and it is less sensitive to nanotube bundling than PL. However, requirement of a tunable laser and a dedicated spectrometer is a strong technical impediment.

Radial Breathing Mode

Radial breathing mode (RBM) corresponds to radial expansion-contraction of the nanotube. Therefore, its frequency ν_{RBM} (in cm^{-1}) depends on the nanotube diameter d as, $\nu_{RBM} = A/d + B$ (where A and B are constants dependent on the environment in which the nanotube is present. For example, B=0 for individual nanotubes.) (in nanometers) and can be estimated as $\nu_{RBM} = 234/d + 10$ for SWNT or $\nu_{RBM} = 248/d$ for DWNT, which is very useful in deducing the CNT diameter from the RBM position. Typical RBM range is 100–350 cm^{-1}. If RBM intensity is particularly strong, its weak second overtone can be observed at double frequency.

Bundling Mode

The bundling mode is a special form of RBM supposedly originating from collective vibration in a bundle of SWCNTs.

G Mode

Another very important mode is the G mode (G from graphite). This mode corresponds to planar vibrations of carbon atoms and is present in most graphite-like materials. G band in SWCNT is shifted to lower frequencies relative to graphite (1580 cm^{-1}) and is split into several peaks. The splitting pattern and intensity depend on the tube structure and excitation energy; they can be used, though with much lower accuracy compared to RBM mode, to estimate the tube diameter and whether the tube is metallic or semiconducting.

D Mode

D mode is present in all graphite-like carbons and originates from structural defects. Therefore, the ratio of the *G*/*D* modes is conventionally used to quantify the structural quality of carbon nanotubes. High-quality nanotubes have this ratio significantly higher than 100. At a lower functionalisation of the nanotube, the *G*/*D* ratio remains almost unchanged. This ratio gives an idea of the functionalisation of a nanotube.

G' Mode

The name of this mode is misleading: it is given because in graphite, this mode is usually the second strongest after the G mode. However, it is actually the second overtone of the defect-induced D mode (and thus should logically be named D'). Its intensity is stronger than that of the D mode due to different selection rules. In particular, D mode is forbidden in the ideal nanotube and requires a structural defect, providing a phonon of certain angular momentum, to be induced. In contrast, G' mode involves a "self-annihilating" pair of phonons and thus does not require defects. The spectral position of G' mode depends on diameter, so it can be used roughly to estimate the SWCNT diameter. In particular, G' mode is a doublet in double-wall carbon nanotubes, but the doublet is often unresolved due to line broadening.

Other overtones, such as a combination of RBM+G mode at ~1750 cm^{-1}, are frequently seen in CNT Raman spectra. However, they are less important and are not considered here.

Anti-Stokes Scattering

All the above Raman modes can be observed both as Stokes and anti-Stokes scattering. As mentioned above, Raman scattering from CNTs is resonant in nature, i.e. only tubes whose band gap energy is similar to the laser energy are excited. The difference between those two energies, and thus the band gap of individual tubes, can be estimated from the intensity ratio of the Stokes/anti-Stokes lines. This estimate however relies on the temperature factor (Boltzmann factor), which is often miscalculated – a focused laser beam is used in the measurement, which can locally heat the nanotubes without changing the overall temperature of the studied sample.

Rayleigh Scattering

Carbon nanotubes have very large aspect ratio, i.e., their length is much larger than their diameter. Consequently, as expected from the classical electromagnetic theory, elastic light scattering (or Rayleigh scattering) by straight CNTs has anisotropic angular dependence, and from its spectrum, the band gaps of individual nanotubes can be deduced.

Another manifestation of Rayleigh scattering is the "antenna effect", an array of nanotubes standing on a substrate has specific angular and spectral distributions of reflected light, and both those distributions depend on the nanotube length.

Mechanical Properties of Carbon Nanotubes

The mechanical properties of carbon nanotubes reveal them as one of the strongest materials in nature. Carbon nanotubes (CNTs) are long hollow cylinders of graphene. Although graphene sheets have 2D symmetry, carbon nanotubes by geometry have different properties in axial and radial directions. It has been shown that CNTs are very strong in the axial direction. Young's modulus on the order of 270 - 950 GPa and tensile strength of 11 - 63 GPa were obtained.

Strength

Carbon nanotubes are the strongest and stiffest materials yet discovered in terms of tensile strength and elastic modulus respectively. This strength results from the covalent sp^2 bonds formed between the individual carbon atoms. In 2000, a multi-walled carbon nanotube was tested to have a tensile strength of 63 gigapascals (9,100,000 psi). (For illustration, this translates into the ability to endure tension of a weight equivalent to 6,422 kilograms-force (62,980 N; 14,160 lbf) on a cable with cross-section of 1 square millimetre (0.0016 sq in).) Further studies, such as one conducted in 2008, revealed that individual CNT shells have strengths of up to ≈100 gigapascals (15,000,000 psi), which is in agreement with quantum/atomistic models. Since carbon nanotubes have a low density for a solid of 1.3 to 1.4 g/cm^3, its specific strength of up to 48,000 $kN{\cdot}m{\cdot}kg^{-1}$ is the best of known materials, compared to high-carbon steel's 154 $kN{\cdot}m{\cdot}kg^{-1}$.

Under excessive tensile strain, the tubes will undergo plastic deformation, which means the deformation is permanent. This deformation begins at strains of approximately 5% and can increase the maximum strain the tubes undergo before fracture by releasing strain energy.

Although the strength of individual CNT shells is extremely high, weak shear interactions between adjacent shells and tubes lead to significant reduction in the effective strength of multi-walled carbon nanotubes and carbon nanotube bundles down to only a few GPa. This limitation has been recently addressed by applying high-energy electron irradiation, which crosslinks inner shells and tubes, and effectively increases the strength of these materials to ≈60 GPa for multi-walled carbon nanotubes and ≈17 GPa for double-walled carbon nanotube bundles.

CNTs are not nearly as strong under compression. Because of their hollow structure and high aspect ratio, they tend to undergo buckling when placed under compressive, torsional, or bending stress.

Comparison of mechanical properties			
Material	**Young's modulus (TPa)**	**Tensile strength (GPa)**	**Elongation at break (%)**
SWNT^E	≈1 (from 1 to 5)	13–53	16
Armchair SWNT^T	0.94	126.2	23.1
Zigzag SWNT^T	0.94	94.5	15.6–17.5
Chiral SWNT	0.92		
MWNT^E	0.2–0.8–0.95	11–63–150	
Stainless steel^E	0.186–0.214	0.38–1.55	15–50
Kevlar–29&149^E	0.06–0.18	3.6–3.8	≈2

^{E}Experimental observation; ^{T}Theoretical prediction

Radial Elasticity

On the other hand, there was evidence that in the radial direction they are rather soft. The first transmission electron microscope observation of radial elasticity suggested that even the van der Waals forces can deform two adjacent nanotubes. Later, nanoindentations with atomic force microscope were performed by several groups to quantitatively measure radial elasticity of multi-walled carbon nanotubes and tapping/contact mode atomic force microscopy was also performed on single-walled carbon nanotubes. Young's modulus of on the order of several GPa showed that CNTs are in fact very soft in the radial direction.

Radial direction elasticity of CNTs is important especially for carbon nanotube composites where the embedded tubes are subjected to large deformation in the transverse direction under the applied load on the composite structure.

One of the main problems in characterizing the radial elasticity of CNTs is the knowledge about the internal radius of the CNT; carbon nanotubes with identical outer diameter may have different internal diameter (or the number of walls). In 2008, a method using an atomic force microscope was introduced to determine the exact number of layers and hence the internal diameter of the CNT. In this way, mechanical characterization is more accurate.

Hardness

Standard single-walled carbon nanotubes can withstand a pressure up to 25 GPa without [plastic/permanent] deformation. They then undergo a transformation to superhard phase nanotubes. Maximum pressures measured using current experimental techniques are around 55 GPa. However, these new superhard phase nanotubes collapse at an even higher, albeit unknown, pressure.

The bulk modulus of superhard phase nanotubes is 462 to 546 GPa, even higher than that of diamond (420 GPa for single diamond crystal).

Wettability

The surface wettability of CNT is of importance for its applications in various settings. Although the intrinsic contact angle of graphite is around 90°, the contact angles of most as-synthesized CNT arrays are over 160°, exhibiting a superhydrophobic property. By applying a voltage as low as 1.3V, the extreme water repellant surface can be switched to a superhydrophilic one.

Kinetic Properties

Multi-walled nanotubes are multiple concentric nanotubes precisely nested within one another. These exhibit a striking telescoping property whereby an inner nanotube core may slide, almost without friction, within its outer nanotube shell, thus creating an atomically perfect linear or rotational bearing. This is one of the first true examples of molecular nanotechnology, the precise positioning of atoms to create useful machines. Already, this property has been utilized to create the world's smallest rotational motor. Future applications such as a gigahertz mechanical oscillator are also envisioned.

Defects

As with any material, the existence of a crystallographic defect affects the material properties. Defects can occur in the form of atomic vacancies. High levels of such defects can lower the tensile strength by up to 85%. An important example is the Stone Wales defect, which creates a pentagon and heptagon pair by rearrangement of the bonds. Because of the very small structure of CNTs, the tensile strength of the tube is dependent on its weakest segment in a similar manner to a chain, where the strength of the weakest link becomes the maximum strength of the chain.

Synthesis of Carbon Nanotubes

Techniques have been developed to produce carbon nanotubes in sizable quantities, including arc discharge, laser ablation, high-pressure carbon monoxide disproportionation, and chemical vapor deposition (CVD). Most of these processes take place in a vacuum or with process gases. CVD growth of CNTs can occur in vacuum or at atmospheric pressure. Large quantities of nanotubes can be synthesized by these methods; advances in catalysis and continuous growth are making CNTs more commercially viable.

powder of carbon nanotubes

Types

Arc Discharge

Nanotubes were observed in 1991 in the carbon soot of graphite electrodes during an arc discharge, by using a current of 100 amps, that was intended to produce fullerenes. However the first macroscopic production of carbon nanotubes was made in 1992 by two researchers at NEC's Fundamental Research Laboratory. The method used was the same as in 1991. During this process, the carbon contained in the negative electrode sublimates because of the high-discharge temperatures.

The yield for this method is up to 30% by weight and it produces both single- and multi-walled nanotubes with lengths of up to 50 micrometers with few structural defects. Arc-discharge technique uses higher temperatures (above 1,700 °C) for CNT synthesis which typically causes the expansion of CNTs with fewer structural defects in comparison with other methods.

Laser Ablation

In laser ablation, a pulsed laser vaporizes a graphite target in a high-temperature reactor while an inert gas is bled into the chamber. Nanotubes develop on the cooler surfaces of the reactor as the vaporized carbon condenses. A water-cooled surface may be included in the system to collect the nanotubes.

This process was developed by Dr. Richard Smalley and co-workers at Rice University, who at the time of the discovery of carbon nanotubes, were blasting metals with a laser to produce various metal molecules. When they heard of the existence of nanotubes they replaced the metals with graphite to create multi-walled carbon nanotubes. Later that year the team used a composite of graphite and metal catalyst particles (the best yield was from a cobalt and nickel mixture) to synthesize single-walled carbon nanotubes.

The laser ablation method yields around 70% and produces primarily single-walled carbon nanotubes with a controllable diameter determined by the reaction temperature. However, it is more expensive than either arc discharge or chemical vapor deposition.

The effective equation for few cycle optical pulse dynamics was obtained by virtue of the Boltzmann collision-less equation solution for conduction band electrons of semiconductor carbon nanotubes in the case when medium with carbon nanotubes has spatially-modulated refractive index.

Plasma Torch

Single-walled carbon nanotubes can also be synthesized by a thermal plasma method, first invented in 2000 at INRS (Institut national de la recherche scientifique) in Varennes, Canada, by Olivier Smiljanic. In this method, the aim is to reproduce the conditions prevailing in the arc discharge and laser ablation approaches, but a carbon-containing gas is used instead of graphite vapors to supply the necessary carbon. Doing so, the growth of SWNT is more efficient (decomposing the gas can be 10 times less energy-consuming than graphite vaporization). The process is also continuous and low cost. A gaseous mixture of argon, ethylene and ferrocene is introduced into a microwave plasma torch, where it is atomized by the atmospheric pressure plasma, which has the form of an intense 'flame'. The fumes created by the flame contain SWNT, metallic and carbon nanoparticles and amorphous carbon.

Another way to produce single-walled carbon nanotubes with a plasma torch is to use the induction thermal plasma method, implemented in 2005 by groups from the University of Sherbrooke and the National Research Council of Canada. The method is similar to arc discharge in that both use ionized gas to reach the high temperature necessary to vaporize carbon-containing substances and the metal catalysts necessary for the ensuing nanotube growth. The thermal plasma is induced by high-frequency oscillating currents in a coil, and is maintained in flowing inert gas. Typically, a feedstock of carbon black and metal catalyst particles is fed into the plasma, and then cooled down to form single-walled carbon nanotubes. Different single-wall carbon nanotube diameter distributions can be synthesized.

The induction thermal plasma method can produce up to 2 grams of nanotube material per minute, which is higher than the arc discharge or the laser ablation methods.

Chemical Vapor Deposition (CVD)

The catalytic vapor phase deposition of carbon was reported in 1952 and 1959, but it was not until 1993 that carbon nanotubes were formed by this process. In 2007, researchers at the University of Cincinnati (UC) developed a process to grow aligned carbon nanotube arrays of length 18 mm on a FirstNano ET3000 carbon nanotube growth system.

During CVD, a substrate is prepared with a layer of metal catalyst particles, most commonly nickel, cobalt, iron, or a combination. The metal nanoparticles can also be produced by other ways, including reduction of oxides or oxides solid solutions. The diameters of the nanotubes that are to be grown are related to the size of the metal particles. This can be controlled by patterned (or masked) deposition of the metal, annealing, or by plasma etching of a metal layer. The substrate is heated to approximately 700 °C. To initiate the growth of nanotubes, two gases are bled into the reactor: a process gas (such as ammonia, nitrogen or hydrogen) and a carbon-containing gas (such as acetylene, ethylene, ethanol or methane). Nanotubes grow at the sites of the metal catalyst; the carbon-containing gas is broken apart at the surface of the catalyst particle, and the carbon is transported to the edges of the particle, where it forms the nanotubes. This mechanism is still being studied. The catalyst particles can stay at the tips of the growing nanotube during growth, or remain at the nanotube base, depending on the adhesion between the catalyst particle and the substrate. Thermal catalytic decomposition of hydrocarbon has become an active area of research and can be a promising route for the bulk production of CNTs. Fluidised bed reactor is the most widely used reactor for CNT preparation. Scale-up of the reactor is the major challenge.

CVD is the most widely used method for the production of carbon nanotubes. For this purpose, the metal nanoparticles are mixed with a catalyst support such as MgO or Al_2O_3 to increase the surface area for higher yield of the catalytic reaction of the carbon feedstock with the metal particles. One issue in this synthesis route is the removal of the catalyst support via an acid treatment, which sometimes could destroy the original structure of the carbon nanotubes. However, alternative catalyst supports that are soluble in water have proven effective for nanotube growth.

nanotubes being grown by plasma enhanced chemical vapor deposition

If a plasma is generated by the application of a strong electric field during growth (plasma-enhanced chemical vapor deposition), then the nanotube growth will follow the direction of the electric field. By adjusting the geometry of the reactor it is possible to synthesize vertically aligned carbon nanotubes (i.e., perpendicular to the substrate), a morphology that has been of interest to researchers interested in electron emission from nanotubes. Without the plasma, the resulting nanotubes are often randomly oriented. Under certain reaction conditions, even in the absence of a plasma, closely spaced nanotubes will maintain a vertical growth direction resulting in a dense array of tubes resembling a carpet or forest.

Of the various means for nanotube synthesis, CVD shows the most promise for industrial-scale deposition, because of its price/unit ratio, and because CVD is capable of growing nanotubes directly on a desired substrate, whereas the nanotubes must be collected in the other growth techniques. The growth sites are controllable by careful deposition of the catalyst. In 2007, a team from Meijo University demonstrated a high-efficiency CVD technique for growing carbon nanotubes from camphor. Researchers at Rice University, until recently led by the late Richard Smalley, have concentrated upon finding methods to produce large, pure amounts of particular types of nanotubes. Their approach grows long fibers from many small seeds cut from a single nanotube; all of the resulting fibers were found to be of the same diameter as the original nanotube and are expected to be of the same type as the original nanotube.

Super-growth CVD

Super-growth CVD (water-assisted chemical vapor deposition) was developed by Kenji Hata, Sumio Iijima and co-workers at AIST, Japan. In this process, the activity and lifetime of the catalyst are enhanced by addition of water into the CVD reactor. Dense millimeter-tall vertically aligned nanotube arrays (VANTAs) or "forests", aligned normal to the substrate, were produced. The forests height could be expressed, as

$$H(t) = \beta\tau_o(1 - e^{-t/\tau_o}).$$

In this equation, β is the initial growth rate and τ_o is the characteristic catalyst lifetime.

Their specific surface exceeds 1,000 m^2/g (capped) or 2,200 m^2/g (uncapped), surpassing the value of 400–1,000 m^2/g for HiPco samples. The synthesis efficiency is about 100 times higher than for the laser ablation method. The time required to make SWNT forests of the height of 2.5 mm by this method was 10 minutes in 2004. Those SWNT forests can be easily separated from the catalyst, yielding clean SWNT material (purity >99.98%) without further purification. For comparison, the as-grown HiPco CNTs contain about 5–35% of metal impurities; it is therefore purified through dispersion and centrifugation that damages the nanotubes. Super-growth avoids this problem. Patterned highly organized single-walled nanotube structures were successfully fabricated using the super-growth technique.

The mass density of super-growth CNTs is about 0.037 g/cm^3. It is much lower than that of conventional CNT powders (~1.34 g/cm^3), probably because the latter contain metals and amorphous carbon.

The super-growth method is basically a variation of CVD. Therefore, it is possible to grow material containing SWNT, DWNTs and MWNTs, and to alter their ratios by tuning the growth conditions. Their ratios change by the thinness of the catalyst. Many MWNTs are included so that the diameter of the tube is wide.

The vertically aligned nanotube forests originate from a "zipping effect" when they are immersed in a solvent and dried. The zipping effect is caused by the surface tension of the solvent and the van der Waals forces between the carbon nanotubes. It aligns the nanotubes into a dense material, which can be formed in various shapes, such as sheets and bars, by applying weak compression during the process. Densification increases the Vickers hardness by about 70 times and density is 0.55 g/cm^3. The packed carbon nanotubes are more than 1 mm long and have a carbon purity of 99.9% or higher; they also retain the desirable alignment properties of the nanotubes forest.

Natural, Incidental, and Controlled Flame Environments

Fullerenes and carbon nanotubes are not necessarily products of high-tech laboratories; they are commonly formed in such mundane places as ordinary flames, produced by burning methane, ethylene, and benzene, and they have been found in soot from both indoor and outdoor air. However, these naturally occurring varieties can be highly irregular in size and quality because the environment in which they are produced is often highly uncontrolled. Thus, although they can be used in some applications, they can lack in the high degree of uniformity necessary to satisfy the many needs of both research and industry. Recent efforts have focused on producing more uniform carbon nanotubes in controlled flame environments. Such methods have promise for large-scale, low-cost nanotube synthesis based on theoretical models, though they must compete with rapidly developing large scale CVD production.

Purification

Removal of Catalysts

Nanoscale metal catalysts are important ingredients for fixed- and fluidized-bed CVD synthesis of CNTs. They allow increasing the growth efficiency of CNTs and may give control over their structure and chirality. During synthesis, catalysts can convert carbon precursors into tubular carbon structures but can also form encapsulating carbon overcoats. Together with metal oxide supports they may therefore attach to or become incorporated into the CNT product. The presence of metal impurities can be problematic for many applications. Especially catalyst metals like nickel, cobalt or yttrium may be of toxicological concern. While unencapsulated catalyst metals may be readily removable by acid washing, encapsulated ones require oxidative treatment for opening their carbon shell. The effective removal of catalysts, especially of encapsulated ones, while preserving the CNT structure is a challenge and has been addressed in many studies. A new approach to break carbonaceous catalyst encapsulations is based on rapid thermal annealing.

Application-related Issues

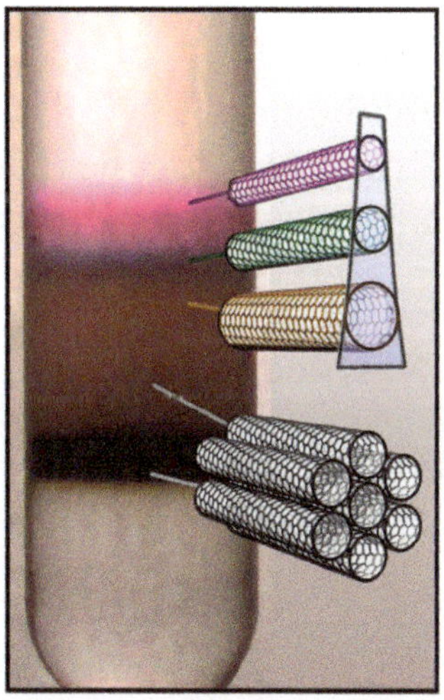

Centrifuge tube with a solution of carbon nanotubes, which were sorted by diameter using density-gradient ultracentrifugation.

Many electronic applications of carbon nanotubes crucially rely on techniques of selectively producing either semiconducting or metallic CNTs, preferably of a certain chirality. Several methods of separating semiconducting and metallic CNTs are known, but most of them are not yet suitable for large-scale technological processes. The most efficient method relies on density-gradient ultracentrifugation, which separates surfactant-wrapped nanotubes by the minute difference in their density. This density difference often translates into difference in the nanotube diameter and (semi)conducting properties. Another method of separation uses a sequence of freezing, thawing, and compression of SWNTs embedded in agarose gel. This process results in a solution containing 70% metallic SWNTs and leaves a gel containing 95% semiconducting SWNTs. The diluted solutions separated by this method show various colors. The separated carbon nanotubes using this method have been applied to electrodes, e.g. electric double-layer capacitor. Moreover, SWNTs can be separated by the column chromatography method. Yield is 95% in semiconductor type SWNT and 90% in metallic type SWNT.

In addition to separation of semiconducting and metallic SWNTs, it is possible to sort SWNTs by length, diameter, and chirality. The highest resolution length sorting, with length variation of <10%, has thus far been achieved by size exclusion chromatography (SEC) of DNA-dispersed carbon nanotubes (DNA-SWNT). SWNT diameter separation has been achieved by density-gradient ultracentrifugation (DGU) using surfactant-dispersed SWNTs and by ion-exchange chromatography (IEC) for DNA-SWNT. Purification of individual chiralities has also been demonstrated with IEC of DNA-SWNT: specific short DNA oligomers can be used to isolate individual SWNT chiralities. Thus far, 12 chiralities have been isolated at purities ranging from 70% for (8,3) and (9,5) SWNTs to 90% for (6,5), (7,5) and (10,5)SWNTs. There have been successful efforts to integrate these purified nanotubes into devices, e. g. FETs.

An alternative to separation is development of a selective growth of semiconducting or metallic CNTs. Recently, a new CVD recipe that involves a combination of ethanol and methanol gases and quartz substrates resulting in horizontally aligned arrays of 95–98% semiconducting nanotubes was announced.

Nanotubes are usually grown on nanoparticles of magnetic metal (Fe, Co), which facilitates production of electronic (spintronic) devices. In particular, control of current through a field-effect transistor by magnetic field has been demonstrated in such a single-tube nanostructure.

Potential Applications of Carbon Nanotubes

Carbon nanotubes (CNTs) are cylinders of one or more layers of graphene (lattice). Diameters of single-walled carbon nanotubes (SWNTs) and multi-walled carbon nanotubes (MWNTs) are typically 0.8 to 2 nm and 5 to 20 nm, respectively, although MWNT diameters can exceed 100 nm. CNT lengths range from less than 100 nm to 0.5 m.

Individual CNT walls can be metallic or semiconducting depending on the orientation of the lattice with respect to the tube axis, which is called chirality. MWNT's cross-sectional area offers an elastic modulus approaching 1 TPa and a tensile strength of 100 GPa, over 10-fold higher than any industrial fiber. MWNTs are typically metallic and can carry currents of up to 10^9 A cm^{-2}. SWNTs can display thermal conductivity of 3500 W m^{-1} K^{-1}, exceeding that of diamond.

As of 2013, carbon nanotube production exceeded several thousand tons per year, used for applications in energy storage, automotive parts, boat hulls, sporting goods, water filters, thin-film electronics, coatings, actuators and electromagnetic shields. CNT-related publications more than tripled in the prior decade, while rates of patent issuance also increased. Most output was of unorganized architecture. Organized CNT architectures such as "forests", yarns and regular sheets were produced in much smaller volumes. CNTs have even been proposed as the tether for a purported space elevator.

Recently, several studies have highlighted the prospect of using carbon nanotubes as building blocks to fabricate three-dimensional macroscopic (>1mm in all three dimensions) all-carbon devices. Lalwani et al. have reported a novel radical initiated thermal crosslinking method to fabricated macroscopic, free-standing, porous, all-carbon scaffolds using single- and multi-walled carbon nanotubes as building blocks. These scaffolds possess macro-, micro-, and nano- structured pores and the porosity can be tailored for specific applications. These 3D all-carbon scaffolds/architectures may be used for the fabrication of the next generation of energy storage, supercapacitors, field emission transistors, high-performance catalysis, photovoltaics, and biomedical devices and implants.

Biological and Biomedical Research

Researchers from Rice University and State University of New York – Stony Brook have shown that the addition of low weight % of carbon nanotubes can lead to significant improvements in the mechanical properties of biodegradable polymeric nanocomposites for applications in tissue engineering including bone, cartilage, muscle and nerve tissue. Dispersion of low weight % of graphene (~0.02 wt.%) results in significant increases in compressive and flexural mechanical properties of polymeric nanocomposites. Researchers at Rice University, Stony Brook University, Radboud University Nijmegen Medical Centre and University of California, Riverside have shown that carbon nanotubes and their polymer nanocomposites are suitable scaffold materials for bone tissue engineering and bone formation.

CNTs exhibit dimensional and chemical compatibility with biomolecules, such as DNA and proteins. CNTs enable fluorescent and photoacoustic imaging, as well as localized heating using near-infrared radiation.

SWNT biosensors exhibit large changes in electrical impedance and optical properties, which is typically modulated by adsorption of a target on the CNT surface. Low detection limits and high selectivity require engineering the CNT surface and field effects, capacitance, Raman spectral shifts and photoluminescence for sensor design. Products under development include printed test strips for estrogen and progesterone detection, microarrays for DNA and protein detection and sensors for NO_2 and cardiac troponin. Similar CNT sensors support food industry, military and environmental applications.

CNTs can be internalized by cells, first by binding their tips to cell membrane receptors. This enables transfection of molecular cargo attached to the CNT walls or encapsulated by CNTs. For example, the cancer drug doxorubicin was loaded at up to 60 wt % on CNTs compared with a maximum of 8 to 10 wt % on liposomes. Cargo release can be triggered by near-infrared radiation. However, limiting the retention of CNTs within the body is critical to prevent undesirable accumulation.

CNT toxicity remains a concern, although CNT biocompatibility may be engineerable. The degree of lung inflammation caused by injection of well-dispersed SWNTs was insignificant compared with asbestos and with particulate matter in air. Medical acceptance of CNTs requires understanding of immune response and appropriate exposure standards for inhalation, injection, ingestion and skin contact. CNT forests immobilized in a polymer did not show elevated inflammatory response in rats relative to controls. CNTs are under consideration as low-impedance neural interface electrodes and for coating of catheters to reduce thrombosis.

CNT enabled x-ray sources for medical imaging are also in development. Relying on the unique properties of the CNTs, researchers have developed field emission cathodes that allow precise x-ray control and close placement of multiple sources. CNT enabled x-ray sources have been demonstrated for pre-clinical, small animal imaging applications, and are currently in clinical trials.

In November 2012 researchers at the American National Institute of Standards and Technology (NIST) proved that single-wall carbon nanotubes may help protect DNA molecules from damage by oxidation.

A highly effective method of delivering carbon nanotubes into cells is Cell squeezing, a high-throughput vector-free microfluidic platform for intracellular delivery developed at the Massachusetts Institute of Technology in the labs of Robert S. Langer.

Carbon nanotubes have furthermore been grown inside microfluidic channels for chemical analysis, based on electrochromatography. Here, the high surface-area-to-volume ratio and high hydrophobicity of CNTs are used in order to greatly decrease the analysis time of small neutral molecules that typically require large bulky equipment for analysis.

Composite Materials

Because of the carbon nanotube's superior mechanical properties, many structures have been proposed ranging from everyday items like clothes and sports gear to combat jackets and space elevators. However, the space elevator will require further efforts in refining carbon nanotube technology, as the practical tensile strength of carbon nanotubes must be greatly improved.

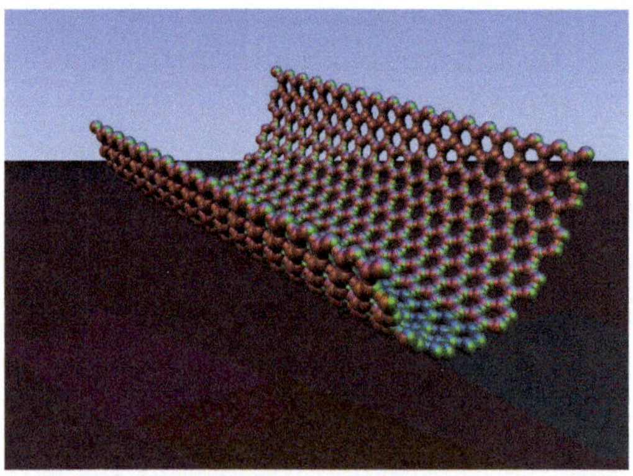

Carbon nanotubes being spun to form a yarn, CSIRO

For perspective, outstanding breakthroughs have already been made. Pioneering work led by Ray H. Baughman at the NanoTech Institute has shown that single and multi-walled nanotubes can produce materials with toughness unmatched in the man-made and natural worlds.

Carbon nanotubes are also a promising material as building blocks in hierarchical composite materials given their exceptional mechanical properties (~1 TPa in modulus, and ~100 GPa in strength). Initial attempts to incorporate CNTs into hierarchical structures (such as yarns, fibres or films) has led to mechanical properties that were significantly lower than these potential limits. The hierarchical integration of multi-walled carbon nanotubes and metal/metal oxides within a single nanostructure can leverage the potentiality of carbon nanotubes composite for water splitting and electrocatalysis. Windle *et al.* have used an *in situ* chemical vapor deposition (CVD) spinning method to produce continuous CNT yarns from CVD-grown CNT aerogels. CNT yarns can also be manufactured by drawing out CNT bundles from a CNT forest and subsequently twisting to form the fibre (draw-twist method). The Windle group have fabricated CNT yarns with strengths as high as ~9 GPa at small gage lengths of ~1 mm, however, strengths of only about ~1 GPa were reported at the longer gage length of 20 mm. The reason why fibre strengths have been low compared to the strength of individual CNTs is due to a failure to effectively transfer load to the constituent (discontinuous) CNTs within the fibre. One potential route for alleviating this problem is via irradiation (or deposition) induced covalent inter-bundle and inter-CNT cross-linking to effectively 'join up' the CNTs, with higher dosage levels leading to the possibility of amorphous carbon/carbon nanotube composite fibres. Espinosa *et al.* developed high performance DWNT-polymer composite yarns by twisting and stretching ribbons of randomly oriented bundles of DWNTs thinly coated with polymeric organic compounds. These DWNT-polymer yarns exhibited an unusually high energy to failure of ~100 $J{\cdot}g^{-1}$ (comparable to one of the toughest natural materials – spider silk), and strength as high as ~1.4 GPa. Effort is ongoing to produce CNT composites that incorporate tougher matrix materials, such as Kevlar, to further improve on the mechanical properties toward those of individual CNTs.

Because of the high mechanical strength of carbon nanotubes, research is being made into weaving them into clothes to create stab-proof and bulletproof clothing. The nanotubes would effectively stop the bullet from penetrating the body, although the bullet's kinetic energy would likely cause broken bones and internal bleeding.

Mixtures

MWNTs were first used as electrically conductive fillers in metals, at concentrations as high as 83.78 percent by weight (wt%). MWNT-polymer composites reach conductivities as high as 10,000 $S\ m^{-1}$ at 10 wt % loading. In the automotive industry, CNT plastics are used in electrostatic-assisted painting of mirror housings, as well as fuel lines and filters that dissipate electrostatic charge. Other products include electromagnetic interference (EMI)–shielding packages and silicon wafer carriers.

For load-bearing applications, CNT powders are mixed with polymers or precursor resins to increase stiffness, strength and toughness. These enhancements depend on CNT diameter, aspect ratio, alignment, dispersion and interfacial interaction. Premixed resins and master batches employ CNT loadings from 0.1 to 20 wt%. Nanoscale stick-slip among CNTs and CNT-polymer contacts can increase material damping, enhancing sporting goods, including tennis racquets, baseball bats and bicycle frames.

CNT resins enhance fiber composites, including wind turbine blades and hulls for maritime security boats that are made by enhancing carbon fiber composites with CNT-enhanced resin. CNTs are deployed as additives in the organic precursors of stronger 1-μm diameter carbon fibers. CNTs influence the arrangement of carbon in pyrolyzed fiber.

Toward the challenge of organizing CNTs at larger scales, hierarchical fiber composites are created by growing aligned forests onto glass, silicon carbide (SiC), alumina and carbon fibers, creating so-called "fuzzy" fibers. Fuzzy epoxy CNT-SiC and CNT-alumina fabric showed 69% improved crack-opening (mode I) and/or in-plane shear interlaminar (mode II) toughness. Applications under investigation include lightning-strike protection, deicing, and structural health monitoring for aircraft.

MWNTs can be used as a flame-retardant additive to plastics due to changes in rheology by nanotube loading. Such additives can replace halogenated flame retardants, which face environmental restrictions.

CNT/Concrete blends offer increased tensile strength and reduced crack propagation.

Buckypaper (nanotube aggregate) can significantly improve fire resistance due to efficient heat reflection.

Textiles

The previous studies on the use of CNTs for textile functionalization were focused on fiber spinning for improving physical and mechanical properties. Recently a great deal of attention has been focused on coating CNTs on textile fabrics. Various methods have been employed for modifying fabrics using CNTs. Shim et al. produced intelligent e-textiles for Human Biomonitoring using a polyelectrolyte-based coating with CNTs. Additionally, Panhuis et al. dyed textile material by immersion in either a poly (2-methoxy aniline-5-sulfonic acid) PMAS polymer solution or PMAS-SWNT dispersion with enhanced conductivity and capacitance with a durable behavior. In another study, Hu and coworkers coated single-walled carbon nanotubes with a simple "dipping and drying" process for wearable electronics and energy storage applications. In the recent study, Li and coworkers using elastomeric separator and almost achieved a fully stretchable supercapacitor based on buckled single-walled carbon nanotube macrofilms. The electrospun polyurethane was used and provided sound mechanical stretchability and the whole cell achieve excellent charge-discharge cycling stability. CNTs have an aligned nanotube structure and a negative surface charge. Therefore, they have similar structures to direct dyes, so the exhaustion method is applied for coating and absorbing CNTs on the fiber surface for preparing multifunctional fabric including antibacterial, electric conductive, flame retardant and electromagnetic absorbance properties.

Later, CNT yarns and laminated sheets made by direct chemical vapor deposition (CVD) or forest spinning or drawing methods may compete with carbon fiber for high-end uses, especially in weight-sensitive applications requiring combined electrical and mechanical functionality. Research yarns made from few-walled CNTs have reached a stiffness of 357 GPa and a strength of 8.8 GPa for a gauge length comparable to the millimeter-long CNTs within the yarn. Centimeter-scale gauge lengths offer only 2-GPa gravimetric strengths, matching that of Kevlar.

Because the probability of a critical flaw increases with volume, yarns may never achieve the strength of individual CNTs. However, CNT's high surface area may provide interfacial coupling that mitigates these deficiencies. CNT yarns can be knotted without loss of strength. Coating forest-drawn CNT sheets with functional powder before inserting twist yields weavable, braidable and sewable yarns containing up to 95 wt % powder. Uses include superconducting wires, battery and fuel cell electrodes and self-cleaning textiles.

As yet impractical fibers of aligned SWNTs can be made by coagulation-based spinning of CNT suspensions. Cheaper SWNTs or spun MWNTs are necessary for commercialization. Carbon nanotubes can be dissolved in superacids such as fluorosulfuric acid and drawn into fibers in dry jet-wet spinning.

DWNT-polymer composite yarns have been made by twisting and stretching ribbons of randomly oriented bundles of DWNTs thinly coated with polymeric organic compounds.

Body armor—combat jackets Cambridge University developed the fibres and licensed a company to make them. In comparison, the bullet-resistant fiber Kevlar fails at 27–33 J/g.

Synthetic muscles offer high contraction/extension ratio given an electric current.

SWNT are in use as an experimental material for removable, structural bridge panels.

In 2015 researchers incorporated CNTs and graphene into spider silk, increasing its strength and toughness to a new record. They sprayed 15 Pholcidae spiders with water containing the nanotubes or flakes.The resulting silk had a fracture strength up to 5.4 GPa, a Young's modulus up to 47.8 GPa and a toughness modulus up to 2.1 GPa, surpassing both synthetic polymeric high performance fibres (e.g. Kevlar49) and knotted fibers.

Carbon Nanotube Springs

"Forests" of stretched, aligned MWNT springs can achieve an energy density 10 times greater than that of steel springs, offering cycling durability, temperature insensitivity, no spontaneous discharge and arbitrary discharge rate. SWNT forests are expected to be able to store far more than MWNTs.

Alloys

Adding small amounts of CNTs to metals increases tensile strength and modulus with potential in aerospace and automotive structures. Commercial aluminum-MWNT composites have strengths comparable to stainless steel (0.7 to 1 GPa) at one-third the density (2.6 g cm^{-3}), comparable to more expensive aluminium-lithium alloys.

Coatings and Films

CNTs can serve as a multifunctional coating material. For example, paint/MWNT mixtures can reduce biofouling of ship hulls by discouraging attachment of algae and barnacles. They are a possible alternative to environmentally hazardous biocide-containing paints. Mixing CNTs into anticorrosion coatings for metals can enhance coating stiffness and strength and provide a path for cathodic protection.

CNTs provide a less expensive alternative to ITO for a range of consumer devices. Besides cost, CNT's flexible, transparent conductors offer an advantage over brittle ITO coatings for flexible displays. CNT conductors can be deposited from solution and patterned by methods such as screen printing. SWNT films offer 90% transparency and a sheet resistivity of 100 ohm per square. Such films are under development for thin-film heaters, such as for defrosting windows or sidewalks.

Carbon nanotubes forests and foams can also be coated with a variety of different materials to change their functionality and performance. Examples include silicon coated CNTs to create flexible energy-dense batteries, graphene coatings to create highly elastic aerogels and silicon carbide coatings to create a strong structural material for robust high-aspect-ratio 3D-micro architectures.

There is a wide range of methods how CNTs can be formed into coatings and films.

Optical Power Detectors

A spray-on mixture of carbon nanotubes and ceramic demonstrates unprecedented ability to resist damage while absorbing laser light. Such coatings that absorb as the energy of high-powered lasers without breaking down are essential for optical power detectors that measure the output of such lasers. These are used, for example, in military equipment for defusing unexploded mines. The composite consists of multiwall carbon nanotubes and a ceramic made of silicon, carbon and nitrogen. Including boron boosts the breakdown temperature. The nanotubes and graphene-like carbon transmit heat well, while the oxidation-resistant ceramic boosts damage resistance. Creating the coating involves dispersing the nanotubes in toluene, to which a clear liquid polymer containing boron was added. The mixture was heated to 1,100 °C (2,010 °F). The result is crushed into a fine powder, dispersed again in toluene and sprayed in a thin coat on a copper surface. The coating absorbed 97.5 percent of the light from a far-infrared laser and tolerated 15 kilowatts per square centimeter for 10 seconds. Damage tolerance is about 50 percent higher than for similar coatings, e.g., nanotubes alone and carbon paint.

Radar Absorption

Radars work in the microwave frequency range, which can be absorbed by MWNTs. Applying the MWNTs to the aircraft would cause the radar to be absorbed and therefore seem to have a smaller radar cross-section. One such application could be to paint the nanotubes onto the plane. Recently there has been some work done at the University of Michigan regarding carbon nanotubes usefulness as stealth technology on aircraft. It has been found that in addition to the radar absorbing properties, the nanotubes neither reflect nor scatter visible light, making it essentially invisible at night, much like painting current stealth aircraft black except much more effective. Current limitations in manufacturing, however, mean that current production of nanotube-coated aircraft is not possible. One theory to overcome these current limitations is to cover small particles with the nanotubes and suspend the nanotube-covered particles in a medium such as paint, which can then be applied to a surface, like a stealth aircraft.

Microelectronics

Nanotube-based transistors, also known as carbon nanotube field-effect transistors (CNTFETs), have been made that operate at room temperature and that are capable of digital switching using a single electron. However, one major obstacle to realization of nanotubes has been the lack of technology for mass production. In 2001 IBM researchers demonstrated how metallic nanotubes can be destroyed, leaving semiconducting ones behind for use as transistors. Their process is called "constructive destruction," which includes the automatic destruction of defective nanotubes on the wafer. This process, however, only gives control over the electrical properties on a statistical scale.

SWNTs are attractive for transistors because of their low electron scattering and their bandgap. SWNTs are compatible with field-effect transistor (FET) architectures and high-k dielectrics. Despite progress following the CNT transistor's appearance in 1998, including a tunneling FET with a subthreshold swing of <60 mV per decade (2004), a radio (2007) and an FET with sub-10-nm channel length and a normalized current density of 2.41 mA μm^{-1} at 0.5 V, greater than those obtained for silicon devices.

However, control of diameter, chirality, density and placement remains insufficient for commercial production. Less demanding devices of tens to thousands of SWNTs are more immediately practical. The use of CNT arrays/transistor increases output current and compensates for defects and chirality differences, improving device uniformity and reproducibility. For example, transistors using horizontally aligned CNT arrays achieved mobilities of 80 $cm^2\ V^{-1}\ s^{-1}$, subthreshold slopes of 140 mV per decade and on/off ratios as high as 10^5. CNT film deposition methods enable conventional semiconductor fabrication of more than 10,000 CNT devices per chip.

Printed CNT thin-film transistors (TFTs) are attractive for driving organic light-emitting diode displays, showing higher mobility than amorphous silicon (~1 $cm^2\ V^{-1}\ s^{-1}$) and can be deposited by low-temperature, nonvacuum methods. Flexible CNT TFTs with a mobility of 35 $cm^2\ V^{-1}\ s^{-1}$ and an on/off ratio of 6×10^6 were demonstrated. A vertical CNT FET showed sufficient current output to drive OLEDs at low voltage, enabling red-green-blue emission through a transparent CNT network. CNTs are under consideration for radio-frequency identification tags. Selective retention of semiconducting SWNTs during spin-coating and reduced sensitivity to adsorbates were demonstrated.

The International Technology Roadmap for Semiconductors suggests that CNTs could replace Cu in microelectronic interconnects, owing to their low scattering, high current-carrying capacity, and resistance to electromigration. For this, vias comprising tightly packed ($>10^{13}\ cm^{-2}$) metallic CNTs with low defect density and low contact resistance are needed. Recently, complementary metal oxide semiconductor (CMOS)–compatible 150-nm-diameter interconnects with a single CNT–contact hole resistance of 2.8 kOhm were demonstrated on full 200-mm-diameter wafers. Also, as a replacement for solder bumps, CNTs can function both as electrical leads and heat dissipaters for use in high-power amplifiers.

Last, a concept for a nonvolatile memory based on individual CNT crossbar electromechanical switches has been adapted for commercialization by patterning tangled CNT thin films as the functional elements. This required development of ultrapure CNT suspensions that can be spin-coated and processed in industrial clean room environments and are therefore compatible with CMOS processing standards.

Transistors

Carbon nanotube field-effect transistors (CNTFETs) can operate at room temperature and are capable of digital switching using a single electron. In 2013, a CNT logic circuit was demonstrated that could perform useful work. Major obstacles to nanotube-based microelectronics include the absence of technology for mass production, circuit density, positioning of individual electrical contacts, sample purity, control over length, chirality and desired alignment, thermal budget and contact resistance.

One of the main challenges was regulating conductivity. Depending on subtle surface features, a nanotube may act as a conductor or as a semiconductor.

Another way to make carbon nanotube transistors has been to use random networks of them. By doing so one averages all of their electrical differences and one can produce devices in large scale at the wafer level. This approach was first patented by Nanomix Inc. (date of original application June 2002). It was first published in the academic literature by the United States Naval Research Laboratory in 2003 through independent research work. This approach also enabled Nanomix to make the first transistor on a flexible and transparent substrate.

Since the electron mean free path in SWCNTs can exceed 1 micrometer, long channel CNTFETs exhibit near-ballistic transport characteristics, resulting in high speeds. CNT devices are projected to operate in the frequency range of hundreds of gigahertz.

Nanotubes can be grown on nanoparticles of magnetic metal (Fe, Co) that facilitates production of electronic (spintronic) devices. In particular control of current through a field-effect transistor by magnetic field has been demonstrated in such a single-tube nanostructure.

History

In 2001 IBM researchers demonstrated how metallic nanotubes can be destroyed, leaving semiconducting nanotubes for use as components. Using "constructive destruction", they destroyed defective nanotubes on the wafer. This process, however, only gives control over the electrical properties on a statistical scale. In 2003 room-temperature ballistic transistors with ohmic metal contacts and high-k gate dielectric were reported, showing 20–30x more current than state-of-the-art siliconMOSFETs. Palladium is a high-work function metal that was shown to exhibit Schottky barrier-free contacts to semiconducting nanotubes with diameters >1.7 nm.

The potential of carbon nanotubes was demonstrated in 2003 when room-temperature ballistic transistors with ohmic metal contacts and high-k gate dielectric were reported, showing 20–30x higher ON current than state-of-the-art Si MOSFETs. This presented an important advance in the field as CNT was shown to potentially outperform Si. At the time, a major challenge was ohmic metal contact formation. In this regard, palladium, which is a high-work function metal was shown to exhibit Schottky barrier-free contacts to semiconducting nanotubes with diameters >1.7 nm.

The first nanotube integrated memory circuit was made in 2004. One of the main challenges has been regulating the conductivity of nanotubes. Depending on subtle surface features a nanotube may act as a plain conductor or as a semiconductor. A fully automated method has however been developed to remove non-semiconductor tubes.

In 2013, researchers demonstrated a Turing-complete prototype micrometer-scale computer. Carbon nanotube transistors as logic-gate circuits with densities comparable to modern CMOS technology has not yet been demonstrated.

In 2014 networks of purified semiconducting carbon nanotubes were used as the active material in p-type thin film transistors. They were created using 3-D printers using inkjet or gravure methods on flexible substrates, including polyimide and polyethylene (PET) and transparent substrates such as glass. These transistors reliably exhibit high-mobilities (> 10 $cm^2 V^{-1} s^{-1}$) and ON/OFF ratios (> 1000) as well as threshold voltages below 5 V. They offer current density and low power consumption as well as environmental stability and mechanical flexibility. Hysterisis in the current-voltage curses as well as variability in the threshold voltage remain to be solved.

In 2015 researchers announced a new way to connect wires to SWNTs that make it possible to continue shrinking the width of the wires without increasing electrical resistance. The advance was expected to shrink the contact point between the two materials to just 40 atoms in width and later less. They tubes align in regularly spaced rows on silicon wafers. Simluations indicated that designs could be optimized either for high performance or for low power consumption. Commercial devices were not expected until the 2020s.

Thermal Management

Large structures of carbon nanotubes can be used for thermal management of electronic circuits. An approximately 1 mm–thick carbon nanotube layer was used as a special material to fabricate coolers, this material has very low density, ~20 times lower weight than a similar copper structure, while the cooling properties are similar for the two materials.

Buckypaper has characteristics appropriate for use as a heat sink for chipboards, a backlight for LCD screens or as a faraday cage.

Solar Cells

One of the promising applications of single-walled carbon nanotubes (SWNTs) is their use in solar panels, due to their strong UV/Vis-NIR absorption characteristics. Research has shown that they can provide a sizable increase in efficiency, even at their current unoptimized state. Solar cells developed at the New Jersey Institute of Technology use a carbon nanotube complex, formed by a mixture of carbon nanotubes and carbon buckyballs (known as fullerenes) to form snake-like structures. Buckyballs trap electrons, but they can't make electrons flow. Add sunlight to excite the polymers, and the buckyballs will grab the electrons. Nanotubes, behaving like copper wires, will then be able to make the electrons or current flow.

Additional research has been conducted on creating SWNT hybrid solar panels to increase the efficiency further. These hybrids are created by combining SWNT's with photo-excitable electron donors to increase the number of electrons generated. It has been found that the interaction between the photo-excited porphyrin and SWNT generates electro-hole pairs at the SWNT surfaces. This phenomenon has been observed experimentally, and contributes practically to an increase in efficiency up to 8.5%.

Nanotubes can potentially replace indium tin oxide in solar cells as a transparent conductive film in solar cells to allow light to pass to the active layers and generate photocurrent.

CNTs in organic solar cells help reduce energy loss (carrier recombination) and enhance resistance to photooxidation. Photovoltaic technologies may someday incorporate CNT-Silicon heterojunctions to leverage efficient multiple-exciton generation at p-n junctions formed within individual CNTs. In the nearer term, commercial photovoltaics may incorporate transparent SWNT electrodes.

Hydrogen Storage

In addition to being able to store electrical energy, there has been some research in using carbon nanotubes to store hydrogen to be used as a fuel source. By taking advantage of the capillary effects of the small carbon nanotubes, it is possible to condense gases in high density inside single-walled nanotubes. This allows for gases, most notably hydrogen (H_2), to be stored at high densities without being condensed into a liquid. Potentially, this storage method could be used on vehicles in place of gas fuel tanks for a hydrogen-powered car. A current issue regarding hydrogen-powered vehicles is the on-board storage of the fuel. Current storage methods involve cooling and condensing the H_2 gas to a liquid state for storage which causes a loss of potential energy (25–45%) when compared to the energy associated with the gaseous state. Storage using SWNTs would allow one to keep the H_2 in its gaseous state, thereby increasing the storage efficiency. This method allows for a volume to energy ratio slightly smaller to that of current gas powered vehicles, allowing for a slightly lower but comparable range.

An area of controversy and frequent experimentation regarding the storage of hydrogen by adsorption in carbon nanotubes is the efficiency by which this process occurs. The effectiveness of hydrogen storage is integral to its use as a primary fuel source since hydrogen only contains about one fourth the energy per unit volume as gasoline. Studies however show that what is the most important is the surface area of the materials used. Hence activated carbon with surface area of 2600 m^2/g can store up to 5,8% w/w. In all these carbonaceous materials, hydrogen is stored by physisorption at 70-90K.

Experimental Capacity

One experiment sought to determine the amount of hydrogen stored in CNTs by utilizing elastic recoil detection analysis (ERDA). CNTs (primarily SWNTs) were synthesized via chemical vapor disposition (CVD) and subjected to a two-stage purification process including air oxidation and acid treatment, then formed into flat, uniform discs and exposed to pure, pressurized hydrogen at various temperatures. When the data was analyzed, it was found that the ability of CNTs to store hydrogen decreased as temperature increased. Moreover, the highest hydrogen concentration measured was ~0.18%; significantly lower than commercially viable hydrogen storage needs to be. A separate experimental work performed by using a gravimetric method also revealed the maximum hydrogen uptake capacity of CNTs to be as low as 0.2%.

In another experiment, CNTs were synthesized via CVD and their structure was characterized using Raman spectroscopy. Utilizing microwave digestion, the samples were exposed to different acid concentrations and different temperatures for various amounts of time in an attempt to find the optimum purification method for SWNTs of the diameter determined earlier. The purified samples were then exposed to hydrogen gas at various high pressures, and their adsorption by weight percent was plotted. The data showed that hydrogen adsorption levels of up to 3.7% are possible with a very pure sample and under the proper conditions. It is thought that microwave digestion helps improve the hydrogen adsorption capacity of the CNTs by opening up the ends, allowing access to the inner cavities of the nanotubes.

Limitations on Efficient Hydrogen Adsorption

The biggest obstacle to efficient hydrogen storage using CNTs is the purity of the nanotubes. To achieve maximum hydrogen adsorption, there must be minimum graphene, amorphous carbon, and metallic deposits in the nanotube sample. Current methods of CNT synthesis require a purification step. However, even with pure nanotubes, the adsorption capacity is only maximized under high pressures, which are undesirable in commercial fuel tanks.

Electronic Components

Various companies are developing transparent, electrically conductive CNT films and nanobuds to replace indium tin oxide (ITO) in LCDs, touch screens and photovoltaic devices. Nanotube films show promise for use in displays for computers, cell phones, Personal digital assistants, and automated teller machines. CNT diodes display a photovoltaic effect.

Multi-walled nanotubes (MWNT coated with magnetite) can generate strong magnetic fields. Recent advances show that MWNT decorated with maghemite nanoparticles can be oriented in a magnetic field and enhance the electrical properties of the composite material in the direction of the field for use in electric motor brushes.

A layer of 29% iron enriched single-walled nanotubes (SWNT) placed on top of a layer of explosive material such as PETN can be ignited with a regular camera flash.

CNTs can be used as electron guns in miniature cathode ray tubes (CRT) in high-brightness, low-energy, low-weight displays. A display would consist of a group of tiny CRTs, each providing the electrons to illuminate the phosphor of one pixel, instead of having one CRT whose electrons are aimed using electric and magnetic fields. These displays are known as field emission displays (FEDs).

CNTs can act as antennas for radios and other electromagnetic devices.

Conductive CNTs are used in brushes for commercial electric motors. They replace traditional carbon black. The nanotubes improve electrical and thermal conductivity because they stretch through the plastic matrix of the brush. This permits the carbon filler to be reduced from 30% down to 3.6%, so that more matrix is present in the brush. Nanotube composite motor brushes are better-lubricated (from the matrix), cooler-running (both from better lubrication and superior thermal conductivity), less brittle (more matrix, and fiber reinforcement), stronger and more accurately moldable (more matrix). Since brushes are a critical failure point in electric motors, and also don't need much material, they became economical before almost any other application.

Wires for carrying electric current may be fabricated from nanotubes and nanotube-polymer composites. Small wires have been fabricated with specific conductivity exceeding copper and aluminum; the highest conductivity non-metallic cables.

CNT are under investigation as an alternative to tungsten filaments in incandescent light bulbs.

Interconnects

Metallic carbon nanotubes have aroused research interest for their applicability as very-large-scale integration (VLSI) interconnects because of their high thermal stability, high thermal conductivity and large current carrying capacity. An isolated CNT can carry current densities in excess of 1000 MA/sq-cm without damage even at an elevated temperature of 250 °C (482 °F), eliminating electromigration reliability concerns that plague Cu interconnects. Recent modeling work comparing the two has shown that CNT bundle interconnects can potentially offer advantages over copper. Recent experiments demonstrated resistances as low as 20 Ohms using different architectures, detailed conductance measurements over a wide temperature range were shown to agree with theory for a strongly disordered quasi-one-dimensional conductor.

Hybrid interconnects that employ CNT vias in tandem with copper interconnects may offer advantages from a reliability/thermal-management perspective. In 2016, the European Union has funded a four million euro project over three years to evaluate manufacturability and performance of composite interconnects employing both CNT and copper interconnects. The project named CONNECT (CarbON Nanotube compositE InterconneCTs) involves the joint efforts of seven European research and industry partners on fabrication techniques and processes to enable reliable Carbon NanoTubes for on-chip interconnects in ULSI microchip production.

Electrical Cables and Wires

Wires for carrying electric current may be fabricated from pure nanotubes and nanotube-polymer composites. It has already been demonstrated that carbon nanotube wires can successfully be used for power or data transmission. Recently small wires have been fabricated with specific conductivity exceeding copper and aluminum; these cables are the highest conductivity carbon nanotube and also highest conductivity non-metal cables. Recently, composite of carbon nanotube and copper have been shown to exhibit nearly one hundred times higher current-carrying-capacity than pure copper or gold. Significantly, the electrical conductivity of such a composite is similar to pure Cu. Thus, this Carbon nanotube-copper (CNT-Cu) composite possesses the highest observed current-carrying capacity among electrical conductors. Thus for a given cross-section of electrical conductor, the CNT-Cu composite can withstand and transport one hundred times higher current compared to metals such as copper and gold.

Energy Storage

The use of CNTs as a catalyst support in fuel cells can potentially reduce platinum usage by 60% compared with carbon black. Doped CNTs may enable the complete elimination of Pt.

Supercapacitor

MIT Research Laboratory of Electronics uses nanotubes to improve supercapacitors. The activated charcoal used in conventional ultracapacitors has many small hollow spaces of various size, which create together a large surface to store electric charge. But as charge is quantized into elementary charges, i.e. electrons, and each such elementary charge needs a minimum space, a significant fraction of the electrode surface is not available for storage because the hollow spaces are not compatible with the charge's requirements. With a nanotube electrode the spaces may be tailored to size—few too large or too small—and consequently the capacity should be increased considerably.

A 40-F supercapacitor with a maximum voltage of 3.5 V that employed forest-grown SWNTs that are binder- and additive-free achieved an energy density of 15.6 Wh kg^{-1} and a power density of 37 kW kg^{-1}. CNTs can be bound to the charge plates of capacitors to dramatically increase the surface area and therefore energy density.

Batteries

Carbon nanotubes' (CNTs) exciting electronic properties have shown promise in the field of batteries, where typically they are being experimented as a new electrode material, particularly the anode for lithium ion batteries. This is due to the fact that the anode requires a relatively high reversible capacity at a potential close to metallic lithium, and a moderate irreversible capacity, observed thus far only by graphite-based composites, such as CNTs. They have shown to greatly improve capacity and cyclability of lithium-ion batteries, as well as the capability to be very effective buffering components, alleviating the degradation of the batteries that is typically due to repeated charging and discharging. Further, electronic transport in the anode can be greatly improved using highly metallic CNTs.

More specifically, CNTs have shown reversible capacities from 300 to 600 $mAhg^{-1}$, with some treatments to them showing these numbers rise to up to 1000 $mAhg^{-1}$. Meanwhile, graphite, which is most widely used as an anode material for these lithium batteries, has shown capacities of only 320 $mAhg^{-1}$. By creating composites out of the CNTs, scientists see much potential in taking advantage of these exceptional capacities, as well as their excellent mechanical strength, conductivities, and low densities.

MWNTs are used in lithium ion batteries. In these batteries, small amounts of MWNT powder are blended with active materials and a polymer binder, such as 1 wt % CNT loading in LiCoO 2 cathodes and graphite anodes. CNTs provide increased electrical connectivity and mechanical integrity, which enhances rate capability and cycle life.

Paper Batteries

A paper battery is a battery engineered to use a paper-thin sheet of cellulose (which is the major constituent of regular paper, among other things) infused with aligned carbon nanotubes. The potential for these devices is great, as they may be manufactured via a roll-to-roll process, which would make it very low-cost, and they would be lightweight, flexible, and thin. In order to productively use paper electronics (or any thin electronic devices), the power source must be equally thin, thus indicating the need for paper batteries. Recently, it has been shown that surfaces coated with CNTs can be used to replace heavy metals in batteries. More recently, functional paper batteries have been demonstrated, where a lithium-ion battery is integrated on a single sheet of paper through a lamination process as a composite with Li4Ti5O12 (LTO) or LiCoO2 (LCO). The paper substrate would function well as the separator for the battery, where the CNT films function as the current collectors for both the anode *and* the cathode. These rechargeable energy devices show potential in RFID tags, functional packaging, or new disposable electronic applications.

Improvements have also been shown in lead-acid batteries, based on research performed by Bar-Ilan University using high quality SWCNT manufactured by OCSiAl. The study demonstrated an increase in the lifetime of lead acid batteries by 4.5 times and a capacity increase of 30% on average and up to 200% at high discharge rates.

Chemical

CNT can be used for desalination. Water molecules can be separated from salt by forcing them through electrochemically robust nanotube networks with controlled nanoscale porosity. This process requires far lower pressures than conventional reverse osmosis methods. Compared to a plain membrane, it operates at a 20 °C lower temperature, and at a 6x greater flow rate. Membranes using aligned, encapsulated CNTs with open ends permit flow through the CNTs' interiors. Very-small-diameter SWNTs are needed to reject salt at seawater concentrations. Portable filters containing CNT meshes can purify contaminated drinking water. Such networks can electrochemically oxidize organic contaminants, bacteria and viruses.

CNT membranes can filter carbon dioxide from power plant emissions.

CNT can be filled with biological molecules, aiding biotechnology.

CNT have the potential to store between 4.2 and 65% hydrogen by weight. If they can be mass-produced economically, 13.2 litres (2.9 imp gal; 3.5 US gal) of CNT could contain the same amount of energy as a 50 litres (11 imp gal; 13 US gal) gasoline tank.

CNTs can be used to produce nanowires of other elements/molecules, such as gold or zinc oxide. Nanowires in turn can be used to cast nanotubes of other materials, such as gallium nitride. These can have very different properties from CNTs—for example, gallium nitride nanotubes are hydrophilic, while CNTs are hydrophobic, giving them possible uses in organic chemistry.

Mechanical

Oscillators based on CNT have achieved speeds of > 50 GHz.

CNT electrical and mechanical properties suggest them as alternatives to traditional electrical actuators.

Actuators

The exceptional electrical and mechanical properties of carbon nanotubes have made them alternatives to the traditional electrical actuators for both microscopic and macroscopic applications. Carbon nanotubes are very good conductors of both electricity and heat, and they are also very strong and elastic molecules in certain directions.

Loudspeaker

Carbon nanotubes have also been applied in the acoustics(such as loudspeaker and earphone). In

2008 it was shown that a sheet of nanotubes can operate as a loudspeaker if an alternating current is applied. The sound is not produced through vibration but thermoacoustically. In 2013, a carbon nanotube (CNT) thin yarn thermoacoustic earphone together with CNT thin yarn thermoacoustic chip was demonstrated by a research group of Tsinghua-Foxconn Nanotechnology Research Center in Tsinghua University, using a Si-based semi-conducting technology compatible fabrication process.

Near-term commercial uses include replacing piezoelectric speakers in greeting cards.

Optical

- Carbon nanotube photoluminescence (fluorescence) can be used to observe semiconducting single-walled carbon nanotube species. Photoluminescence maps, made by acquiring the emission and scanning the excitation energy, can facilitate sample characterization.
- Nanotube fluorescence is under investigation for biomedical imaging and sensors.
- The reflectivity of buckypaper produced with "super-growth" chemical vapor deposition is 0.03 or less, potentially enabling performance gains for pyroelectric infrared detectors.

Environmental

Environmental Remediation

A CNT nano-structured sponge (nanosponge) containing sulfur and iron is more effective at soaking up water contaminants such as oil, fertilizers, pesticides and pharmaceuticals. Their magnetic properties make them easier to retrieve once the clean-up job is done. The sulfur and iron increases sponge size to around 2 centimetres (0.79 in). It also increases porosity due to beneficial defects, creating buoyancy and reusability. Iron, in the form of ferrocene makes the structure easier to control and enables recovery using magnets. Such nanosponges increase the absorption of the toxic organic solvent dichlorobenzene from water by 3.5 times. The sponges can absorb vegetable oil up to 150 times their initial weight and can absorb engine oil as well.

Earlier, a magnetic boron-doped MWNT nanosponge that could absorb oil from water. The sponge was grown as a forest on a substrate via chemical vapor disposition. Boron puts kinks and elbows into the tubes as they grow and promotes the formation of covalent bonds. The nanosponges retain their elastic property after 10,000 compressions in the lab. The sponges are both superhydrophobic, forcing them to remain at the water's surface and oleophilic, drawing oil to them.

Water Treatment

It has been shown that carbon nanotubes exhibit strong adsorption affinities to a wide range of aromatic and aliphatic contaminants in water, due to their large and hydrophobic surface areas. They also showed similar adsorption capacities as activated carbons in the presence of natural organic matter. As a result, they have been suggested as promising adsorbents for removal of contaminant in water and wastewater treatment systems.

Moreover, membranes made out of carbon nanotube arrays have been suggested as switchable molecular sieves, with sieving and permeation features that can be dynamically activated/deactivated by either pore size distribution (passive control) or external electrostatic fields (active control).

Other Applications

Carbon nanotubes have been implemented in nanoelectromechanical systems, including mechanical memory elements (NRAM being developed by Nantero Inc.) and nanoscale electric motors.

Carboxyl-modified single-walled carbon nanotubes (so called zig-zag, armchair type) can act as sensors of atoms and ions of alkali metals Na, Li, K. In May 2005, Nanomix Inc. placed on the market a hydrogen sensor that integrated carbon nanotubes on a silicon platform. Since then, Nanomix has been patenting many such sensor applications, such as in the field of carbon dioxide, nitrous oxide, glucose, DNA detection, etc. End of 2014, Tulane University researchers have tested Nanomix's fast and fully automated point of care diagnostic system in Sierra Leone to help for rapid testing for Ebola. Nanomix announced that a product could be launched within three to six months.

Eikos Inc of Franklin, Massachusetts and Unidym Inc. of Silicon Valley, California are developing transparent, electrically conductive films of carbon nanotubes to replace indium tin oxide (ITO). Carbon nanotube films are substantially more mechanically robust than ITO films, making them ideal for high-reliability touchscreens and flexible displays. Printable water-based inks of carbon nanotubes are desired to enable the production of these films to replace ITO. Nanotube films show promise for use in displays for computers, cell phones, PDAs, and ATMs.

A nanoradio, a radio receiver consisting of a single nanotube, was demonstrated in 2007.

The use in tensile stress or toxic gas sensors was proposed by Tsagarakis.

A flywheel made of carbon nanotubes could be spun at extremely high velocity on a floating magnetic axis in a vacuum, and potentially store energy at a density approaching that of conventional fossil fuels. Since energy can be added to and removed from flywheels very efficiently in the form of electricity, this might offer a way of storing electricity, making the electrical grid more efficient and variable power suppliers (like wind turbines) more useful in meeting energy needs. The practicality of this depends heavily upon the cost of making massive, unbroken nanotube structures, and their failure rate under stress.

Carbon nanotube springs have the potential to indefinitely store elastic potential energy at ten times the density of lithium-ion batteries with flexible charge and discharge rates and extremely high cycling durability.

Ultra-short SWNTs (US-tubes) have been used as nanoscaled capsules for delivering MRI contrast agents in vivo.

Carbon nanotubes provide a certain potential for metal-free catalysis of inorganic and organic reactions. For instance, oxygen groups attached to the surface of carbon nanotubes have the potential to catalyze oxidative dehydrogenations or selective oxidations. Nitrogen-doped carbon nanotubes may replace platinum catalysts used to reduce oxygen in fuel cells. A forest of vertically aligned nanotubes can reduce oxygen in alkaline solution more effectively than platinum, which has been used in such applications since the 1960s. Here, the nanotubes have the added benefit of not being subject to carbon monoxide poisoning.

Wake Forest University engineers are using multiwalled carbon nanotubes to enhance the brightness of field-induced polymer electroluminescent technology, potentially offering a step forward in the search for safe, pleasing, high-efficiency lighting. In this technology, moldable polymer matrix emits light when exposed to an electric current. It could eventually yield high-efficiency lights without the mercury vapor of compact fluorescent lamps or the bluish tint of some fluorescents and LEDs, which has been linked with circadian rhythm disruption.

Candida albicans has been used in combination with carbon nanotubes (CNT) to produce stable electrically conductive bio-nano-composite tissue materials that have been used as temperature sensing elements.

The SWNT production company OCSiAl developed a series of masterbatches for industrial use of single-wall CNTs in multiple types of rubber blends and tires, with initial trials showing increases in hardness, viscosity, tensile strain resistance and resistance to abrasion while reducing elongation and compression In tires the three primary characteristics of durability, fuel efficiency and traction were improved using SWNTs. The development of rubber masterbatches built on earlier work by the Japanese National Institute of Advanced Industrial Science & Technology showing rubber to be a viable candidate for improvement with SWNTs.

Introducing MWNTs to polymers can improve flame retardancy and retard thermal degradation of polymer. The results confirmed that combination of MWNTs and ammonium polyphosphates show a synergistic effect for improving flame retardancy.

References

- Ma, K.L. (2011). "Electronic transport properties of junctions between carbon nanotubes and graphene nanoribbons". European Physical Journal B. 83: 487–492. doi:10.1140/epjb/e2011-20313-9
- Schulz, Mark J.; Shanov, Vesselin N.; Yun, Yeoheung (2009). Nanomedicine Design of Particles, Sensors, Motors, Implants, Robots, and Devices. Artech House. ISBN 9781596932807
- Mintmire, J.W.; Dunlap, B.I.; White, C.T. (1992). "Are Fullerene Tubules Metallic?". Phys. Rev. Lett. 68 (5): 631–634. Bibcode:1992PhRvL..68..631M. PMID 10045950. doi:10.1103/PhysRevLett.68.631
- Oberlin, A.; Endo, M.; Koyama, T. (1976). "Filamentous growth of carbon through benzene decomposition" (PDF). Journal of Crystal Growth. 32 (3): 335–349. Bibcode:1976JCrGr..32..335O. doi:10.1016/0022-0248(76)90115-9
- Pacios Pujadó, Mercè (2012). Carbon Nanotubes as Platforms for Biosensors with Electrochemical and Electronic Transduction. Springer Heidelberg. pp. XX,208. ISBN 978-3-642-31421-6. doi:10.1007/978-3-642-31421-6
- Su, H.; Goddard, W.A.; Zhao, Y. (2006). "Dynamic friction force in a carbon peapod oscillator". Nanotechnology. 17 (22): 5691–5695. Bibcode:2006Nanot..17.5691S. arXiv:cond-mat/0611671. doi:10.1088/0957-4484/17/22/026
- N. Ishigami; et al. (2008). "Crystal Plane Dependent Growth of Aligned Single-Walled Carbon Nanotubes on Sapphire". Journal of the American Chemical Society. 130 (30): 9918–9924. PMID 18597459. doi:10.1021/ja8024752
- Banerjee, K.; Srivastava, N. (2006). "Are carbon nanotubes the future of VLSI interconnections?". 2006 43rd ACM/IEEE Design Automation Conference. p. 809. ISBN 1-59593-381-6. doi:10.1109/DAC.2006.229330

- Iijima, Sumio; Ichihashi, Toshinari (1993). "Single-shell carbon nanotubes of 1-nm diameter". Nature. 363 (6430): 603–605. Bibcode:1993Natur.363..603I. doi:10.1038/363603a0
- Hesami, Mahdis; Bagheri, Rouhollah; Masoomi, Mahmood (15 February 2014). "Flammability and thermal properties of epoxy/glass/MWNT Composites". Journal of Applied Polymer Science. 131 (4): n/a. doi:10.1002/app.39849
- Naeemi, A.; Meindl, J. D. (2007). "Carbon nanotube interconnects". Proceedings of the 2007 international symposium on Physical design – ISPD '07. p. 77. ISBN 9781595936134. doi:10.1145/1231996.1232014
- Ebbesen TW, Ajayan PM, Hiura H, Tanigaki K; Ajayan; Hiura; Tanigaki (1994). "Purification of nanotubes". Nature. 367 (6463): 519. Bibcode:1994Natur.367..519E. doi:10.1038/367519a0
- "European Commission : CORDIS : Projects & Results Service : CarbON Nanotube compositE InterconneCTs". Retrieved 11 July 2016
- Hesami, Mahdis; Bagheri, Rouhollah; Masoomi, Mahmood (8 April 2014). "Combination effects of carbon nanotubes, MMT and phosphorus flame retardant on fire and thermal resistance of fiber-reinforced epoxy composites". Iranian Polymer Journal. 23 (6): 469–476. doi:10.1007/s13726-014-0241-z
- High electromigration-resistant copper/carbon nanotube composite for interconnect application. IEEE Electron Devices Meeting, 2008. 2008. p. 1. ISBN 978-1-4244-2377-4. doi:10.1109/IEDM.2008.4796764
- Janas, D.; Koziol, K. K. (2014). "A review of production methods of carbon nanotube and graphene thin films for electrothermal applications". Nanoscale. 6 (6): 3037. doi:10.1039/C3NR05636H
- Energy Exploration Energy Efficiency Energy Management. "Five World Records. One Cool Technology.". FastCap Systems. Retrieved 2013-12-18

5

An Overview of Nanotechnology

Nanotechnology is the technology that is used in the manipulation of matter that exist at the nanoscale. Nanotechnology is used in creating new material which can have various applications in fields such as tissue engineering and nanoelectronics. This chapter is an overview of the subject matter incorporating all the major aspects of nanotechnology.

Nanotechnology

Nanotechnology ("nanotech") is manipulation of matter on an atomic, molecular, and supramolecular scale. The earliest, widespread description of nanotechnology referred to the particular technological goal of precisely manipulating atoms and molecules for fabrication of macroscale products, also now referred to as molecular nanotechnology. A more generalized description of nanotechnology was subsequently established by the National Nanotechnology Initiative, which defines nanotechnology as the manipulation of matter with at least one dimension sized from 1 to 100 nanometers. This definition reflects the fact that quantum mechanical effects are important at this quantum-realm scale, and so the definition shifted from a particular technological goal to a research category inclusive of all types of research and technologies that deal with the special properties of matter which occur below the given size threshold. It is therefore common to see the plural form "nanotechnologies" as well as "nanoscale technologies" to refer to the broad range of research and applications whose common trait is size. Because of the variety of potential applications (including industrial and military), governments have invested billions of dollars in nanotechnology research. Until 2012, through its National Nanotechnology Initiative, the USA has invested 3.7 billion dollars, the European Union has invested 1.2 billion and Japan 750 million dollars.

Nanotechnology as defined by size is naturally very broad, including fields of science as diverse as surface science, organic chemistry, molecular biology, semiconductor physics, microfabrication, molecular engineering, etc. The associated research and applications are equally diverse, ranging from extensions of conventional device physics to completely new approaches based upon molecular self-assembly, from developing new materials with dimensions on the nanoscale to direct control of matter on the atomic scale.

Scientists currently debate the future implications of nanotechnology. Nanotechnology may be able to create many new materials and devices with a vast range of applications, such as in nanomedicine, nanoelectronics, biomaterials energy production, and consumer products. On the other hand, nanotechnology raises many of the same issues as any new technology, including concerns about the toxicity and environmental impact of nanomaterials, and their potential effects on global economics, as well as speculation about various doomsday scenarios. These concerns have led to a debate among advocacy groups and governments on whether special regulation of nanotechnology is warranted.

The concepts that seeded nanotechnology were first discussed in 1959 by renowned physicist Richard Feynman in his talk *There's Plenty of Room at the Bottom*, in which he described the possibility of synthesis via direct manipulation of atoms. The term "nano-technology" was first used by Norio Taniguchi in 1974, though it was not widely known.

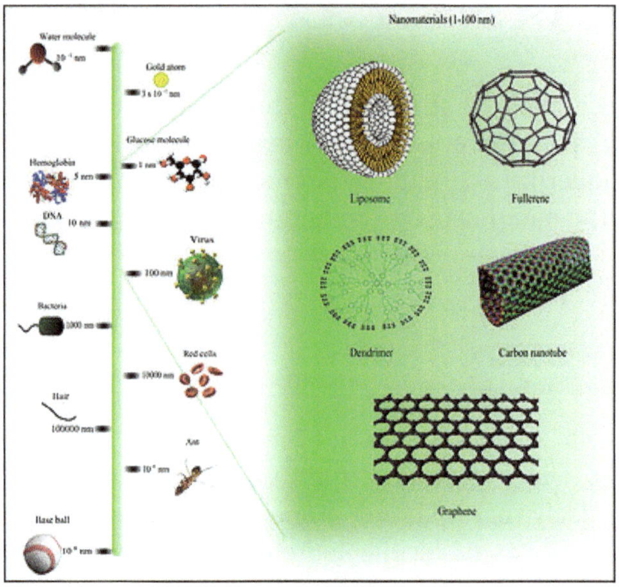

Comparison of Nanomaterials Sizes

Inspired by Feynman's concepts, K. Eric Drexler used the term "nanotechnology" in his 1986 book *Engines of Creation: The Coming Era of Nanotechnology*, which proposed the idea of a nanoscale "assembler" which would be able to build a copy of itself and of other items of arbitrary complexity with atomic control. Also in 1986, Drexler co-founded The Foresight Institute (with which he is no longer affiliated) to help increase public awareness and understanding of nanotechnology concepts and implications.

Thus, emergence of nanotechnology as a field in the 1980s occurred through convergence of Drexler's theoretical and public work, which developed and popularized a conceptual framework for nanotechnology, and high-visibility experimental advances that drew additional wide-scale attention to the prospects of atomic control of matter. In the 1980s, two major breakthroughs sparked the growth of nanotechnology in modern era.

First, the invention of the scanning tunneling microscope in 1981 which provided unprecedented visualization of individual atoms and bonds, and was successfully used to manipulate individual atoms in 1989. The microscope's developers Gerd Binnig and Heinrich Rohrer at IBM Zurich Research Laboratory received a Nobel Prize in Physics in 1986. Binnig, Quate and Gerber also invented the analogous atomic force microscope that year.

Second, Fullerenes were discovered in 1985 by Harry Kroto, Richard Smalley, and Robert Curl, who together won the 1996 Nobel Prize in Chemistry. C_{60} was not initially described as nanotechnology; the term was used regarding subsequent work with related graphene tubes (called carbon nanotubes and sometimes called Bucky tubes) which suggested potential applications for nanoscale electronics and devices.

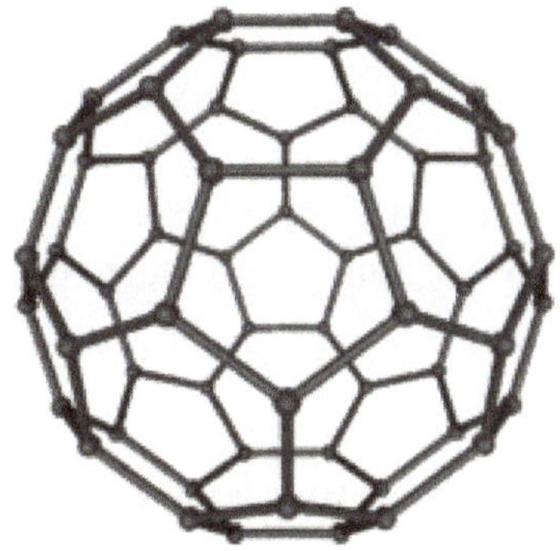

Buckminsterfullerene C_{60}, also known as the buckyball, is a representative member of the carbon structures known as fullerenes. Members of the fullerene family are a major subject of research falling under the nanotechnology umbrella.

In the early 2000s, the field garnered increased scientific, political, and commercial attention that led to both controversy and progress. Controversies emerged regarding the definitions and potential implications of nanotechnologies, exemplified by the Royal Society's report on nanotechnology. Challenges were raised regarding the feasibility of applications envisioned by advocates of molecular nanotechnology, which culminated in a public debate between Drexler and Smalley in 2001 and 2003.

Meanwhile, commercialization of products based on advancements in nanoscale technologies began emerging. These products are limited to bulk applications of nanomaterials and do not involve atomic control of matter. Some examples include the Silver Nano platform for using silver nanoparticles as an antibacterial agent, nanoparticle-based transparent sunscreens, carbon fiber strengthening using silica nanoparticles, and carbon nanotubes for stain-resistant textiles.

Governments moved to promote and fund research into nanotechnology, such as in the U.S. with the National Nanotechnology Initiative, which formalized a size-based definition of nanotechnology and established funding for research on the nanoscale, and in Europe via the European Framework Programmes for Research and Technological Development.

By the mid-2000s new and serious scientific attention began to flourish. Projects emerged to produce nanotechnology roadmaps which center on atomically precise manipulation of matter and discuss existing and projected capabilities, goals, and applications.

Fundamental Concepts

Nanotechnology is the engineering of functional systems at the molecular scale. This covers both current work and concepts that are more advanced. In its original sense, nanotechnology refers to the projected ability to construct items from the bottom up, using techniques and tools being developed today to make complete, high performance products.

One nanometer (nm) is one billionth, or 10^{-9}, of a meter. By comparison, typical carbon-carbon bond lengths, or the spacing between these atoms in a molecule, are in the range 0.12–0.15 nm, and a DNA double-helix has a diameter around 2 nm. On the other hand, the smallest cellular life-forms, the bacteria of the genus *Mycoplasma*, are around 200 nm in length. By convention, nanotechnology is taken as the scale range 1 to 100 nm following the definition used by the National Nanotechnology Initiative in the US. The lower limit is set by the size of atoms (hydrogen has the smallest atoms, which are approximately a quarter of a nm diameter) since nanotechnology must build its devices from atoms and molecules. The upper limit is more or less arbitrary but is around the size below which phenomena not observed in larger structures start to become apparent and can be made use of in the nano device. These new phenomena make nanotechnology distinct from devices which are merely miniaturised versions of an equivalent macroscopic device; such devices are on a larger scale and come under the description of microtechnology.

To put that scale in another context, the comparative size of a nanometer to a meter is the same as that of a marble to the size of the earth. Or another way of putting it: a nanometer is the amount an average man's beard grows in the time it takes him to raise the razor to his face.

Two main approaches are used in nanotechnology. In the "bottom-up" approach, materials and devices are built from molecular components which assemble themselves chemically by principles of molecular recognition. In the "top-down" approach, nano-objects are constructed from larger entities without atomic-level control.

Areas of physics such as nanoelectronics, nanomechanics, nanophotonics and nanoionics have evolved during the last few decades to provide a basic scientific foundation of nanotechnology.

Larger to Smaller: A Materials Perspective

Image of reconstruction on a clean Gold(100) surface, as visualized using scanning tunneling microscopy. The positions of the individual atoms composing the surface are visible.

Several phenomena become pronounced as the size of the system decreases. These include statistical mechanical effects, as well as quantum mechanical effects, for example the "quantum size effect" where the electronic properties of solids are altered with great reductions in particle size. This effect does not come into play by going from macro to micro dimensions. However, quantum effects can become significant when the nanometer size range is reached, typically at distances of 100 nanometers or less, the so-called quantum realm. Additionally, a number of physical (mechanical, electrical, optical, etc.) properties change when compared to macroscopic systems. One example is the increase in surface area to volume ratio altering mechanical, thermal and catalytic properties of materials. Diffusion and reactions at nanoscale, nanostructures materials and nanodevices with fast ion transport are generally referred to nanoionics. *Mechanical* properties of nanosystems are of interest in the nanomechanics research. The catalytic activity of nanomaterials also opens potential risks in their interaction with biomaterials.

Materials reduced to the nanoscale can show different properties compared to what they exhibit on a macroscale, enabling unique applications. For instance, opaque substances can become transparent (copper); stable materials can turn combustible (aluminium); insoluble materials may become soluble (gold). A material such as gold, which is chemically inert at normal scales, can serve as a potent chemical catalyst at nanoscales. Much of the fascination with nanotechnology stems from these quantum and surface phenomena that matter exhibits at the nanoscale.

Simple to Complex: A Molecular Perspective

Modern synthetic chemistry has reached the point where it is possible to prepare small molecules to almost any structure. These methods are used today to manufacture a wide variety of useful chemicals such as pharmaceuticals or commercial polymers. This ability raises the question of extending this kind of control to the next-larger level, seeking methods to assemble these single molecules into supramolecular assemblies consisting of many molecules arranged in a well defined manner.

These approaches utilize the concepts of molecular self-assembly and/or supramolecular chemistry to automatically arrange themselves into some useful conformation through a bottom-up approach. The concept of molecular recognition is especially important: molecules can be designed so that a specific configuration or arrangement is favored due to non-covalent intermolecular forces. The Watson–Crick basepairing rules are a direct result of this, as is the specificity of an enzyme being targeted to a single substrate, or the specific folding of the protein itself. Thus, two or more components can be designed to be complementary and mutually attractive so that they make a more complex and useful whole.

Such bottom-up approaches should be capable of producing devices in parallel and be much cheaper than top-down methods, but could potentially be overwhelmed as the size and complexity of the desired assembly increases. Most useful structures require complex and thermodynamically unlikely arrangements of atoms. Nevertheless, there are many examples of self-assembly based on molecular recognition in biology, most notably Watson–Crick basepairing and enzyme-substrate interactions. The challenge for nanotechnology is whether these principles can be used to engineer new constructs in addition to natural ones.

Molecular Nanotechnology: A Long-term View

Molecular nanotechnology, sometimes called molecular manufacturing, describes engineered nanosystems (nanoscale machines) operating on the molecular scale. Molecular nanotechnology is especially associated with the molecular assembler, a machine that can produce a desired structure or device atom-by-atom using the principles of mechanosynthesis. Manufacturing in the context of productive nanosystems is not related to, and should be clearly distinguished from, the conventional technologies used to manufacture nanomaterials such as carbon nanotubes and nanoparticles.

When the term "nanotechnology" was independently coined and popularized by Eric Drexler (who at the time was unaware of an earlier usage by Norio Taniguchi) it referred to a future manufacturing technology based on molecular machine systems. The premise was that molecular scale biological analogies of traditional machine components demonstrated molecular machines were possible: by the countless examples found in biology, it is known that sophisticated, stochastically optimised biological machines can be produced.

It is hoped that developments in nanotechnology will make possible their construction by some other means, perhaps using biomimetic principles. However, Drexler and other researchers have proposed that advanced nanotechnology, although perhaps initially implemented by biomimetic means, ultimately could be based on mechanical engineering principles, namely, a manufacturing technology based on the mechanical functionality of these components (such as gears, bearings, motors, and structural members) that would enable programmable, positional assembly to atomic specification. The physics and engineering performance of exemplar designs were analyzed in Drexler's book *Nanosystems*.

In general it is very difficult to assemble devices on the atomic scale, as one has to position atoms on other atoms of comparable size and stickiness. Another view, put forth by Carlo Montemagno, is that future nanosystems will be hybrids of silicon technology and biological molecular machines. Richard Smalley argued that mechanosynthesis are impossible due to the difficulties in mechanically manipulating individual molecules.

This led to an exchange of letters in the ACS publication Chemical & Engineering News in 2003. Though biology clearly demonstrates that molecular machine systems are possible, non-biological molecular machines are today only in their infancy. Leaders in research on non-biological molecular machines are Dr. Alex Zettl and his colleagues at Lawrence Berkeley Laboratories and UC Berkeley. They have constructed at least three distinct molecular devices whose motion is controlled from the desktop with changing voltage: a nanotube nanomotor, a molecular actuator, and a nanoelectromechanical relaxation oscillator.

An experiment indicating that positional molecular assembly is possible was performed by Ho and Lee at Cornell University in 1999. They used a scanning tunneling microscope to move an individual carbon monoxide molecule (CO) to an individual iron atom (Fe) sitting on a flat silver crystal, and chemically bound the CO to the Fe by applying a voltage.

Current Research

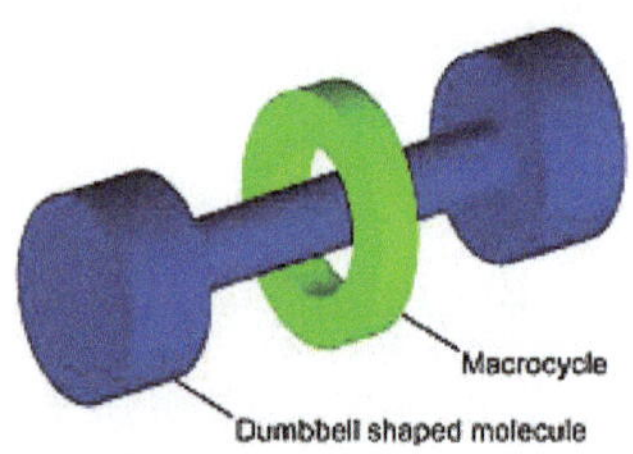

Graphical representation of a rotaxane, useful as a molecular switch.

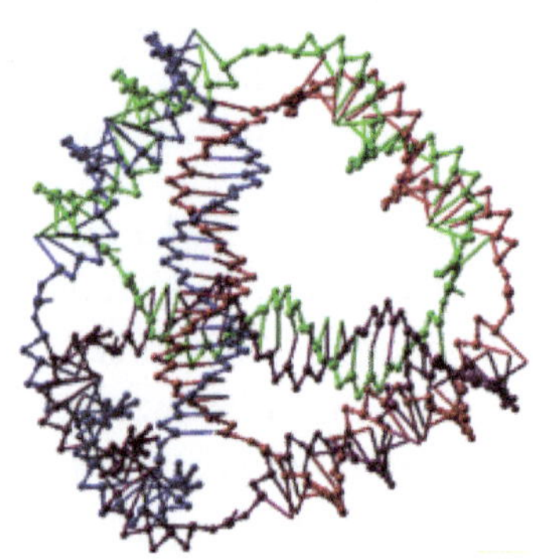
This DNA tetrahedron is an artificially designed nanostructure of the type made in the field of DNA nanotechnology. Each edge of the tetrahedron is a 20 base pair DNA double helix, and each vertex is a three-arm junction.

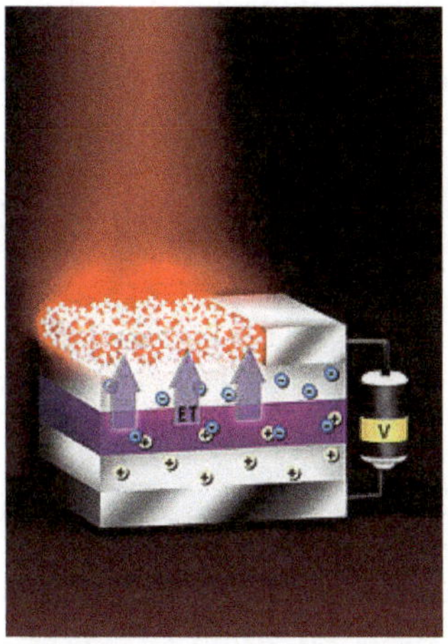

This device transfers energy from nano-thin layers of quantum wells to nanocrystals above them, causing the nanocrystals to emit visible light.

Nanomaterials

The nanomaterials field includes subfields which develop or study materials having unique properties arising from their nanoscale dimensions.

- Interface and colloid science has given rise to many materials which may be useful in nanotechnology, such as carbon nanotubes and other fullerenes, and various nanoparticles and nanorods. Nanomaterials with fast ion transport are related also to nanoionics and nanoelectronics.
- Nanoscale materials can also be used for bulk applications; most present commercial applications of nanotechnology are of this flavor.
- Progress has been made in using these materials for medical applications.
- Nanoscale materials such as nanopillars are sometimes used in solar cells which combats the cost of traditional silicon solar cells.
- Development of applications incorporating semiconductor nanoparticles to be used in the next generation of products, such as display technology, lighting, solar cells and biological imaging.
- Recent application of nanomaterials include a range of biomedical applications, such as tissue engineering, drug delivery, and biosensors.

Bottom-up Approaches

These seek to arrange smaller components into more complex assemblies.

- DNA nanotechnology utilizes the specificity of Watson–Crick basepairing to construct well-defined structures out of DNA and other nucleic acids.
- Approaches from the field of "classical" chemical synthesis (Inorganic and organic synthesis) also aim at designing molecules with well-defined shape (e.g. bis-peptides).

- More generally, molecular self-assembly seeks to use concepts of supramolecular chemistry, and molecular recognition in particular, to cause single-molecule components to automatically arrange themselves into some useful conformation.
- Atomic force microscope tips can be used as a nanoscale "write head" to deposit a chemical upon a surface in a desired pattern in a process called dip pen nanolithography. This technique fits into the larger subfield of nanolithography.

Top-down Approaches

These seek to create smaller devices by using larger ones to direct their assembly.

- Many technologies that descended from conventional solid-state silicon methods for fabricating microprocessors are now capable of creating features smaller than 100 nm, falling under the definition of nanotechnology. Giant magnetoresistance-based hard drives already on the market fit this description, as do atomic layer deposition (ALD) techniques. Peter Grünberg and Albert Fert received the Nobel Prize in Physics in 2007 for their discovery of Giant magnetoresistance and contributions to the field of spintronics.
- Solid-state techniques can also be used to create devices known as nanoelectromechanical systems or NEMS, which are related to microelectromechanical systems or MEMS.
- Focused ion beams can directly remove material, or even deposit material when suitable precursor gasses are applied at the same time. For example, this technique is used routinely to create sub-100 nm sections of material for analysis in Transmission electron microscopy.
- Atomic force microscope tips can be used as a nanoscale "write head" to deposit a resist, which is then followed by an etching process to remove material in a top-down method.

Functional Approaches

These seek to develop components of a desired functionality without regard to how they might be assembled.

- Magnetic assembly for the synthesis of anisotropic superparamagnetic materials such as recently presented magnetic nanochains.
- Molecular scale electronics seeks to develop molecules with useful electronic properties. These could then be used as single-molecule components in a nanoelectronic device.
- Synthetic chemical methods can also be used to create synthetic molecular motors, such as in a so-called nanocar.

Biomimetic Approaches

- Bionics or biomimicry seeks to apply biological methods and systems found in nature, to the study and design of engineering systems and modern technology. Biomineralization is one example of the systems studied.
- Bionanotechnology is the use of biomolecules for applications in nanotechnology, including use of viruses and lipid assemblies. Nanocellulose is a potential bulk-scale application.

Speculative

These subfields seek to anticipate what inventions nanotechnology might yield, or attempt to propose an agenda along which inquiry might progress. These often take a big-picture view of nanotechnology, with more emphasis on its societal implications than the details of how such inventions could actually be created.

- Molecular nanotechnology is a proposed approach which involves manipulating single molecules in finely controlled, deterministic ways. This is more theoretical than the other subfields, and many of its proposed techniques are beyond current capabilities.
- Nanorobotics centers on self-sufficient machines of some functionality operating at the nanoscale. There are hopes for applying nanorobots in medicine, but it may not be easy to do such a thing because of several drawbacks of such devices. Nevertheless, progress on innovative materials and methodologies has been demonstrated with some patents granted about new nanomanufacturing devices for future commercial applications, which also progressively helps in the development towards nanorobots with the use of embedded nanobioelectronics concepts.
- Productive nanosystems are "systems of nanosystems" which will be complex nanosystems that produce atomically precise parts for other nanosystems, not necessarily using novel nanoscale-emergent properties, but well-understood fundamentals of manufacturing. Because of the discrete (i.e. atomic) nature of matter and the possibility of exponential growth, this stage is seen as the basis of another industrial revolution. Mihail Roco, one of the architects of the USA's National Nanotechnology Initiative, has proposed four states of nanotechnology that seem to parallel the technical progress of the Industrial Revolution, progressing from passive nanostructures to active nanodevices to complex nanomachines and ultimately to productive nanosystems.
- Programmable matter seeks to design materials whose properties can be easily, reversibly and externally controlled though a fusion of information science and materials science.
- Due to the popularity and media exposure of the term nanotechnology, the words picotechnology and femtotechnology have been coined in analogy to it, although these are only used rarely and informally.

Dimensionality in Nanomaterials

Nanomaterials can be classified in 0D, 1D, 2D and 3D nanomaterials. The dimensionality play a major role in determining the characteristic of nanomaterials including physical, chemical and biological characteristics. With the decrease in dimensionality, an increase in surface-to-volume ratio is observed. This indicate that smaller dimensional nanomaterials have higher surface area compared to 3D nanomaterials. Recently, two dimensional (2D) nanomaterials are extensively investigated for electronic, biomedical, drug delivery and biosensor applications.

Tools and Techniques

There are several important modern developments. The atomic force microscope (AFM) and the Scanning Tunneling Microscope (STM) are two early versions of scanning probes that launched nanotechnology. There are other types of scanning probe microscopy. Although conceptually similar to the scanning confocal microscope developed by Marvin Minsky in 1961 and the scanning acoustic microscope (SAM) developed by Calvin Quate and coworkers in the 1970s, newer scanning probe microscopes have much higher resolution, since they are not limited by the wavelength of sound or light.

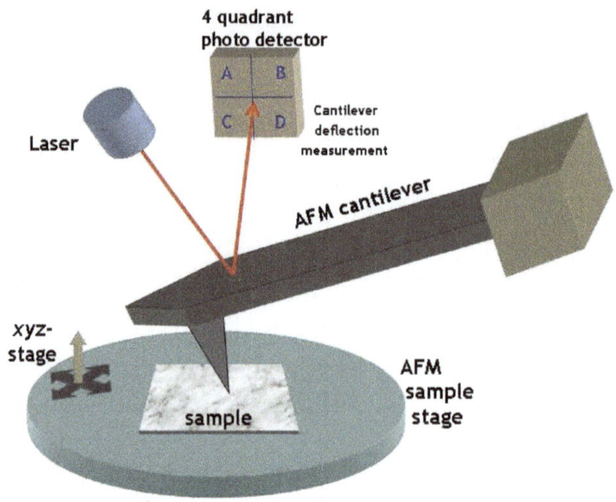

Typical AFM setup. A microfabricated cantilever with a sharp tip is deflected by features on a sample surface, much like in a phonograph but on a much smaller scale. A laser beam reflects off the backside of the cantilever into a set of photodetectors, allowing the deflection to be measured and assembled into an image of the surface.

The tip of a scanning probe can also be used to manipulate nanostructures (a process called positional assembly). Feature-oriented scanning methodology may be a promising way to implement these nanomanipulations in automatic mode. However, this is still a slow process because of low scanning velocity of the microscope.

Various techniques of nanolithography such as optical lithography, X-ray lithography dip pen nanolithography, electron beam lithography or nanoimprint lithography were also developed. Lithography is a top-down fabrication technique where a bulk material is reduced in size to nanoscale pattern.

Another group of nanotechnological techniques include those used for fabrication of nanotubes and nanowires, those used in semiconductor fabrication such as deep ultraviolet lithography, electron beam lithography, focused ion beam machining, nanoimprint lithography, atomic layer deposition, and molecular vapor deposition, and further including molecular self-assembly techniques such as those employing di-block copolymers. The precursors of these techniques preceded the nanotech era, and are extensions in the development of scientific advancements rather than techniques which were devised with the sole purpose of creating nanotechnology and which were results of nanotechnology research.

The top-down approach anticipates nanodevices that must be built piece by piece in stages, much as manufactured items are made. Scanning probe microscopy is an important technique both for characterization and synthesis of nanomaterials. Atomic force microscopes and scanning tunneling microscopes can be used to look at surfaces and to move atoms around. By designing different tips for these microscopes, they can be used for carving out structures on surfaces and to help guide self-assembling structures. By using, for example, feature-oriented scanning approach, atoms or molecules can be moved around on a surface with scanning probe microscopy techniques. At present, it is expensive and time-consuming for mass production but very suitable for laboratory experimentation.

In contrast, bottom-up techniques build or grow larger structures atom by atom or molecule by molecule. These techniques include chemical synthesis, self-assembly and positional assembly. Dual polarisation interferometry is one tool suitable for characterisation of self assembled thin films. Another variation of the bottom-up approach is molecular beam epitaxy or MBE. Researchers at Bell Telephone Laboratories like John R. Arthur. Alfred Y. Cho, and Art C. Gossard developed and implemented MBE as a research tool in the late 1960s and 1970s. Samples made by MBE were key to the discovery of the fractional quantum Hall effect for which the 1998 Nobel Prize in Physics was awarded. MBE allows scientists to lay down atomically precise layers of atoms and, in the process, build up complex structures. Important for research on semiconductors, MBE is also widely used to make samples and devices for the newly emerging field of spintronics.

However, new therapeutic products, based on responsive nanomaterials, such as the ultradeformable, stress-sensitive Transfersome vesicles, are under development and already approved for human use in some countries.

Applications

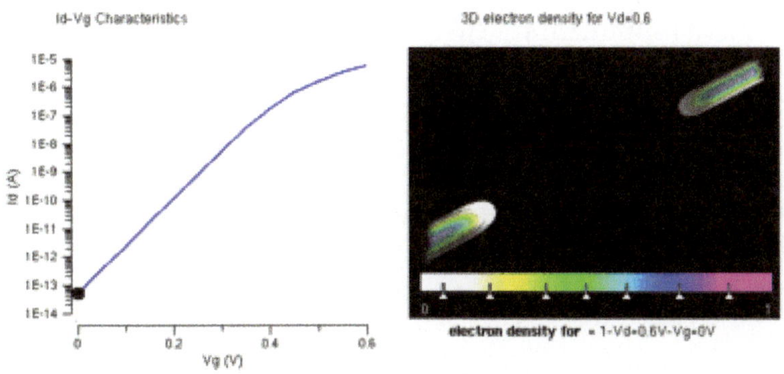

One of the major applications of nanotechnology is in the area of nanoelectronics with MOSFET's being made of small nanowires ≈10 nm in length. Here is a simulation of such a nanowire.

As of August 21, 2008, the Project on Emerging Nanotechnologies estimates that over 800 manufacturer-identified nanotech products are publicly available, with new ones hitting the market at a pace of 3–4 per week. The project lists all of the products in a publicly accessible online database. Most applications are limited to the use of "first generation" passive nanomaterials which includes titanium dioxide in sunscreen, cosmetics, surface coatings, and some food products; Carbon allotropes used to produce gecko tape; silver in food packaging, clothing, disinfectants and household appliances; zinc oxide in sunscreens and cosmetics, surface coatings, paints and outdoor furniture varnishes; and cerium oxide as a fuel catalyst.

Further applications allow tennis balls to last longer, golf balls to fly straighter, and even bowling balls to become more durable and have a harder surface. Trousers and socks have been infused with nanotechnology so that they will last longer and keep people cool in the summer. Bandages are being infused with silver nanoparticles to heal cuts faster. Video game consoles and personal computers may become cheaper, faster, and contain more memory thanks to nanotechnology. Nanotechnology may have the ability to make existing medical applications cheaper and easier to use in places like the general practitioner's office and at home. Cars are being manufactured with nanomaterials so they may need fewer metals and less fuel to operate in the future.

Scientists are now turning to nanotechnology in an attempt to develop diesel engines with cleaner exhaust fumes. Platinum is currently used as the diesel engine catalyst in these engines. The catalyst is what cleans the exhaust fume particles. First a reduction catalyst is employed to take nitrogen atoms from NOx molecules in order to free oxygen. Next the oxidation catalyst oxidizes the hydrocarbons and carbon monoxide to form carbon dioxide and water. Platinum is used in both the reduction and the oxidation catalysts. Using platinum though, is inefficient in that it is expensive and unsustainable. Danish company InnovationsFonden invested DKK 15 million in a search for new catalyst substitutes using nanotechnology. The goal of the project, launched in the autumn of 2014, is to maximize surface area and minimize the amount of material required. Objects tend to minimize their surface energy; two drops of water, for example, will join to form one drop and decrease surface area. If the catalyst's surface area that is exposed to the exhaust fumes is maximized, efficiency of the catalyst is maximized. The team working on this project aims to create nanoparticles that will not merge. Every time the surface is optimized, material is saved. Thus, creating these nanoparticles will increase the effectiveness of the resulting diesel engine catalyst—in turn leading to cleaner exhaust fumes—and will decrease cost. If successful, the team hopes to reduce platinum use by 25%.

Nanotechnology also has a prominent role in the fast developing field of Tissue Engineering. When designing scaffolds, researchers attempt to the mimic the nanoscale features of a Cell's microenvironment to direct its differentiation down a suitable lineage. For example, when creating scaffolds to support the growth of bone, researchers may mimic osteoclast resorption pits.

Researchers have successfully used DNA origami-based nanobots capable of carrying out logic functions to achieve targeted drug delivery in cockroaches. It is said that the computational power of these nanobots can be scaled up to that of a Commodore 64.

Implications

An area of concern is the effect that industrial-scale manufacturing and use of nanomaterials would have on human health and the environment, as suggested by nanotoxicology research. For these reasons, some groups advocate that nanotechnology be regulated by governments. Others counter that overregulation would stifle scientific research and the development of beneficial innovations. Public health research agencies, such as the National Institute for Occupational Safety and Health are actively conducting research on potential health effects stemming from exposures to nanoparticles.

Some nanoparticle products may have unintended consequences. Researchers have discovered that bacteriostatic silver nanoparticles used in socks to reduce foot odor are being released in the wash. These particles are then flushed into the waste water stream and may destroy bacteria which are critical components of natural ecosystems, farms, and waste treatment processes.

Public deliberations on risk perception in the US and UK carried out by the Center for Nanotechnology in Society found that participants were more positive about nanotechnologies for energy applications than for health applications, with health applications raising moral and ethical dilemmas such as cost and availability.

Experts, including director of the Woodrow Wilson Center's Project on Emerging Nanotechnologies David Rejeski, have testified that successful commercialization depends on adequate oversight, risk research strategy, and public engagement. Berkeley, California is currently the only city in the United States to regulate nanotechnology; Cambridge, Massachusetts in 2008 considered enacting a similar law, but ultimately rejected it. Relevant for both research on and application of nanotechnologies, the insurability of nanotechnology is contested. Without state regulation of nanotechnology, the availability of private insurance for potential damages is seen as necessary to ensure that burdens are not socialised implicitly.

Health and Environmental Concerns

Nanofibers are used in several areas and in different products, in everything from aircraft wings to tennis rackets. Inhaling airborne nanoparticles and nanofibers may lead to a number of pulmonary diseases, e.g. fibrosis. Researchers have found that when rats breathed in nanoparticles, the particles settled in the brain and lungs, which led to significant increases in biomarkers for inflammation and stress response and that nanoparticles induce skin aging through oxidative stress in hairless mice.

A two-year study at UCLA's School of Public Health found lab mice consuming nano-titanium dioxide showed DNA and chromosome damage to a degree "linked to all the big killers of man, namely cancer, heart disease, neurological disease and aging".

A major study published more recently in Nature Nanotechnology suggests some forms of carbon nanotubes – a poster child for the "nanotechnology revolution" – could be as harmful as asbestos if inhaled in sufficient quantities. Anthony Seaton of the Institute of Occupational Medicine in Edinburgh, Scotland, who contributed to the article on carbon nanotubes said "We know that some of them probably have the potential to cause mesothelioma. So those sorts of materials need to be handled very carefully." In the absence of specific regulation forthcoming from governments, Paull and Lyons (2008) have called for an exclusion of engineered nanoparticles in food. A newspaper article reports that workers in a paint factory developed serious lung disease and nanoparticles were found in their lungs.

Regulation

Calls for tighter regulation of nanotechnology have occurred alongside a growing debate related to the human health and safety risks of nanotechnology. There is significant debate about who is responsible for the regulation of nanotechnology. Some regulatory agencies currently cover some nanotechnology products and processes (to varying degrees) – by "bolting on" nanotechnology to existing regulations – there are clear gaps in these regimes. Davies (2008) has proposed a regulatory road map describing steps to deal with these shortcomings.

Stakeholders concerned by the lack of a regulatory framework to assess and control risks associated with the release of nanoparticles and nanotubes have drawn parallels with bovine spongiform encephalopathy ("mad cow" disease), thalidomide, genetically modified food, nuclear energy, reproductive technologies, biotechnology, and asbestosis. Dr. Andrew Maynard, chief science advisor to the Woodrow Wilson Center's Project on Emerging Nanotechnologies, concludes that there is insufficient funding for human health and safety research, and as a result there is currently limited understanding of the human health and safety risks associated with nanotechnology. As a result, some academics have called for stricter application of the precautionary principle, with delayed marketing approval, enhanced labelling and additional safety data development requirements in relation to certain forms of nanotechnology.

The Royal Society report identified a risk of nanoparticles or nanotubes being released during disposal, destruction and recycling, and recommended that "manufacturers of products that fall under extended producer responsibility regimes such as end-of-life regulations publish procedures outlining how these materials will be managed to minimize possible human and environmental exposure".

The Center for Nanotechnology in Society has found that people respond to nanotechnologies differently, depending on application – with participants in public deliberations more positive about nanotechnologies for energy than health applications – suggesting that any public calls for nano regulations may differ by technology sector.

Nanomedicine

Nanomedicine is the medical application of nanotechnology. Nanomedicine ranges from the medical applications of nanomaterials and biological devices, to nanoelectronic biosensors, and even possible future applications of molecular nanotechnology such as biological machines. Current problems for nanomedicine involve understanding the issues related to toxicity and environmental impact of nanoscale materials (materials whose structure is on the scale of nanometers, i.e. billionths of a meter).

Functionalities can be added to nanomaterials by interfacing them with biological molecules or structures. The size of nanomaterials is similar to that of most biological molecules and structures; therefore, nanomaterials can be useful for both in vivo and in vitro biomedical research and applications. Thus far, the integration of nanomaterials with biology has led to the development of diagnostic devices, contrast agents, analytical tools, physical therapy applications, and drug delivery vehicles.

Nanomedicine seeks to deliver a valuable set of research tools and clinically useful devices in the near future. The National Nanotechnology Initiative expects new commercial applications in the pharmaceutical industry that may include advanced drug delivery systems, new therapies, and in vivo imaging. Nanomedicine research is receiving funding from the US National Institutes of Health Common Fund program, supporting four nanomedicine development centers.

Nanomedicine sales reached $16 billion in 2015, with a minimum of $3.8 billion in nanotechnology R&D being invested every year. Global funding for emerging nanotechnology increased by 45% per year in recent years, with product sales exceeding $1 trillion in 2013. As the nanomedicine industry continues to grow, it is expected to have a significant impact on the economy.

Drug Delivery

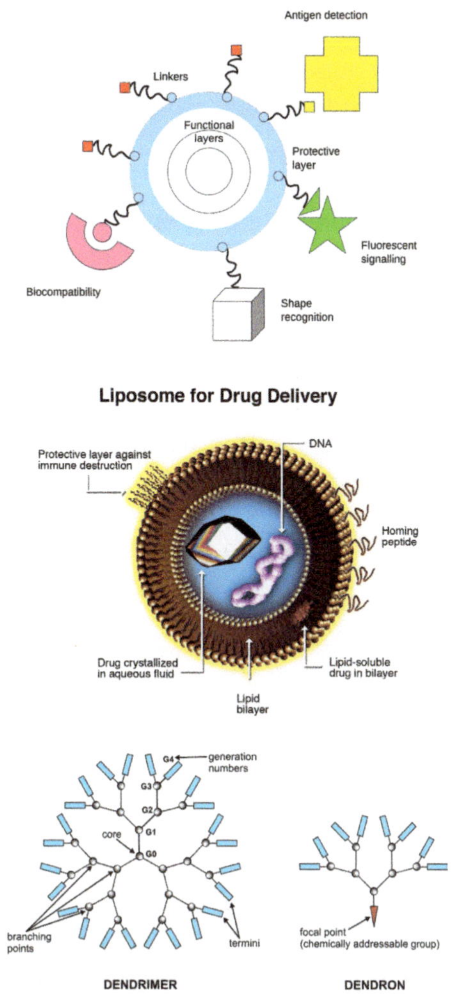

Nanoparticles *(top)*, liposomes *(middle)*, and dendrimers *(bottom)* are some nanomaterials being investigated for use in nanomedicine.

Nanotechnology has provided the possibility of delivering drugs to specific cells using nanoparticles. The overall drug consumption and side-effects may be lowered significantly by depositing the active agent in the morbid region only and in no higher dose than needed. Targeted drug delivery is intended to reduce the side effects of drugs with concomitant decreases in consumption and treatment expenses. Drug delivery focuses on maximizing bioavailability both at specific places in the body and over a period of time. This can potentially be achieved by molecular targeting by nanoengineered devices. A benefit of using nanoscale for medical technologies is that smaller devices are less invasive and can possibly be implanted inside the body, plus biochemical reaction times are much shorter. These devices are faster and more sensitive than typical drug delivery. The efficacy of drug delivery through nanomedicine is largely based upon: a) efficient encapsulation of the drugs, b) successful delivery of drug to the targeted region of the body, and c) successful release of the drug.

Drug delivery systems, lipid- or polymer-based nanoparticles, can be designed to improve the pharmacokinetics and biodistribution of the drug. However, the pharmacokinetics and pharmacodynamics of nanomedicine is highly variable among different patients. When designed to avoid the body's defence mechanisms, nanoparticles have beneficial properties that can be used to improve drug delivery. Complex drug delivery mechanisms are being developed, including the ability to get drugs through cell membranes and into cell cytoplasm. Triggered response is one way for drug molecules to be used more efficiently. Drugs are placed in the body and only activate on encountering a particular signal. For example, a drug with poor solubility will be replaced by a drug delivery system where both hydrophilic and hydrophobic environments exist, improving the solubility. Drug delivery systems may also be able to prevent tissue damage through regulated drug release; reduce drug clearance rates; or lower the volume of distribution and reduce the effect on non-target tissue. However, the biodistribution of these nanoparticles is still imperfect due to the complex host's reactions to nano- and microsized materials and the difficulty in targeting specific organs in the body. Nevertheless, a lot of work is still ongoing to optimize and better understand the potential and limitations of nanoparticulate systems. While advancement of research proves that targeting and distribution can be augmented by nanoparticles, the dangers of nanotoxicity become an important next step in further understanding of their medical uses.

Nanoparticles are under research for their potential to decrease antibiotic resistance or for various antimicrobial uses. Nanoparticles might also used to circumvent multidrug resistance (MDR) mechanisms.

Systems under Research

Two forms of nanomedicine that have already been tested in mice and are awaiting human testing will use gold nanoshells to help diagnose and treat cancer, along with liposomes as vaccine adjuvants and drug transport vehicles. Similarly, drug detoxification is also another application for nanomedicine which has shown promising results in rats. Advances in Lipid nanotechnology was also instrumental in engineering medical nanodevices and novel drug delivery systems as well as in developing sensing applications. Another example can be found in dendrimers and nanoporous materials. Another example is to use block co-polymers, which form micelles for drug encapsulation.

Polymeric nanoparticles are a competing technology to lipidic (based mainly on Phospholipids) nanoparticles. There is an additional risk of toxicity associated with polymers not widely studied or understood. The major advantages of polymers is stability, lower cost and predictable characterisation. However, in the patient's body this very stability (slow degradation) is a negative factor. Phospholipids on the other hand are membrane lipids (already present in the body and surrounding each cell), have a GRAS (Generally Recognised As Safe) status from FDA and are derived from natural sources without any complex chemistry involved. They are not metabolised but rather absorbed by the body and the degradation products are themselves nutrients (fats or micronutrients).

Protein and peptides exert multiple biological actions in the human body and they have been identified as showing great promise for treatment of various diseases and disorders. These macromolecules are called biopharmaceuticals. Targeted and/or controlled delivery of these biopharmaceuticals using nanomaterials like nanoparticles and Dendrimers is an emerging field called nanobiopharmaceutics, and these products are called nanobiopharmaceuticals.

Another highly efficient system for microRNA delivery for example are nanoparticles formed by the self-assembly of two different microRNAs deregulated in cancer.

Another vision is based on small electromechanical systems; nanoelectromechanical systems are being investigated for the active release of drugs. Some potentially important applications include cancer treatment with iron nanoparticles or gold shells.Nanotechnology is also opening up new opportunities in implantable delivery systems, which are often preferable to the use of injectable drugs, because the latter frequently display first-order kinetics (the blood concentration goes up rapidly, but drops exponentially over time). This rapid rise may cause difficulties with toxicity, and drug efficacy can diminish as the drug concentration falls below the targeted range.

Applications

Some nanotechnology-based drugs that are commercially available or in human clinical trials include:

- Abraxane, approved by the U.S. Food and Drug Administration (FDA) to treat breast cancer, non-small- cell lung cancer (NSCLC) and pancreatic cancer, is the nanoparticle albumin bound paclitaxel.
- Doxil was originally approved by the FDA for the use on HIV-related Kaposi's sarcoma. It is now being used to also treat ovarian cancer and multiple myeloma. The drug is encased in liposomes, which helps to extend the life of the drug that is being distributed. Liposomes are self-assembling, spherical, closed colloidal structures that are composed of lipid bilayers that surround an aqueous space. The liposomes also help to increase the functionality and it helps to decrease the damage that the drug does to the heart muscles specifically.
- Onivyde, liposome encapsulated irinotecan to treat metastatic pancreatic cancer, was approved by FDA in October 2015.
- C-dots (Cornell dots) are the smallest silica-based nanoparticles with the size <10 nm. The particles are infused with organic dye which will light up with fluorescence. Clinical trial is underway since 2011 to use the C-dots as diagnostic tool to assist surgeons to identify the location of tumor cells.
- An early phase clinical trial using the platform of 'Minicell' nanoparticle for drug delivery have been tested on patients with advanced and untreatable cancer. Built from the membranes of mutant bacteria, the minicells were loaded with paclitaxel and coated with cetuximab, antibodies that bind the epidermal growth factor receptor (EGFR) which is often overexpressed in a number of cancers, as a 'homing' device to the tumor cells. The tumor cells recognize the bacteria from which the minicells have been derived, regard it as invading microorganism and engulf it. Once inside, the payload of anti-cancer drug kills the tumor cells. Measured at 400 nanometers, the minicell is bigger than synthetic particles developed for drug delivery. The researchers indicated that this larger size gives the minicells a better profile in side-effects because the minicells will preferentially leak out of the porous blood vessels around the tumor cells and do not reach the liver, digestive system and skin. This Phase 1 clinical trial demonstrated that this treatment is well tolerated by the

patients. As a platform technology, the minicell drug delivery system can be used to treat a number of different cancers with different anti-cancer drugs with the benefit of lower dose and less side-effects.

- In 2014, a Phase 3 clinical trial for treating inflammation and pain after cataract surgery, and a Phase 2 trial for treating dry eye disease were initiated using nanoparticle loteprednol etabonate. In 2015, the product, KPI-121 was found to produce statistically significant positive results for the post-surgery treatment.

Cancer

Existing and potential drug nanocarriers have been reviewed.

Nanoparticles have high surface area to volume ratio. This allows for many functional groups to be attached to a nanoparticle, which can seek out and bind to certain tumor cells. Additionally, the small size of nanoparticles (10 to 100 nanometers), allows them to preferentially accumulate at tumor sites (because tumors lack an effective lymphatic drainage system). Limitations to conventional cancer chemotherapy include drug resistance, lack of selectivity, and lack of solubility. Nanoparticles have the potential to overcome these problems.

In photodynamic therapy, a particle is placed within the body and is illuminated with light from the outside. The light gets absorbed by the particle and if the particle is metal, energy from the light will heat the particle and surrounding tissue. Light may also be used to produce high energy oxygen molecules which will chemically react with and destroy most organic molecules that are next to them (like tumors). This therapy is appealing for many reasons. It does not leave a "toxic trail" of reactive molecules throughout the body (chemotherapy) because it is directed where only the light is shined and the particles exist. Photodynamic therapy has potential for a noninvasive procedure for dealing with diseases, growth and tumors. Kanzius RF therapy is one example of such therapy (nanoparticle hyperthermia). Also, gold nanoparticles have the potential to join numerous therapeutic functions into a single platform, by targeting specific tumor cells, tissues and organs.

Imaging

In vivo imaging is another area where tools and devices are being developed. Using nanoparticle contrast agents, images such as ultrasound and MRI have a favorable distribution and improved contrast. In cardiovascular imaging, nanoparticles have potential to aid visualization of blood pooling, ischemia, angiogenesis, atherosclerosis, and focal areas where inflammation is present.

The small size of nanoparticles endows them with properties that can be very useful in oncology, particularly in imaging. Quantum dots (nanoparticles with quantum confinement properties, such as size-tunable light emission), when used in conjunction with MRI (magnetic resonance imaging), can produce exceptional images of tumor sites. Nanoparticles of cadmium selenide (quantum dots) glow when exposed to ultraviolet light. When injected, they seep into cancer tumors. The surgeon can see the glowing tumor, and use it as a guide for more accurate tumor removal.These nanoparticles are much brighter than organic dyes and only need one light source for excitation. This means that the use of fluorescent quantum dots could produce a higher contrast image and at a lower cost than today's organic dyes used as contrast media. The downside, however, is that quantum dots are usually made of quite toxic elements, but this concern may be addressed by use of fluorescent dopants.

Tracking movement can help determine how well drugs are being distributed or how substances are metabolized. It is difficult to track a small group of cells throughout the body, so scientists used to dye the cells. These dyes needed to be excited by light of a certain wavelength in order for them to light up. While different color dyes absorb different frequencies of light, there was a need for as many light sources as cells. A way around this problem is with luminescent tags. These tags are quantum dots attached to proteins that penetrate cell membranes. The dots can be random in size, can be made of bio-inert material, and they demonstrate the nanoscale property that color is size-dependent. As a result, sizes are selected so that the frequency of light used to make a group of quantum dots fluoresce is an even multiple of the frequency required to make another group incandesce. Then both groups can be lit with a single light source. They have also found a way to insert nanoparticles into the affected parts of the body so that those parts of the body will glow showing the tumor growth or shrinkage or also organ trouble.

Sensing

Nanotechnology-on-a-chip is one more dimension of lab-on-a-chip technology. Magnetic nanoparticles, bound to a suitable antibody, are used to label specific molecules, structures or microorganisms. Gold nanoparticles tagged with short segments of DNA can be used for detection of genetic sequence in a sample. Multicolor optical coding for biological assays has been achieved by embedding different-sized quantum dots into polymeric microbeads. Nanopore technology for analysis of nucleic acids converts strings of nucleotides directly into electronic signatures.

Sensor test chips containing thousands of nanowires, able to detect proteins and other biomarkers left behind by cancer cells, could enable the detection and diagnosis of cancer in the early stages from a few drops of a patient's blood. Nanotechnology is helping to advance the use of arthroscopes, which are pencil-sized devices that are used in surgeries with lights and cameras so surgeons can do the surgeries with smaller incisions. The smaller the incisions the faster the healing time which is better for the patients. It is also helping to find a way to make an arthroscope smaller than a strand of hair.

Research on nanoelectronics-based cancer diagnostics could lead to tests that can be done in pharmacies. The results promise to be highly accurate and the product promises to be inexpensive. They could take a very small amount of blood and detect cancer anywhere in the body in about five minutes, with a sensitivity that is a thousand times better than in a conventional laboratory test. These devices that are built with nanowires to detect cancer proteins; each nanowire detector is primed to be sensitive to a different cancer marker. The biggest advantage of the nanowire detectors is that they could test for anywhere from ten to one hundred similar medical conditions without adding cost to the testing device. Nanotechnology has also helped to personalize oncology for the detection, diagnosis, and treatment of cancer. It is now able to be tailored to each individual's tumor for better performance. They have found ways that they will be able to target a specific part of the body that is being affected by cancer.

Blood Purification

Magnetic micro particles are proven research instruments for the separation of cells and proteins from complex media. The technology is available under the name Magnetic-activated cell sorting or Dynabeads among others. More recently it was shown in animal models that magnetic nanoparticles can be used for the removal of various noxious compounds including toxins, pathogens, and proteins from whole blood in an extracorporeal circuit similar to dialysis. In contrast to dialysis, which works on the principle of the size related diffusion of solutes and ultrafiltration of fluid across a semi-permeable membrane, the purification with nanoparticles allows specific targeting of substances. Additionally larger compounds which are commonly not dialyzable can be removed.

The purification process is based on functionalized iron oxide or carbon coated metal nanoparticles with ferromagnetic or superparamagnetic properties. Binding agents such as proteins, antibodies, antibiotics, or synthetic ligands are covalently linked to the particle surface. These binding agents are able to interact with target species forming an agglomerate. Applying an external magnetic field gradient allows exerting a force on the nanoparticles. Hence the particles can be separated from the bulk fluid, thereby cleaning it from the contaminants.

The small size (< 100 nm) and large surface area of functionalized nanomagnets leads to advantageous properties compared to hemoperfusion, which is a clinically used technique for the purification of blood and is based on surface adsorption. These advantages are high loading and accessibility of the binding agents, high selectivity towards the target compound, fast diffusion, small hydrodynamic resistance, and low dosage.

This approach offers new therapeutic possibilities for the treatment of systemic infections such as sepsis by directly removing the pathogen. It can also be used to selectively remove cytokines or endotoxins or for the dialysis of compounds which are not accessible by traditional dialysis methods. However the technology is still in a preclinical phase and first clinical trials are not expected before 2017.

Tissue Engineering

Nanotechnology may be used as part of tissue engineering to help reproduce or repair or reshape damaged tissue using suitable nanomaterial-based scaffolds and growth factors. Tissue engineering if successful may replace conventional treatments like organ transplants or artificial implants. Nanoparticles such as graphene, carbon nanotubes, molybdenum disulfide and tungsten disulfide are being used as reinforcing agents to fabricate mechanically strong biodegradable polymeric nanocomposites for bone tissue engineering applications. The addition of these nanoparticles in the polymer matrix at low concentrations (~0.2 weight %) leads to significant improvements in the compressive and flexural mechanical properties of polymeric nanocomposites. Potentially, these nanocomposites may be used as a novel, mechanically strong, light weight composite as bone implants.

For example, a flesh welder was demonstrated to fuse two pieces of chicken meat into a single piece using a suspension of gold-coated nanoshells activated by an infrared laser. This could be used to weld arteries during surgery. Another example is nanonephrology, the use of nanomedicine on the kidney.

Medical Devices

Neuro-electronic interfacing is a visionary goal dealing with the construction of nanodevices that will permit computers to be joined and linked to the nervous system. This idea requires the building of a molecular structure that will permit control and detection of nerve impulses by an external computer.

A refuelable strategy implies energy is refilled continuously or periodically with external sonic, chemical, tethered, magnetic, or biological electrical sources, while a nonrefuelable strategy implies that all power is drawn from internal energy storage which would stop when all energy is drained. A nanoscale enzymatic biofuel cell for self-powered nanodevices have been developed that uses glucose from biofluids including human blood and watermelons. One limitation to this innovation is the fact that electrical interference or leakage or overheating from power consumption is possible. The wiring of the structure is extremely difficult because they must be positioned precisely in the nervous system. The structures that will provide the interface must also be compatible with the body's immune system.

Molecular nanotechnology is a speculative subfield of nanotechnology regarding the possibility of engineering molecular assemblers, machines which could re-order matter at a molecular or atomic scale. Nanomedicine would make use of these nanorobots, introduced into the body, to repair or detect damages and infections. Molecular nanotechnology is highly theoretical, seeking to anticipate what inventions nanotechnology might yield and to propose an agenda for future inquiry. The proposed elements of molecular nanotechnology, such as molecular assemblers and nanorobots are far beyond current capabilities. Future advances in nanomedicine could give rise to life extension through the repair of many processes thought to be responsible for aging. K. Eric Drexler, one of the founders of nanotechnology, postulated cell repair machines, including ones operating within cells and utilizing as yet hypothetical molecular machines, in his 1986 book Engines of Creation, with the first technical discussion of medical nanorobots by Robert Freitas appearing in 1999. Raymond Kurzweil, a futurist and transhumanist, stated in his book *The Singularity Is Near* that he believes that advanced medical nanorobotics could completely remedy the effects of aging by 2030. According to Richard Feynman, it was his former graduate student and collaborator Albert Hibbs who originally suggested to him (circa 1959) the idea of a *medical* use for Feynman's theoretical micromachines. Hibbs suggested that certain repair machines might one day be reduced in size to the point that it would, in theory, be possible to (as Feynman put it) "swallow the doctor". The idea was incorporated into Feynman's 1959 essay *There's Plenty of Room at the Bottom.*

Nanobiotechnology

Nanobiotechnology, bionanotechnology, and nanobiology are terms that refer to the intersection of nanotechnology and biology. Given that the subject is one that has only emerged very recently, bionanotechnology and nanobiotechnology serve as blanket terms for various related technologies.

This discipline helps to indicate the merger of biological research with various fields of nanotechnology. Concepts that are enhanced through nanobiology include: nanodevices (such as biological machines), nanoparticles, and nanoscale phenomena that occurs within the discipline of nanotechnology. This technical approach to biology allows scientists to imagine and create systems that can be used for biological research. Biologically inspired nanotechnology uses biological systems as the inspirations for technologies not yet created. However, as with nanotechnology and biotechnology, bionanotechnology does have many potential ethical issues associated with it.

The most important objectives that are frequently found in nanobiology involve applying nanotools to relevant medical/biological problems and refining these applications. Developing new tools, such as peptoid nanosheets, for medical and biological purposes is another primary objective in nanotechnology. New nanotools are often made by refining the applications of the nanotools that are already being used. The imaging of native biomolecules, biological membranes, and tissues is also a major topic for the nanobiology researchers. Other topics concerning nanobiology include the use of cantilever array sensors and the application of nanophotonics for manipulating molecular processes in living cells.

Recently, the use of microorganisms to synthesize functional nanoparticles has been of great interest. Microorganisms can change the oxidation state of metals. These microbial processes have opened up new opportunities for us to explore novel applications, for example, the biosynthesis of metal nanomaterials. In contrast to chemical and physical methods, microbial processes for synthesizing nanomaterials can be achieved in aqueous phase under gentle and environmentally benign conditions. This approach has become an attractive focus in current green bionanotechnology research towards sustainable development.

Terminology

The terms are often used interchangeably. When a distinction is intended, though, it is based on whether the focus is on applying biological ideas or on studying biology with nanotechnology. Bionanotechnology generally refers to the study of how the goals of nanotechnology can be guided by studying how biological "machines" work and adapting these biological motifs into improving existing nanotechnologies or creating new ones. Nanobiotechnology, on the other hand, refers to the ways that nanotechnology is used to create devices to study biological systems.

In other words, nanobiotechnology is essentially miniaturized biotechnology, whereas bionanotechnology is a specific application of nanotechnology. For example, DNA nanotechnology or cellular engineering would be classified as bionanotechnology because they involve working with biomolecules on the nanoscale. Conversely, many new medical technologies involving nanoparticles as delivery systems or as sensors would be examples of nanobiotechnology since they involve using nanotechnology to advance the goals of biology.

The definitions enumerated above will be utilized whenever a distinction between nanobio and bionano is made. However, given the overlapping usage of the terms in modern parlance, individual technologies may need to be evaluated to determine which term is more fitting. As such, they are best discussed in parallel.

Concepts

Most of the scientific concepts in bionanotechnology are derived from other fields. Biochemical principles that are used to understand the material properties of biological systems are central in bionanotechnology because those same principles are to be used to create new technologies. Material properties and applications studied in bionanoscience include mechanical properties (e.g. deformation, adhesion, failure), electrical/electronic (e.g. electromechanical stimulation, capacitors, energy storage/batteries), optical (e.g. absorption, luminescence, photochemistry), thermal (e.g. thermomutability, thermal management), biological (e.g. how cells interact with nanomaterials, molecular flaws/defects, biosensing, biological mechanisms s.a. mechanosensing), nanoscience of disease (e.g. genetic disease, cancer, organ/tissue failure), as well as computing (e.g. DNA computing) and agriculture (target delivery of pesticides, hormones and fertilizers. The impact of bionanoscience, achieved through structural and mechanistic analyses of biological processes at nanoscale, is their translation into synthetic and technological applications through nanotechnology.

Nano-biotechnology takes most of its fundamentals from nanotechnology. Most of the devices designed for nano-biotechnological use are directly based on other existing nanotechnologies. Nano-biotechnology is often used to describe the overlapping multidisciplinary activities associated with biosensors, particularly where photonics, chemistry, biology, biophysics, nano-medicine, and engineering converge. Measurement in biology using wave guide techniques, such as dual polarization interferometry, are another example.

Applications

Applications of bionanotechnology are extremely widespread. Insofar as the distinction holds, nanobiotechnology is much more commonplace in that it simply provides more tools for the study of biology. Bionanotechnology, on the other hand, promises to recreate biological mechanisms and pathways in a form that is useful in other ways.

Nanomedicine

Nanomedicine is a field of medical science whose applications are increasing more and more thanks to nanorobots and biological machines, which constitute a very useful tool to develop this area of knowledge. In the past years, researchers have done many improvements in the different devices and systems required to develop nanorobots. This supposes a new way of treating and dealing with diseases such as cancer; thanks to nanorobots, side effects of chemotherapy have been controlled, reduced and even eliminated, so some years from now, cancer patients will be offered an alternative to treat this disease instead of chemotherapy, which causes secondary effects such as hair loss, fatigue or nausea killing not only cancerous cells but also the healthy ones. At a clinical level, cancer treatment with nanomedicine will consist on the supply of nanorobots to the patient through an injection that will seek for cancerous cells leaving untouched the healthy ones. Patients that will be treated through nanomedicine will not notice the presence of this nanomachines inside them; the only thing that is going to be noticeable is the progressive improvement of their health.

Nanobiotechnology

Nanobiotechnology (sometimes referred to as nanobiology) is best described as helping modern medicine progress from treating symptoms to generating cures and regenerating biological tissues. Three American patients have received whole cultured bladders with the help of doctors who use nanobiology techniques in their practice. Also, it has been demonstrated in animal studies that a uterus can be grown outside the body and then placed in the body in order to produce a baby. Stem cell treatments have been used to fix diseases that are found in the human heart and are in clinical trials in the United States. There is also funding for research into allowing people to have new limbs without having to resort to prosthesis. Artificial proteins might also become available to manufacture without the need for harsh chemicals and expensive machines. It has even been surmised that by the year 2055, computers may be made out of biochemicals and organic salts.

Another example of current nanobiotechnological research involves nanospheres coated with fluorescent polymers. Researchers are seeking to design polymers whose fluorescence is quenched when they encounter specific molecules. Different polymers would detect different metabolites. The polymer-coated spheres could become part of new biological assays, and the technology might someday lead to particles which could be introduced into the human body to track down metabolites associated with tumors and other health problems. Another example, from a different perspective, would be evaluation and therapy at the nanoscopic level, i.e. the treatment of Nanobacteria (25-200 nm sized) as is done by NanoBiotech Pharma.

While nanobiology is in its infancy, there are a lot of promising methods that will rely on nanobiology in the future. Biological systems are inherently nano in scale; nanoscience must merge with biology in order to deliver biomacromolecules and molecular machines that are similar to nature. Controlling and mimicking the devices and processes that are constructed from molecules is a tremendous challenge to face the converging disciplines of nanotechnology. All living things, including humans, can be considered to be nanofoundries. Natural evolution has optimized the "natural" form of nanobiology over millions of years. In the 21st century, humans have developed the technology to artificially tap into nanobiology. This process is best described as "organic merging with synthetic." Colonies of live neurons can live together on a biochip device; according to research from Dr. Gunther Gross at the University of North Texas. Self-assembling nanotubes have the ability to be used as a structural system. They would be composed together with rhodopsins; which would facilitate the optical computing process and help with the storage of biological materials. DNA (as the software for all living things) can be used as a structural proteomic system - a logical component for molecular computing. Ned Seeman - a researcher at New York University - along with other researchers are currently researching concepts that are similar to each other.

Bionanotechnology

DNA nanotechnology is one important example of bionanotechnology. The utilization of the inherent properties of nucleic acids like DNA to create useful materials is a promising area of modern research. Another important area of research involves taking advantage of membrane properties to generate synthetic membranes. Proteins that self-assemble to generate functional materials could be used as a novel approach for the large-scale production of programmable nanomaterials. One example is the development of amyloids found in bacterial biofilms as engineered nanomaterials that can be programmed genetically to have different properties. Protein folding studies provide a third important avenue of research, but one that has been largely inhibited by our inability to predict protein folding with a sufficiently high degree of accuracy. Given the myriad uses that biological systems have for proteins, though, research into understanding protein folding is of high importance and could prove fruitful for bionanotechnology in the future.

Lipid nanotechnology is another major area of research in bionanotechnology, where physico-chemical properties of lipids such as their antifouling and self-assembly is exploited to build nanodevices with applications in medicine and engineering.

Agriculture

Meanwhile, nanotechnology application to biotechnology will also leave no field untouched by its groundbreaking scientific innovations for human wellness; the agricultural industry is no exception. Basically, nanomaterials are distinguished depending on the origin: natural, incidental and engineered nanoparticles. Among these, engineered nanoparticles have received wide attention in all fields of science, including medical, materials and agriculture technology with significant socio-economical growth. In the agriculture industry, engineered nanoparticles have been serving as nano carrier, containing herbicides, chemicals, or genes, which target particular plant parts to release their content. Previously nanocapsules containing herbicides have been reported to effectively penetrate through cuticles and tissues, allowing the slow and constant release of the active substances. Likewise, other literature describes that nano-encapsulated slow release of fertilizers has also become a trend to save fertilizer consumption and to minimize environmental pollution through precision farming. These are only a few examples from numerous research works which might open up exciting opportunities for nanobiotechnology application in agriculture. Also, application of this kind of engineered nanoparticles to plants should be considered the level of amicability before it is employed in agriculture practices. Based on a thorough literature survey, it was understood that there is only limited authentic information available to explain the biological consequence of engineered nanoparticles on treated plants. Certain reports underline the phytotoxicity of various origin of engineered nanoparticles to the plant caused by the subject of concentrations and sizes. At the same time, however, an equal number of studies were reported with a positive outcome of nanoparticles, which facilitate growth promoting nature to treat plant. In particular, compared to other nanoparticles, silver and gold nanoparticles based applications elicited beneficial results on various plant species with less and/or no toxicity. Silver nanoparticles (AgNPs) treated leaves of Asparagus showed the increased content of ascorbate and chlorophyll. Similarly, AgNPs-treated common bean and corn has increased shoot and root length, leaf surface area, chlorophyll, carbohydrate and protein contents reported earlier. The gold nanoparticle has been used to induce growth and seed yield in Brassica juncea.

Tools

This field relies on a variety of research methods, including experimental tools (e.g. imaging, characterization via AFM/optical tweezers etc.), x-ray diffraction based tools, synthesis via self-assembly, characterization of self-assembly (using e.g. MP-SPR, DPI, recombinant DNA methods, etc.), theory (e.g. statistical mechanics, nanomechanics, etc.), as well as computational approaches (bottom-up multi-scale simulation, supercomputing).

Nanoreactor

Nanoreactors are a form of chemical reactor that are particularly in the disciplines of nanotechnology and nanobiotechnology. These special reactors are crucial in maintaining a working nanofoundry; which is essentially a foundry that manufactures products on a nanotechnological scale.

General Information

Researchers in the Netherlands have succeeded in building nanoreactors that can perform one-pot multistep reactions - the next step towards artificial cell-like devices in addition for applications involving the screening and diagnosis of a disease or illness. A biochemical nanoreactor is created simply by unwrapping a biological virus through scientific methods, eliminating its harmful contents, and re-assembling its protein coat around a single molecule of enzyme. The kinetic isotope effect is trapped in a single molecule within a membrane-based nanoreactor. This is a phenomenon that has been found by researchers in the United Kingdom during experiments done on September 2010. The kinetic isotope effect, where the rate of a reaction is influenced by the presence of an isotopic atom in solution, is an important principle for elucidating reaction mechanisms. This recent finding could open up new methods to study chemical reactions. They may even aid in the process of creating new (and even more powerful) nanoreactors.

Using nanocrystals, a scalable and inexpensive process can ultimately create nanoreactors. Researchers at the Lawrence Berkeley National Laboratory in Berkeley have the ability to take advantage of the large difference in select components to create these nanocrystals and nanoreactors. Nanocrystals are easier to use and less expensive than methods that employ sacrificial templates in the creation process of hollow particles. Catalyst particles are separated into shells in order to prevent particle aggregation. Selective entry into the catalysis chamber reduces the likelihood of desired products undergoing secondary reactions.

Nanoreactors can also be built by controlling the positioning of two different enzymes in the central water reservoir or the plastic membrane of synthetic nanoscopic bubbles. Once the third enzyme is added into the surrounding solution, it becomes possible for three different enzymatic reactions to occur at once without interfering with each other (resulting in a "one-pot" reaction). The potential for nanoreactors can be demonstrated by binding the enzyme horseradish peroxidase into the membrane itself; trapping the enzyme glucose oxidase. The surrounding solution would end up containing the enzyme lipase B with the glucose molecules containing four acetyl groups as the substrate. The resulting glucose would cross the membrane, become oxidized, and the horseradish peroxidase would convert the sample substrate ABTS (2,2'-azinobis(3-ethylbenzthiazoline-6-sulfonic acid)) into its radical cation.

Abilities

Nanoreactors can also be used to emulisfy water, create hydrofuels (which essentially blends 15% water into the refined diesel product), play a helpful role in the chemical industry by allowing multiple streams of raw materials to exists in a single nanoreactor, manufacture personal care products (i.e., lotions, pharmaceutical creams, shampoos, conditioners, shower gels, deodorants), and improve the food and beverage industries (by processing sauces, purées, cooking bases for soup, emulsifying non-alcoholic beverages, and salad dressings).

Personal care goods can be enhanced by companies feeding multiple phases of material, using a mixing device with water, and creating instant emulsions. These emulsions would come with smaller particles, are expected to have a longer shelf life and an give off an enhanced appearance when sold at retailers. The needs of the food and beverage industry can result in lower processing costs, more space, better efficiency, and lower equipment costs. This may bring down the cost of food and beverages for consumers; even alcoholic beverages that are subject to hidden sin taxes.

Hydrofuel can be used to move heavy duty transports, trains, earth-moving equipment (including bulldozers), in addition to providing fuel to most boats and ships. Reduced pollution and increased fuel efficiency may come out of nanoreactor-produced hydrofuel. The increased usage of renewable energy may also help to improve the world's environment thanks to nanoreactors.

Green Nanotechnology

Green nanotechnology refers to the use of nanotechnology to enhance the environmental sustainability of processes producing negative externalities. It also refers to the use of the products of nanotechnology to enhance sustainability. It includes making green nano-products and using nano-products in support of sustainability.

Green nanotechnology has been described as the development of clean technologies, "to minimize potential environmental and human health risks associated with the manufacture and use of nanotechnology products, and to encourage replacement of existing products with new nano-products that are more environmentally friendly throughout their lifecycle."

Goals

Green nanotechnology has two goals: producing nanomaterials and products without harming the environment or human health, and producing nano-products that provide solutions to environmental problems. It uses existing principles of green chemistry and green engineering to make nanomaterials and nano-products without toxic ingredients, at low temperatures using less energy and renewable inputs wherever possible, and using lifecycle thinking in all design and engineering stages.

In addition to making nanomaterials and products with less impact to the environment, green nanotechnology also means using nanotechnology to make current manufacturing processes for non-nano materials and products more environmentally friendly. For example, nanoscale membranes can help separate desired chemical reaction products from waste materials. Nanoscale catalysts can make chemical reactions more efficient and less wasteful. Sensors at the nanoscale can form a part of process control systems, working with nano-enabled information systems. Using alternative energy systems, made possible by nanotechnology, is another way to "green" manufacturing processes.

The second goal of green nanotechnology involves developing products that benefit the environment either directly or indirectly. Nanomaterials or products directly can clean hazardous waste sites, desalinate water, treat pollutants, or sense and monitor environmental pollutants. Indirectly, lightweight nanocomposites for automobiles and other means of transportation could save fuel and reduce materials used for production; nanotechnology-enabled fuel cells and light-emitting diodes (LEDs) could reduce pollution from energy generation and help conserve fossil fuels; self-cleaning nanoscale surface coatings could reduce or eliminate many cleaning chemicals used in regular maintenance routines; and enhanced battery life could lead to less material use and less waste. Green Nanotechnology takes a broad systems view of nanomaterials and products, ensuring that unforeseen consequences are minimized and that impacts are anticipated throughout the full life cycle.

Current Research

Solar Cells

Research is underway to use nanomaterials for purposes including more efficient solar cells, practical fuel cells, and environmentally friendly batteries. The most advanced nanotechnology projects related to energy are: storage, conversion, manufacturing improvements by reducing materials and process rates, energy saving (by better thermal insulation for example), and enhanced renewable energy sources.

One major project that is being worked on is the development of nanotechnology in solar cells. Solar cells are more efficient as they get tinier and solar energy is a renewable resource. The price per watt of solar energy is lower than one dollar.

Research is ongoing to use nanowires and other nanostructured materials with the hope of to create cheaper and more efficient solar cells than are possible with conventional planar silicon solar cells. Another example is the use of fuel cells powered by hydrogen, potentially using a catalyst consisting of carbon supported noble metal particles with diameters of 1-5 nm. Materials with small nanosized pores may be suitable for hydrogen storage. Nanotechnology may also find applications in batteries, where the use of nanomaterials may enable batteries with higher energy content or supercapacitors with a higher rate of recharging.

Nanotechnology is already used to provide improved performance coatings for photovoltaic (PV) and solar thermal panels. Hydrophobic and self-cleaning properties combine to create more efficient solar panels, especially during inclement weather. PV covered with nanotechnology coatings are said to stay cleaner for longer to ensure maximum energy efficiency is maintained.

Nanoremediation and Water Treatment

Nanotechnology offers the potential of novel nanomaterials for the treatment of surface water, groundwater, wastewater, and other environmental materials contaminated by toxic metal ions, organic and inorganic solutes, and microorganisms. Due to their unique activity toward recalcitrant contaminants, many nanomaterials are under active research and development for use in the treatment of water and contaminated sites.

The present market of nanotech-based technologies applied in water treatment consists of reverse osmosis, nanofiltration, ultrafiltration membranes. Indeed, among emerging products one can name nanofiber filters, carbon nanotubes and various nanoparticles. Nanotechnology is expected to deal more efficiently with contaminants which convectional water treatment systems struggle to treat, including bacteria, viruses and heavy metals. This efficiency generally stems from the very high specific surface area of nanomaterials which increases dissolution, reactivity and sorption of contaminants.

Some potential applications include:

- To maintain public health, pathogens in water need to be identified rapidly and reliably. Unfortunately, traditional laboratory culture tests take days to complete. Faster methods involving enzymes, immunological or genetic tests are under development.
- Water filtration may be improved with the use of nanofiber membranes and the use of nanobiocides, which appear promisingly effective.
- Biofilms are mats of bacteria wrapped in natural polymers. These can be difficult to treat with antimicrobials or other chemicals. They can be cleaned up mechanically, but at the

cost of substantial down-time and labour. Work is in progress to develop enzyme treatments that may be able to break down such biofilms.

Environmental Remediation

Nanoremediation is the use of nanoparticles for environmental remediation. Nanoremediation has been most widely used for groundwater treatment, with additional extensive research in wastewater treatment. Nanoremediation has also been tested for soil and sediment cleanup. Even more preliminary research is exploring the use of nanoparticles to remove toxic materials from gases.

Some nanoremediation methods, particularly the use of nano zero-valent iron for groundwater cleanup, have been deployed at full-scale cleanup sites. Nanoremediation is an emerging industry; by 2009, nanoremediation technologies had been documented in at least 44 cleanup sites around the world, predominantly in the United States. During nanoremediation, a nanoparticle agent must be brought into contact with the target contaminant under conditions that allow a detoxifying or immobilizing reaction. This process typically involves a pump-and-treat process or *in situ* application. Other methods remain in research phases.

Scientists have been researching the capabilities of buckminsterfullerene in controlling pollution, as it may be able to control certain chemical reactions. Buckminsterfullerene has been demonstrated as having the ability of inducing the protection of reactive oxygen species and causing lipid peroxidation. This material may allow for hydrogen fuel to be more accessible to consumers.

Water Cleaning Technology

In 2017 the RingwooditE Co Ltd was formed in order to explore Thermonuclear Trap Technology (TTT) for the purpose of cleaning all sources of water from pollution and toxic contents. This patented nanotechnology uses a high pressure and temperature chamber to separate isotopes that should by nature not be in drinking water to pure drinking water, as to the by the WHO´s established classification. This methode has been developed by among others, by professor Vladimir Afanasiew, at the Moscow Nuclear Institution. This technology is targeted to clean Sea, river, lake and landfill weist waters. It even removes radioactive isothopes from the sea water, after Nuclear Power Stations catastrophes and cooling water plant towers. By this technology pharmaca rests are being removed as well as narcotics and tranquilizers. Bottom layers and sides at lake and rivers can be returned, after being cleaned. Machinery used for this purpose are much similar to those of deep sea mining. Removed waste items are being sorted by the process, and can be re used as raw material for other industrial production.

Water Filtration

Nanofiltration is a relatively recent membrane filtration process used most often with low total dissolved solids water such as surface water and fresh groundwater, with the purpose of softening (polyvalent cation removal) and removal of disinfection by-product precursors such as natural organic matter and synthetic organic matter. Nanofiltration is also becoming more widely used in food processing applications such as dairy, for simultaneous concentration and partial (monovalent ion) demineralisation.

Nanofiltration is a membrane filtration based method that uses nanometer sized cylindrical through-pores that pass through the membrane at a 90°. Nanofiltration membranes have pore sizes from 1-10 Angstrom, smaller than that used in microfiltration and ultrafiltration, but just larger than that in reverse osmosis. Membranes used are predominantly created from polymer thin films. Materials that are commonly used include polyethylene terephthalate or metals such as aluminum. Pore dimensions are controlled by pH, temperature and time during development with pore densities ranging from 1 to 106 pores per cm^2. Membranes made from polyethylene terephthalate and other similar materials, are referred to as "track-etch" membranes, named after the way the pores on the membranes are made. "Tracking" involves bombarding the polymer thin film with high energy particles. This results in making tracks that are chemically developed into the membrane, or "etched" into the membrane, which are the pores. Membranes created from metal such as alumina membranes, are made by electrochemically growing a thin layer of aluminum oxide from aluminum metal in an acidic medium.

Some water-treatment devices incorporating nanotechnology are already on the market, with more in development. Low-cost nanostructured separation membranes methods have been shown to be effective in producing potable water in a recent study.

Nanoelectronics

Nanoelectronics refer to the use of nanotechnology in electronic components. The term covers a diverse set of devices and materials, with the common characteristic that they are so small that inter-atomic interactions and quantum mechanical properties need to be studied extensively. Some of these candidates include: hybrid molecular/semiconductor electronics, one-dimensional nanotubes/nanowires (e.g. Silicon nanowires or Carbon nanotubes) or advanced molecular electronics. Recent silicon CMOS technology generations, such as the 22 nanometer node, are already within this regime. Nanoelectronics are sometimes considered as disruptive technology because present candidates are significantly different from traditional transistors.

Fundamental Concepts

In 1965 Gordon Moore observed that silicon transistors were undergoing a continual process of scaling downward, an observation which was later codified as Moore's law. Since his observation transistor minimum feature sizes have decreased from 10 micrometers to the 28-22 nm range in 2011. The field of nanoelectronics aims to enable the continued realization of this law by using new methods and materials to build electronic devices with feature sizes on the nanoscale.

The volume of an object decreases as the third power of its linear dimensions, but the surface area only decreases as its second power. This somewhat subtle and unavoidable principle has huge ramifications. For example, the power of a drill (or any other machine) is proportional to the volume, while the friction of the drill's bearings and gears is proportional to their surface area. For a normal-sized drill, the power of the device is enough to handily overcome any friction. However, scaling its length down by a factor of 1000, for example, decreases its power by 1000^3 (a factor of a billion) while reducing the friction by only 1000^2 (a factor of only a million). Proportionally it has 1000 times less power per unit friction than the original drill. If the original friction-to-power ratio was, say, 1%, that implies the smaller drill will have 10 times as much friction as power; the drill is useless.

For this reason, while super-miniature electronic integrated circuits are fully functional, the same technology cannot be used to make working mechanical devices beyond the scales where frictional forces start to exceed the available power. So even though you may see microphotographs of delicately etched silicon gears, such devices are currently little more than curiosities with limited real world applications, for example, in moving mirrors and shutters. Surface tension increases in much the same way, thus magnifying the tendency for very small objects to stick together. This could possibly make any kind of "micro factory" impractical: even if robotic arms and hands could be scaled down, anything they pick up will tend to be impossible to put down. The above being said, molecular evolution has resulted in working cilia, flagella, muscle fibers and rotary motors in aqueous environments, all on the nanoscale. These machines exploit the increased frictional forces found at the micro or nanoscale. Unlike a paddle or a propeller which depends on normal frictional forces (the frictional forces perpendicular to the surface) to achieve propulsion, cilia develop motion from the exaggerated drag or laminar forces (frictional forces parallel to the surface) present at micro and nano dimensions. To build meaningful "machines" at the nanoscale, the relevant forces need to be considered. We are faced with the development and design of intrinsically pertinent machines rather than the simple reproductions of macroscopic ones.

All scaling issues therefore need to be assessed thoroughly when evaluating nanotechnology for practical applications.

Approaches to Nanoelectronics

Nanofabrication

For example, single electron transistors, which involve transistor operation based on a single electron. Nanoelectromechanical systems also fall under this category. Nanofabrication can be used to construct ultradense parallel arrays of nanowires, as an alternative to synthesizing nanowires individually.

Nanomaterials Electronics

Besides being small and allowing more transistors to be packed into a single chip, the uniform and symmetrical structure of nanowires and/or nanotubes allows a higher electron mobility (faster electron movement in the material), a higher dielectric constant (faster frequency), and a symmetrical electron/hole characteristic.

Also, nanoparticles can be used as quantum dots.

Molecular Electronics

Single molecule devices are another possibility. These schemes would make heavy use of molecular self-assembly, designing the device components to construct a larger structure or even a complete system on their own. This can be very useful for reconfigurable computing, and may even completely replace present FPGA technology.

Molecular electronics is a new technology which is still in its infancy, but also brings hope for truly atomic scale electronic systems in the future. One of the more promising applications of molecular electronics was proposed by the IBM researcher Ari Aviram and the theoretical chemist Mark Ratner in their 1974 and 1988 papers *Molecules for Memory, Logic and Amplification.*

This is one of many possible ways in which a molecular level diode / transistor might be synthesized by organic chemistry. A model system was proposed with a spiro carbon structure giving a molecular diode about half a nanometre across which could be connected by polythiophene molecular wires. Theoretical calculations showed the design to be sound in principle and there is still hope that such a system can be made to work.

Other Approaches

Nanoionics studies the transport of ions rather than electrons in nanoscale systems.

Nanophotonics studies the behavior of light on the nanoscale, and has the goal of developing devices that take advantage of this behavior.

Nanoelectronic Devices

Current high-technology production processes are based on traditional top down strategies, where nanotechnology has already been introduced silently. The critical length scale of integrated circuits is already at the nanoscale (50 nm and below) regarding the gate length of transistors in CPUs or DRAM devices.

Computers

Nanoelectronics holds the promise of making computer processors more powerful than are possible with conventional semiconductor fabrication techniques. A number of approaches are currently being researched, including new forms of nanolithography, as well as the use of nanomaterials such as nanowires or small molecules in place of traditional CMOS components. Field effect transistors have been made using both semiconducting carbon nanotubes and with heterostructured semiconductor nanowires (SiNWs).

In 1999, the CMOS transistor developed at the Laboratory for Electronics and Information Technology in Grenoble, France, tested the limits of the principles of the MOSFET transistor with a diameter of 18 nm (approximately 70 atoms placed side by side). This was almost one tenth the size of the smallest industrial transistor in 2003 (130 nm in 2003, 90 nm in 2004, 65 nm in 2005 and 45 nm in 2007). It enabled the theoretical integration of seven billion junctions on a €1 coin. However, the CMOS transistor, which was created in 1999, was not a simple research experiment to study how CMOS technology functions, but rather a demonstration of how this technology functions now that we ourselves are getting ever closer to working on a molecular scale. Today it would be impossible to master the coordinated assembly of a large number of these transistors on a circuit and it would also be impossible to create this on an industrial level.

Memory Storage

Electronic memory designs in the past have largely relied on the formation of transistors. However, research into crossbar switch based electronics have offered an alternative using reconfigurable interconnections between vertical and horizontal wiring arrays to create ultra high density memories. Two leaders in this area are Nantero which has developed a carbon nanotube based crossbar memory called Nano-RAM and Hewlett-Packard which has proposed the use of memristor material as a future replacement of Flash memory.

An example of such novel devices is based on spintronics.The dependence of the resistance of a material (due to the spin of the electrons) on an external field is called magnetoresistance. This effect can be significantly amplified (GMR - Giant Magneto-Resistance) for nanosized objects, for example when two ferromagnetic layers are separated by a nonmagnetic layer, which is several nanometers thick (e.g. Co-Cu-Co). The GMR effect has led to a strong increase in the data storage density of hard disks and made the gigabyte range possible. The so-called tunneling magnetoresistance (TMR) is very similar to GMR and based on the spin dependent tunneling of electrons through adjacent ferromagnetic layers. Both GMR and TMR effects can be used to create a non-volatile main memory for computers, such as the so-called magnetic random access memory or MRAM.

Novel Optoelectronic Devices

In the modern communication technology traditional analog electrical devices are increasingly replaced by optical or optoelectronic devices due to their enormous bandwidth and capacity, respectively. Two promising examples are photonic crystals and quantum dots. Photonic crystals are materials with a periodic variation in the refractive index with a lattice constant that is half the wavelength of the light used. They offer a selectable band gap for the propagation of a certain wavelength, thus they resemble a semiconductor, but for light or photons instead of electrons. Quantum dots are nanoscaled objects, which can be used, among many other things, for the construction of lasers. The advantage of a quantum dot laser over the traditional semiconductor laser is that their emitted wavelength depends on the diameter of the dot. Quantum dot lasers are cheaper and offer a higher beam quality than conventional laser diodes.

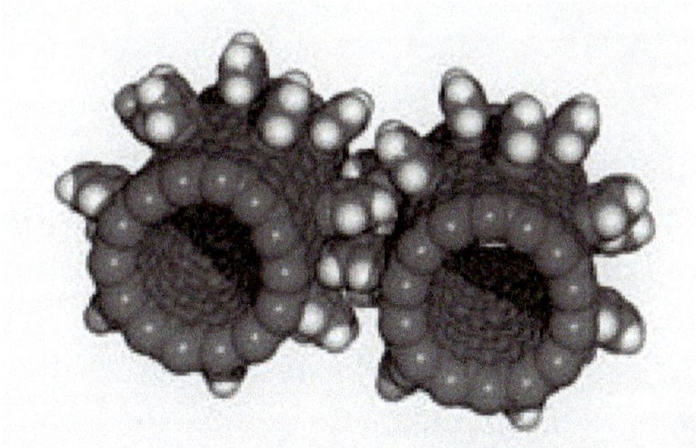

Displays

The production of displays with low energy consumption might be accomplished using carbon nanotubes (CNT) and/or Silicon nanowires. Such nanostructures are electrically conductive and due to their small diameter of several nanometers, they can be used as field emitters with extremely high efficiency for field emission displays (FED). The principle of operation resembles that of the cathode ray tube, but on a much smaller length scale.

Quantum Computers

Entirely new approaches for computing exploit the laws of quantum mechanics for novel quantum computers, which enable the use of fast quantum algorithms. The Quantum computer has quantum bit memory space termed "Qubit" for several computations at the same time. This facility may improve the performance of the older systems.

Radios

Nanoradios have been developed structured around carbon nanotubes.

Energy Production

Research is ongoing to use nanowires and other nanostructured materials with the hope to create cheaper and more efficient solar cells than are possible with conventional planar silicon solar cells. It is believed that the invention of more efficient solar energy would have a great effect on satisfying global energy needs.

There is also research into energy production for devices that would operate *in vivo*, called bio-nano generators. A bio-nano generator is a nanoscale electrochemical device, like a fuel cell or galvanic cell, but drawing power from blood glucose in a living body, much the same as how the body generates energy from food. To achieve the effect, an enzyme is used that is capable of stripping glucose of its electrons, freeing them for use in electrical devices. The average person's body could, theoretically, generate 100 watts of electricity (about 2000 food calories per day) using a bio-nano generator. However, this estimate is only true if all food was converted to electricity, and the human body needs some energy consistently, so possible power generated is likely much lower. The electricity generated by such a device could power devices embedded in the body (such as pacemakers), or sugar-fed nanorobots. Much of the research done on bio-nano generators is still experimental, with Panasonic's Nanotechnology Research Laboratory among those at the forefront.

Medical Diagnostics

There is great interest in constructing nanoelectronic devices that could detect the concentrations of biomolecules in real time for use as medical diagnostics, thus falling into the category of nanomedicine. A parallel line of research seeks to create nanoelectronic devices which could interact with single cells for use in basic biological research. These devices are called nanosensors. Such miniaturization on nanoelectronics towards in vivo proteomic sensing should enable new approaches for health monitoring, surveillance, and defense technology.

Nanotoxicology

Nanotoxicology is the study of the toxicity of nanomaterials. Because of quantum size effects and large surface area to volume ratio, nanomaterials have unique properties compared with their larger counterparts.

Nanotoxicology is a branch of bionanoscience which deals with the study and application of toxicity of nanomaterials. Nanomaterials, even when made of inert elements like gold, become highly active at nanometer dimensions. Nanotoxicological studies are intended to determine whether and to what extent these properties may pose a threat to the environment and to human beings. For instance, Diesel nanoparticles have been found to damage the cardiovascular system in a mouse model.

Background

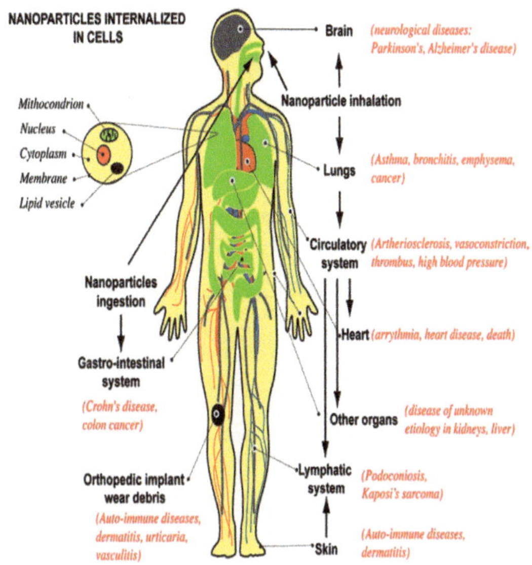

Pathways of exposure to nanoparticles and associated diseases as suggested by epidemiological, in vivo and in vitro studies.

Nanotoxicology is a sub-specialty of particle toxicology. It addresses the toxicology of nanoparticles (particles <100 nm diameter) which appear to have toxicity effects that are unusual and not seen with larger particles. Nanoparticles can be divided into combustion-derived nanoparticles (like diesel soot), manufactured nanoparticles like carbon nanotubes and naturally occurring nanoparticles from volcanic eruptions, atmospheric chemistry etc. Typical nanoparticles that have been studied are titanium dioxide, alumina, zinc oxide, carbon black, and carbon nanotubes, and "nano-C_{60}". Nanoparticles have much larger surface area to unit mass ratios which in some cases may lead to greater pro-inflammatory effects (in, for example, lung tissue). In addition, some nanoparticles seem to be able to translocate from their site of deposition to distant sites such as the blood and the brain. This has resulted in a sea-change in how particle toxicology is viewed- instead of being confined to the lungs, nanoparticle toxicologists study the brain, blood, liver, skin and gut.

Calls for tighter regulation of nanotechnology have arisen alongside a growing debate related to the human health and safety risks associated with nanotechnology. From a large-scale literature review, Yaobo Ding et al. found that release of airborne engineered nanoparticles and associated worker exposure from various production and handling activities at different workplaces are very probable. The Royal Society identifies the potential for nanoparticles to penetrate the skin, and recommends that the use of nanoparticles in cosmetics be conditional upon a favorable assessment by the relevant European Commission safety advisory committee. Andrew Maynard also reports that 'certain nanoparticles may move easily into sensitive lung tissues after inhalation, and cause damage that can lead to chronic breathing problems'.

Carbon nanotubes – characterized by their microscopic size and incredible tensile strength – are frequently likened to asbestos, due to their needle-like fiber shape. In a recent study that introduced carbon nanotubes into the abdominal cavity of mice, results demonstrated that long thin carbon nanotubes showed the same effects as long thin asbestos fibers, raising concerns that exposure to carbon nanotubes may lead to pleural abnormalities such as mesothelioma (cancer of the lining of the lungs caused by exposure to asbestos). Given these risks, effective and rigorous regulation has been called for to determine if, and under what circumstances, carbon nanotubes are manufactured, as well as ensuring their safe handling and disposal.

The Woodrow Wilson Centre's Project on Emerging Technologies conclude that there is insufficient funding for human health and safety research, and as a result there is currently limited understanding of the human health and safety risks associated with nanotechnology. While the US National Nanotechnology Initiative reports that around four percent (about $40 million) is dedicated to risk related research and development, the Woodrow Wilson Centre estimate that only around $11 million is actually directed towards risk related research. They argued in 2007 that it would be necessary to increase funding to a minimum of $50 million in the following two years so as to fill the gaps in knowledge in these areas.

The potential for workplace exposure was highlighted by the 2004 Royal Society report which recommended a review of existing regulations to assess and control workplace exposure to nanoparticles and nanotubes. The report expressed particular concern for the inhalation of large quantities of nanoparticles by workers involved in the manufacturing process.

Stakeholders concerned by the lack of a regulatory framework to assess and control risks associated with the release of nanoparticles and nanotubes have drawn parallels with bovine spongiform encephalopathy ('mad cow's disease'), thalidomide, genetically modified food, nuclear energy, reproductive technologies, biotechnology, and asbestosis. In light of such concerns, the Canadian-based ETC Group have called for a moratorium on nano-related research until comprehensive regulatory frameworks are developed that will ensure workplace safety.

Reactive Oxygen Species

For some types of particles, the smaller they are, the greater their surface area to volume ratio and the higher their chemical reactivity and biological activity. The greater chemical reactivity of nanomaterials can result in increased production of reactive oxygen species (ROS), including free radicals. ROS production has been found in a diverse range of nanomaterials including carbon fullerenes, carbon nanotubes and nanoparticle metal oxides. ROS and free radical production is one of the primary mechanisms of nanoparticle toxicity; it may result in oxidative stress, inflammation, and consequent damage to proteins, membranes and DNA.

Biodistribution

The extremely small size of nanomaterials also means that they much more readily gain entry into the human body than larger sized particles. How these nanoparticles behave inside the body is still a major question that needs to be resolved. The behavior of nanoparticles is a function of their size, shape and surface reactivity with the surrounding tissue. In principle, a large number of particles could overload the body's phagocytes, cells that ingest and destroy foreign matter, thereby triggering stress reactions that lead to inflammation and weaken the body's defense against other pathogens. In addition to questions about what happens if non-degradable or slowly degradable nanoparticles accumulate in bodily organs, another concern is their potential interaction or interference with biological processes inside the body. Because of their large surface area, nanoparticles will, on exposure to tissue and fluids, immediately adsorb onto their surface some of the macromolecules they encounter. This may, for instance, affect the regulatory mechanisms of enzymes and other proteins.

Nanomaterials are able to cross biological membranes and access cells, tissues and organs that larger-sized particles normally cannot. Nanomaterials can gain access to the blood stream via inhalation or ingestion. At least some nanomaterials can penetrate the skin; even larger microparticles may penetrate skin when it is flexed. Broken skin is an ineffective particle barrier, suggesting that acne, eczema, shaving wounds or severe sunburn may accelerate skin uptake of nanomaterials. Then, once in the blood stream, nanomaterials can be transported around the body and be taken up by organs and tissues, including the brain, heart, liver, kidneys, spleen, bone marrow and nervous system. Nanomaterials have proved toxic to human tissue and cell cultures, resulting in increased oxidative stress, inflammatory cytokine production and cell death. Unlike larger particles, nanomaterials may be taken up by cell mitochondria and the cell nucleus. Studies demonstrate the potential for nanomaterials to cause DNA mutation and induce major structural damage to mitochondria, even resulting in cell death.

Nanotoxicity Studies

There is presently no authority to specifically regulate nanotech-based products. Scientific research has indicated the potential for some nanomaterials to be toxic to humans or the environment. In March 2004 tests conducted by environmental toxicologist Eva Oberdörster, Ph.D. working with Southern Methodist University in Texas, found extensive brain damage to fish exposed to fullerenes for a period of just 48 hours at a relatively moderate dose of 0.5 parts per million (commensurate with levels of other kinds of pollution found in bays). The fish also exhibited changed gene markers in their livers, indicating their entire physiology was affected. In a concurrent test, the fullerenes killed water fleas, an important link in the marine food chain. The extremely small size of fabricated nanomaterials also means that they are much more readily taken up by living tissue than presently known toxins. Nanoparticles can be inhaled, swallowed, absorbed through skin and deliberately or accidentally injected during medical procedures. They might be accidentally or inadvertently released from materials implanted into living tissue.

Researcher Shosaku Kashiwada of the National Institute for Environmental Studies in Tsukuba, Japan, in a more recent study, intended to further investigate the effects of nanoparticles on soft-bodied organisms. His study allowed him to explore the distribution of water-suspended fluorescent nanoparticles throughout the eggs and adult bodies of a species of fish, known as the see-through medaka (Oryzias latipes). See-through medaka were used because of their small size, wide temperature and salinity tolerances, and short generation time. Moreover, small fish like the see-through medaka have been popular test subjects for human diseases and organogenesis for other reasons as well, including their transparent embryos, rapid embryo development, and the functional equivalence of their organs and tissue material to that of mammals. Because the see-through medaka have transparent bodies, analyzing the deposition of fluorescent nanoparticles throughout the body is quite simple. For his study, Dr. Kashiwada evaluated four aspects of nanoparticle accumulation. These included the overall accumulation and the size-dependent accumulation of nanoparticles by medaka eggs, the effects of salinity on the aggregation of nanoparticles in solution and on their accumulation by medaka eggs, and the distribution of nanoparticles in the blood and organs of adult medaka. It was also noted that nanoparticles were in fact taken up into the bloodstream and deposited throughout the body. In the medaka eggs, there was a high accumulation of nanoparticles in the yolk; most often bioavailibility was dependent on specific sizes of the particles. Adult samples of medaka had accumulated nanoparticles in the gills, intestine, brain, testis, liver, and bloodstream. One major result from this study was the fact that salinity may have a large influence on the bioavailibility and toxicity of nanoparticles to penetrate membranes and eventually kill the specimen.

As the use of nanomaterials increases worldwide, concerns for worker and user safety are mounting. To address such concerns, the Swedish Karolinska Institute conducted a study in which various nanoparticles were introduced to human lung epithelial cells. The results, released in 2008, showed that iron oxide nanoparticles caused little DNA damage and were non-toxic. Zinc oxide nanoparticles were slightly worse. Titanium dioxide caused only DNA damage. Carbon nanotubes caused DNA damage at low levels. Copper oxide was found to be the worst offender, and was the only nanomaterial identified by the researchers as a clear health risk. The latest toxicology studies on mice involving exposure to carbon nanotubes (CNT) showed a limited pulmonary inflammatory potential of MWCNT at levels corresponding to the average inhalable elemental carbon concentrations observed in U.S.-based CNT facilities. The study estimated that considerable years of exposure are necessary for significant pathology to occur.

No Fullerene Toxicity Reported

Nanoparticles can also be made of C_{60}, as is the case with almost any room temperature solid, and several groups have done this and studied toxicity of such particles. The results in the work of Oberdörster at Southern Methodist University, published in "Environmental Health Perspectives" in July 2004, in which questions were raised of potential cytotoxicity, has now been shown by several sources to be likely caused by the tetrahydrofuran used in preparing the 30 nm–100 nm particles of C_{60} used in the research. Isakovic, et al., 2006, who review this phenomenon, gives results showing that removal of THF from the C_{60} particles resulted in a loss of toxicity. Sayes, et al., 2007, also show that particles prepared as in Oberdorster caused no detectable inflammatory response when instilled intratracheally in rats after observation for 3 months, suggesting that even the particles prepared by Oberdorster do not exhibit markers of toxicity in mammalian models. This work used as a benchmark quartz particles, which did give an inflammatory response.

A comprehensive and recent review of work on fullerene toxicity is available in "Toxicity Studies of Fullerenes and Derivatives," a chapter from the book "Bio-applications of Nanoparticles". In this work, the authors review the work on fullerene toxicity beginning in the early 1990s to present, and conclude that the evidence gathered since the discovery of fullerenes overwhelmingly points to C_{60} being non-toxic. As is the case for toxicity profile with any chemical modification of a structural moiety, the authors suggest that individual molecules be assessed individually.

Toxicity of Metal based Nanoparticles

Metal based nanoparticles (NPs) are a prominent class of NPs synthesized for their functions as semiconductors, electroluminescents, and thermoelectric materials. Biomedically, these antibacterial NPs have been utilized in drug delivery systems to access areas previously inaccessible to conventional medicine. With the recent increase in interest and development of nanotechnology, many studies have been performed to assess whether the unique characteristics of these NPs, namely their small surface area to volume ratio, might negatively impact the environment upon which they were introduced. Researchers have since found that many metal and metal oxide NPs have detrimental effects on the cells with which they come into contact including but not limited to DNA breakage and oxidation, mutations, reduced cell viability, warped morphology, induced apoptosis and necrosis, and decreased proliferation.

Cytotoxicity

A primary marker for the damaging effects of NPs has been cell viability as determined by state and exposed surface area of the cell membrane. Cells exposed to metallic NPs have, in the case of copper oxide, had up to 60% of their cells rendered unviable. When diluted, the positively charged metal ions often experience an electrostatic attraction to the cell membrane of nearby cells, covering the membrane and preventing it from permeating the necessary fuels and wastes. With less exposed membrane for transportation and communication, the cells are often rendered inactive.

NPs have been found to induce apoptosis in certain cells primarily due to the mitochondrial damage and oxidative stress brought on by the foreign NPs electrostatic reactions.

Genotoxicity

Many methods ranging from comet assay to the HPRT gene mutation test have found that metal based NPs disrupt DNA and its replication process in a variety of cells. In a study examining the effects of nanosilver on DNA, AgNPs were introduced to lymphocyte cell DNA which was then examined for abnormalities. The exposure of the NPs correlated to a significant increase in micronuclei indicative of genetic fragmentation. Metal Oxides such as copper oxide, uraninite, and cobalt oxide have also been found to exert significant stress on exposed DNA. The damage done to the DNA will often result in mutated cells and colonies as found with the HPRT gene test.

Coatings and Charges

NPs, in their implementation, are covered with coatings and sometimes given positive or negative charges depending upon the intended function. Studies have found that these external factors affect the degree of toxicity of NPs. Positive charges are usually found to amplify and cause of cellular damage much more noticeably than negative charges do. In a study in which b- and c-polyethylenimine coated AgNPs were attached to strands of Lambda DNA, the cationic b-polyethylenimine coated AgNP was found to lower the melting point of the DNA 50 °C lower than its anionic counterpart.

Immunogenicity of Nanoparticles

Very little attention has been directed towards the potential immunogenicity of nanostructures. Nanostructures can activate the immune system, inducing inflammation, immune responses, allergy, or even affect to the immune cells in a deleterious or beneficial way (immunosuppression in autoimmune diseases, improving immune responses in vaccines). More studies are needed in order to know the potential deleterious or beneficial effects of nanostructures in the immune system. In comparison to conventional pharmaceutical agents, nanostructures have very large sizes, and immune cells, especially phagocytic cells, recognize and try to destroy them.

Complications with Nanotoxicity Studies

Size is therefore a key factor in determining the potential toxicity of a particle. However it is not the only important factor. Other properties of nanomaterials that influence toxicity include: chemical composition, shape, surface structure, surface charge, aggregation and solubility, and the presence or absence of functional groups of other chemicals. The large number of variables influencing toxicity means that it is difficult to generalise about health risks associated with exposure to nanomaterials – each new nanomaterial must be assessed individually and all material properties must be taken into account.

In addition, standarization of toxicology tests between laboratories are needed. Díaz, B. *et al.* from the University of Vigo (Spain) has shown (Small, 2008) that many different cell lines should be studied in order to know if a nanostructure induces toxicity, and human cells can internalize aggregated nanoparticles. Moreover, it is important to take into account that many nanostructures aggregate in biological fluids, but groups manufacturing nanostructures do not care much about this matter. Many efforts of interdisciplinary groups are strongly needed in order to progress in this field.

Effect of Aggregation or Agglomeration of Nanoparticles

Many nanoparticles agglomerate or aggregate when they are placed in environmental or biological fluids. The terms agglomeration and aggregation have distinct definitions according to the standards organizations ISO and ASTM, where agglomeration signifies more loosely bound particles and aggregation signifies very tightly bound or fused particles (typically occurring during synthesis or drying). Nanoparticles frequently agglomerate due to the high ionic strength of environmental and biological fluids, which shields the repulsion due to charges on the nanoparticles. Unfortunately, agglomeration has frequently been ignored in nanotoxicity studies, even though agglomeration would be expected to affect nanotoxicity since it changes the size, surface area, and sedimentation properties of the nanoparticles. In addition, many nanoparticles will agglomerate to some extent in the environment or in the body before they reach their target, so it is desirable to study how toxicity is affected by agglomeration.

A method was published that can be used to produce different mean sizes of stable agglomerates of several metal, metal oxide, and polymer nanoparticles in cell culture media for cell toxicity studies. Different mean sizes of agglomerates are produced by allowing the nanoparticles to agglomerate to a particular size in cell culture media without protein, and then adding protein to coat the agglomerates and "freeze" them at that size. By waiting different amounts of time before adding protein, different mean sizes of agglomerates of a single type of nanoparticle can be produced in an otherwise identical solution, allowing one to study how agglomerate size affects toxicity. In addition, it was found that vortexing while adding a high concentration of nanoparticles to the cell culture media produces much less agglomerated nanoparticles than if the dispersed solution is only mixed after adding the nanoparticles.

The agglomeration/deagglomeration (mechanical stability) potentials of airborne engineered nanoparticle clusters also have significant influences on their size distribution profiles at the end-point of their environmental transport routes. Different aerosolization and deagglomeration systems have been established to test stability of nanoparticle agglomerates. For example, laboratory setups based on critical orifices have been used to apply a wide range of external shear forces onto airborne nanoparticles. After applying shear forces, the particle mean size decreased while the particle generation rate increased. In another pioneering study, four powder aerosolization systems (dustiness testing systems) were compared for the first time for their characteristics linked to aerosol generation.

Challenges of the Nano-visualisation and Related unknowns in Nanotoxicology

With comparison to more conventional toxicology studies, the nanotoxicology field is however suffering from a lack of easy characterisation of the potential contaminants, the "nano" scale being a scale difficult to comprehend. The biological systems are themselves still not completely known at this scale. Ultimate Atomic visualisation methods such as Electron microscopy (SEM and TEM) and Atomic force microscopy (AFM) analysis allow visualisation of the nano world. Further nanotoxicology studies will require precise characterisation of the specificities of a given nano-element : size, chemical composition, detailed shape, level of aggregation, combination with other vectors, etc. Above all, these properties would have to be determined not only on the nanocomponent before its introduction in the living environment but also in the (mostly aqueous) biological environment.

There is a need for new methodologies to quickly assess the presence and reactivity of nanoparticles in commercial, environmental, and biological samples since current detection techniques require expensive and complex analytical instrumentation. There have been recent attempts to address these issues by developing and investigating sensitive, simple and portable colorimetric detection assays that assess for the surface reactivity of NPs, which can be used to detect the presence of NPs, in environmental and biological relevant samples. Surface redox reactivity is a key emerging property related to potential toxicity of NPs with living cells, and can be used as a key surrogate for determine for the presence of NPs and a first tier analytical strategy toward assessing NP exposures.

It is difficult to determine the degree of effect of a specific nanoparticle when compared to those of comparable nanoparticles already present in our natural environment.

AEM - Analytical Electron Microscopy was used over 40 years ago to investigate amphibole asbestos bodies in Lake Superior from the Reserve Mining operations. This could non-destructively characterise sub-micron particles. Today AEM can fully characterise to atom dimensions.

Nanomaterials Pollution

Nanopollution is a generic name for waste generated by nanodevices or during the nanomaterials manufacturing process. Ecotoxicological impacts of nanoparticles and the potential for bioaccumulation in plants and microorganisms is a subject of current research, as nanoparticles are considered to present novel environmental impacts. Of the US$710 million spent in 2002 by the U.S. government on nanotechnology research, $500,000 was spent on environmental impact assessments.

Overview

Nanowaste is mainly the group of particles that are released into the environment, or the particles that are thrown away when still on their products. The thrown away nanoparticles are usually still functioning how they are supposed to (still have their individual properties), they are just not being properly used anymore. Most of the time, they are lost due to contact with different environments. Silver nanoparticles, for example, they are used a lot in clothes to control odor, those particles are lost when washing them. The fact that they are still functioning and are so small is what makes nanowaste a concern. It can float in the air and might easily penetrate animal and plant cells causing unknown effects. Due to its small size, nanoparticles can have different properties than their own material when on a bigger size, and they are also functioning more efficiently because of its greater surface area. Most human-made nanoparticles do not appear in nature, so living organisms may not have appropriate means to deal with nanowaste.

The capacity for nanoparticles to function as a transport mechanism also raises concern about the transport of heavy metals and other environmental contaminants. Two areas of concern can be identified. First, in their free form nanoparticles can be released into the air or water during production, or production accidents, or as waste by-product of production, and ultimately accumulate in the soil, water, or plant life. Second, in fixed form, where they are part of a manufactured substance or product, they will ultimately have to be recycled or disposed of as waste.

Scrinis raises concerns about nano-pollution, and argues that it is not currently possible to "precisely predict or control the ecological impacts of the release of these nano-products into the environment." A May 2007 Report to the UK Department for Environment, Food and Rural Affairs noted concerns about the toxicological impacts of nanoparticles in relation to both hazard and exposure. The report recommended comprehensive toxicological testing and independent performance tests of fuel additives. Risks have been identified by Uskokovic in 2007. Concerns have also been raised about Silver Nano technology used by Samsung in a range of appliances such as washing machines and air purifiers.

One already known consequences to metals exposure is shown by silver, if exposed to humans in a certain concentration, it can cause illnesses such as argyria and argyrosis. Silver can also cause some environmental problems. Due to its antimicrobial properties (antibacterial), when encountered in the soil it can kill beneficial bacteria that are important to keep the soil healthy. Environmental assessment is justified as nanoparticles present novel environmental impacts. Scrinis raises concerns about nano-pollution, and argues that it is not currently possible to "precisely predict or control the ecological impacts of the release of these nano-products into the environment."

Metals, in particular, have a really strong bonds. Their properties follow up to the nanoscale as well. Metals can stay and damage the environment for a long time, since they hardly degrade or get destroyed. With the increase in use of nanotechnology, it is predicted that the nanowaste of metals will keep increasing, and until a solution is found for that problem, that waste will keep accumulating in the environment. On the other hand, some possible future applications of nanotechnology have the potential to benefit the environment. Nanofiltration, based on the use of membranes with extremely small pores smaller than 10 nm (perhaps composed of nanotubes) are suitable for a mechanical filtration for the removal of ions or the separation of different fluids. A couple of studies have found a solution to filtrate and extract those nanoparticles from water. The process is still being studied but simulations have been giving a total of about 90% to 99% removal of nanowaste particles from the water at an upgraded waste water treatment plant. Once the particles are separated from the water, they go to the landfill with the rest of the solids. Furthermore, magnetic nanoparticles offer an effective and reliable method to remove heavy metal contaminants from waste water. Using nanoscale particles increases the efficiency to absorb the contaminants and is comparatively inexpensive compared to traditional precipitation and filtration methods. One current method to recover nanoparticles is the Cloud Point Extraction. With this technique, gold nanoparticles and some other types of particles that are heat conductors are able to be extracted from aqueous solutions. The process consists of a heating section of the solution that contains the nanoparticles, and then centrifuged in order to separate the layers and then separate the nanoparticles.

Life Cycle Responsibility

To properly assess the health hazards of engineered nanoparticles the whole life cycle of these particles needs to be evaluated, including their fabrication, storage and distribution, application and potential abuse, and disposal. The impact on humans or the environment may vary at different stages of the life cycle.

The Royal Society report identified a risk of nanoparticles or nanotubes being released during disposal, destruction and recycling, and recommended that "manufacturers of products that fall under extended producer responsibility regimes such as end-of-life regulations publish procedures outlining how these materials will be managed to minimize possible human and environmental exposure". Reflecting the challenges for ensuring responsible life cycle regulation, the Institute for Food and Agricultural Standards has proposed standards for nanotechnology research and development should be integrated across consumer, worker and environmental standards. They also propose that NGOs and other citizen groups play a meaningful role in the development of these standards.

References

- Saini, Rajiv; Saini, Santosh; Sharma, Sugandha (2010). "Nanotechnology: The Future Medicine". Journal of Cutaneous and Aesthetic Surgery. 3 (1): 32–33. PMC 2890134. PMID 20606992. doi:10.4103/0974-2077.63301
- Drexler, K. Eric (1992). Nanosystems: Molecular Machinery, Manufacturing, and Computation. New York: John Wiley & Sons. ISBN 0-471-57547-X
- "Nanotechnology Information Center: Properties, Applications, Research, and Safety Guidelines". American Elements. Retrieved 13 May 2011
- Cassidy (2014). "Nanotechnology in the Regeneration of Complex Tissues". Bone and Tissue Regeneration Insights: 25. doi:10.4137/BTRI.S12331
- Paull, J. & Lyons, K. (2008). "Nanotechnology: The Next Challenge for Organics" (PDF). Journal of Organic Systems. 3: 3–22

- Allhoff, Fritz; Lin, Patrick; Moore, Daniel (2010). What is nanotechnology and why does it matter?: from science to ethics. John Wiley and Sons. pp. 3–5. ISBN 1-4051-7545-1
- "Analysis: This is the first publicly available on-line inventory of nanotechnology-based consumer products". The Project on Emerging Nanotechnologies. 2008. Retrieved 13 May 2011
- Boisseau, P.; Loubaton, B. (2011). "Nanomedicine, nanotechnology in medicine". Comptes Rendus Physique. 12 (7): 620–636. doi:10.1016/j.crhy.2011.06.001
- Bertrand N; Leroux JC (2011). "The journey of a drug carrier in the body: an anatomo-physiological perspective". Journal of Controlled Release. 161 (2): 152–63. PMID 22001607. doi:10.1016/j.jconrel.2011.09.098
- Ehud Gazit, Plenty of room for biology at the bottom: An introduction to bionanotechnology. Imperial College Press, 2007, ISBN 978-1-86094-677-6
- "CDC – Nanotechnology – NIOSH Workplace Safety and Health Topic". National Institute for Occupational Safety and Health. June 15, 2012. Retrieved 2012-08-24
- Mashaghi S.; Jadidi T.; Koenderink G.; Mashaghi A. (2013). "Lipid Nanotechnology". Int. J. Mol. Sci. 14: 4242–4282. PMC 3588097. PMID 23429269. doi:10.3390/ijms14024242
- Chidambaram, M.; Manavalan, R.; Kathiresan, K. (2011). "Nanotherapeutics to overcome conventional cancer chemotherapy limitations". Journal of Pharmacy & Pharmaceutical Sciences. 14 (1): 67–77. PMID 21501554
- Hanft, Susan (2011). Market Research Report Nanotechnology in water treatment. Wellesley, MA USA: BCC Research. p. 16. ISBN 1596237090
- Kevin Rollins (Nems Mems Works, LLC). "Nanobiotechnology Regulation: A Proposal for Self-Regulation with Limited Oversight". Volume 6 – Issue 2. Retrieved 2 September 2010
- Kim, Y. (2014). "Nanowastes Treatment In Environmental Media". Environmental Health and Toxicology. 29: e2014015. PMC 4224314. PMID 25381912. doi:10.5620/eht.e2014015
- Freitas, Robert A., Jr. (2005). "Current Status of Nanomedicine and Medical Nanorobotics" (PDF). Journal of Computational and Theoretical Nanoscience. 2 (4): 1–25. doi:10.1166/jctn.2005.001
- Raymond D. Letterman (ed.)(1999). "Water Quality and Treatment." 5th Ed. (New York: American Water Works Association and McGraw-Hill.) ISBN 0-07-001659-3
- Royal Society and Royal Academy of Engineering (2004). "Nanoscience and nanotechnologies: opportunities and uncertainties". Retrieved 2008-05-18

Permissions

All chapters in this book are published with permission under the Creative Commons Attribution Share Alike License or equivalent. Every chapter published in this book has been scrutinized by our experts. Their significance has been extensively debated. The topics covered herein carry significant information for a comprehensive understanding. They may even be implemented as practical applications or may be referred to as a beginning point for further studies.

We would like to thank the editorial team for lending their expertise to make the book truly unique. They have played a crucial role in the development of this book. Without their invaluable contributions this book wouldn't have been possible. They have made vital efforts to compile up to date information on the varied aspects of this subject to make this book a valuable addition to the collection of many professionals and students.

This book was conceptualized with the vision of imparting up-to-date and integrated information in this field. To ensure the same, a matchless editorial board was set up. Every individual on the board went through rigorous rounds of assessment to prove their worth. After which they invested a large part of their time researching and compiling the most relevant data for our readers.

The editorial board has been involved in producing this book since its inception. They have spent rigorous hours researching and exploring the diverse topics which have resulted in the successful publishing of this book. They have passed on their knowledge of decades through this book. To expedite this challenging task, the publisher supported the team at every step. A small team of assistant editors was also appointed to further simplify the editing procedure and attain best results for the readers.

Apart from the editorial board, the designing team has also invested a significant amount of their time in understanding the subject and creating the most relevant covers. They scrutinized every image to scout for the most suitable representation of the subject and create an appropriate cover for the book.

The publishing team has been an ardent support to the editorial, designing and production team. Their endless efforts to recruit the best for this project, has resulted in the accomplishment of this book. They are a veteran in the field of academics and their pool of knowledge is as vast as their experience in printing. Their expertise and guidance has proved useful at every step. Their uncompromising quality standards have made this book an exceptional effort. Their encouragement from time to time has been an inspiration for everyone.

The publisher and the editorial board hope that this book will prove to be a valuable piece of knowledge for students, practitioners and scholars across the globe.

Index